Modular Representations
of Finite Groups

Pure and Applied Mathematics

A Series of Monographs and Textbooks

Editors **Samuel Eilenberg and Hyman Bass**

Columbia University, New York

RECENT TITLES

E. R. KOLCHIN. Differential Algebra and Algebraic Groups

GERALD J. JANUSZ. Algebraic Number Fields

A. S. B. HOLLAND. Introduction to the Theory of Entire Functions

WAYNE ROBERTS AND DALE VARBERG. Convex Functions

A. M. OSTROWSKI. Solution of Equations in Euclidean and Banach Spaces, Third Edition of Solution of Equations and Systems of Equations

H. M. EDWARDS. Riemann's Zeta Function

SAMUEL EILENBERG. Automata, Languages, and Machines: Volumes A and B

MORRIS HIRSCH AND STEPHEN SMALE. Differential Equations, Dynamical Systems, and Linear Algebra

WILHELM MAGNUS. Noneuclidean Tesselations and Their Groups

FRANÇOIS TREVES. Basic Linear Partial Differential Equations

WILLIAM M. BOOTHBY. An Introduction to Differentiable Manifolds and Riemannian Geometry

BRAYTON GRAY. Homotopy Theory: An Introduction to Algebraic Topology

ROBERT A. ADAMS. Sobolev Spaces

JOHN J. BENEDETTO. Spectral Synthesis

D. V. WIDDER. The Heat Equation

IRVING EZRA SEGAL. Mathematical Cosmology and Extragalactic Astronomy

J. DIEUDONNÉ. Treatise on Analysis: Volume II, enlarged and corrected printing; Volume IV; Volume V. *In preparation*

WERNER GREUB, STEPHEN HALPERIN, AND RAY VANSTONE. Connections, Curvature, and Cohomology: Volume III, Cohomology of Principal Bundles and Homogeneous Spaces

I. MARTIN ISAACS. Character Theory of Finite Groups

JAMES R. BROWN. Ergodic Theory and Topological Dynamics

K. D. STROYAN AND W. A. J. LUXEMBURG. Introduction to the Theory of Infinitesimals

B. M. PUTTASWAMAIAH AND JOHN D. DIXON. Modular Representations of Finite Groups

In preparation

CLIFFORD A. TRUESDELL. A First Course in Rational Continuum Mechanics: Volume 1, General Concepts

MELVYN BERGER. Nonlinearity and Functional Analysis: Lectures on Nonlinear Problems in Mathematical Analysis

GEORGE GRATZER. Lattice Theory

Modular Representations of Finite Groups

B. M. Puttaswamaiah
John D. Dixon

Department of Mathematics
Carleton University
Ottawa, Canada

ACADEMIC PRESS   New York   San Francisco   London   1977
A Subsidiary of Harcourt Brace Jovanovich, Publishers

COPYRIGHT © 1977, BY ACADEMIC PRESS, INC.
ALL RIGHTS RESERVED.
NO PART OF THIS PUBLICATION MAY BE REPRODUCED OR
TRANSMITTED IN ANY FORM OR BY ANY MEANS, ELECTRONIC
OR MECHANICAL, INCLUDING PHOTOCOPY, RECORDING, OR ANY
INFORMATION STORAGE AND RETRIEVAL SYSTEM, WITHOUT
PERMISSION IN WRITING FROM THE PUBLISHER.

ACADEMIC PRESS, INC.
111 Fifth Avenue, New York, New York 10003

United Kingdom Edition published by
ACADEMIC PRESS, INC. (LONDON) LTD.
24/28 Oval Road, London NW1

Library of Congress Cataloging in Publication Data

Puttaswamaiah, B.M.
    Modular representations of finite groups.

    (Pure and applied mathematics, a series of
monographs and textbooks ;   )
    Bibliography: p.
    Includes index.
    1.   Modular representations of groups.
2.   Finite groups.   I.   Dixon, John D., joint author.
II.   Title.   III.   Series.
QA3.P8   [QA171]   510'.8s   [512'.22]   76-21038
ISBN 0-12-568650-1
AMS (MOS) 1970 Subject Classifications: 20C05, 20C15,
20C20

PRINTED IN THE UNITED STATES OF AMERICA

Contents

Preface ix

Note to the Reader xi

Notation xiii

Chapter I Representation Modules — 1

1.1 Group algebras and modules — 1
1.2 Reducible and irreducible modules — 5
1.3 Semisimple rings and the Wedderburn structure theorem — 8
1.4 Tensor products — 13
1.5 The number of irreducible KG-modules — 15
1.6 Indecomposable modules — 20
1.7 Absolutely indecomposable and absolutely irreducible modules — 24
1.8 Principal indecomposable modules — 26
1.9 Composition factors and intertwining numbers — 29
1.10 Notes and comments — 32

Chapter II Induced Modules and Characters — 33

2.1 Induced modules — 33
2.2 Clifford's theorem — 37
2.3 Group characters — 39
2.4 The theory of ordinary characters — 43

2.5	Induced characters	48
2.6	Brauer's theorem on induced characters	52
2.7	Splitting fields	56
2.8	Notes and comments	60

Chapter III Modular Representations and Characters 62

3.1	The p-adic integers	62
3.2	p-adic algebras	65
3.3	Ordinary and modular representations	69
3.4	Lifting idempotents	73
3.5	The case where p does not divide $|G|$	75
3.6	Modular characters	77
3.7	Cartan invariants, decomposition numbers, and orthogonality relations	79
3.8	Modular characters of p-solvable groups	84
3.9	Notes and comments	88

Chapter IV Blocks of Group Algebras 89

4.1	Blocks	89
4.2	Classifying modules, characters, and idempotents into blocks	93
4.3	Defect groups	100
4.4	Further analysis of the Cartan matrix and decomposition matrix	104
4.5	The characters in a block of given defect	107
4.6	Blocks of small defect	112
4.7	Notes and comments	114

Chapter V The Theory of Indecomposable Modules 116

5.1	Relatively projective modules	116
5.2	Vertices and sources	120
5.3	Green's theorem	123
5.4	The degrees of indecomposable modules	127
5.5	Vertices and defect groups	129
5.6	Restriction of indecomposable modules	130
5.7	Jordan's theorem in characteristic p	133
5.8	Notes and comments	140

Chapter VI The Main Theorems of Brauer 142

6.1	The Brauer homomorphism	142
6.2	Blocks with normal p-subgroups	144
6.3	The Brauer correspondence: The First Main Theorem	149
6.4	Extension of the First Main Theorem	152
6.5	Generalized decomposition numbers: The Second Main Theorem	157
6.6	Principal blocks: The Third Main Theorem	162
6.7	The characters in the principal block	164
6.8	Notes and comments	168

CONTENTS

Chapter VII Fusion of 2-Groups — 169

7.1 Further results on generalized decomposition numbers — 169
7.2 Some technical lemmas — 174
7.3 Groups with Sylow 2-subgroups of type $(2^m, 2^m)$ — 178
7.4 Groups with quaternion Sylow 2-subgroups — 184
7.5 Glauberman's Z^*-theorem — 191
7.6 Notes and comments — 196

Chapter VIII Blocks with Cyclic Defect Groups — 197

8.1 Extending characters from normal subgroups — 197
8.2 Blocks with normal cyclic defect groups — 199
8.3 Groups with cyclic Sylow p-subgroups — 205
8.4 Some technical lemmas — 206
8.5 Groups of order $g = pg_0$ with $p \nmid g_0$ — 208
8.6 Groups with a faithful representation of degree $d < \frac{1}{2}(p - 1)$ — 215
8.7 Criteria for normal Sylow p-groups — 220
8.8 Notes and comments — 226

References — 229

Index — 239

Preface

Our purpose in this book is to give a direct and readable account of the basic concepts and techniques of the theory of modular representations of finite groups and some significant applications. If we have succeeded in our aim, then the book will be useful both as a textbook for students and as an introductory guide to the rather extensive literature in the subject.

The theory of modular representations has a reputation of being much more difficult than the theory of ordinary representations, and to some extent this reputation is justified. At least at the beginning, the modular theory requires a more sophisticated background in algebra, and it takes a little longer to reach the stage where we can apply the theory. On the other hand, a knowledge of elementary properties of rings and modules up to, say, the Artin–Wedderburn structure theorem for Artinian rings together with a reasonable background in the theory of finite groups will be adequate prerequisite for almost all that we shall do in this book. At the same time we shall see that there are serious applications of the theory at quite an early stage, and before the end of the book we shall be using the theory to prove some of the deep and important results on finite groups that have been obtained only in this way.

In presenting this material we have chosen to work throughout with representations over splitting fields; indeed, much of the work is carried out over a specially constructed field described in Chapter III. This seems appropriate in an introduction such as this since the great majority of applications of the theory so far can be given in this context. In fact, at all points where we have had to make a choice we have preferred to treat a particularly important case as clearly as possible rather than give the most general result. To balance this we have included notes and comments at the end of each chapter; in these sections we offer guides to pertinent literature. It is our hope that anyone using this book will thus be able to go much further into the subject than we have been able to do within the confines of this book.

Throughout the volume we have included examples, and at the ends of most sections there are exercises for the reader to test his understanding. Many of the latter are essentially corollaries and extensions of the main results, and some are counterexamples to show how our theorems are limited.

The basic foundations of modular representation theory were laid by Richard Brauer, and his monumental contribution to the subject spreads over the past five decades, through his numerous papers and the work of his students and collaborators. His influence is evident in every chapter of this book.

In writing this book we are indebted to Dr. Roderick Gow who read a draft of the manuscript and whose suggestions led to a number of important corrections and clarifications. We also wish to thank Gillian Murray who typed the final version of the manuscript.

Note to the Reader

The first chapter is essentially introductory, detailing the basic algebraic concepts and setting the notation. Much of this material will be quite familiar and the reader should pass over this chapter quite quickly. To a lesser extent the same may be true of the second chapter; parts of it will be familiar to someone who has studied the theory of ordinary representations, although some will not. The real differences between the ordinary and modular theories become apparent in Chapter III. Each chapter ends with a section of comments and references. These should be read to put our results into context and also for suggestions for further reading.

Our use of notation follows current mathematical usage (such as, for example, Lang [1]), and a list of special notation has been included. One symbol which may not be familiar is the sign ":=" which may be read as "is defined by." The reader should be warned that certain conventions are introduced and used throughout the book. In particular, *group* always means finite group, and *ring* means a ring with unity so the unity is mapped onto unity under ring homomorphisms. Each "module" or "algebra" is finitely generated and unital. The letter p always denotes a prime. Following §3.3 further general conventions are made.

Notation

Relations and operations

$\triangleleft$	normal subgroup of
$:=$	defined to be equal to
$p\mid m$	p divides m
$p \nmid m$	p does not divide m
$U\mid V$	U is isomorphic to a direct summand of V
$\langle S \rangle$	the subgroup generated by the subset S
$\otimes$	tensor product
$\oplus$	direct sum
$\times$	direct product
f_S	restriction of f to S
$\setminus$	set difference

Alphabetic

$[\alpha_{ij}]$	matrix with (i,j)th entry α_{ij}
Δ	decomposition matrix
Δ°	reduced decomposition matrix
Δ_H	decomposition matrix of H
Γ_H	Cartan matrix of kH
χ^G	induced character

$\mathrm{Aut}(G)$	automorphism group of G
$\mathrm{Aut}_R(V)$	automorphism group of R-module V
A_n	alternating group on n symbols
$A[X]$	ring of polynomials over A
B_0	principal p-block
$\mathrm{Char}(G)$	ring of ordinary generalized characters
$\mathrm{Char}(G°)$	ring of modular generalized characters
$\mathrm{Class}_k(G)$	algebra of class functions over k
$C_G(S)$	centralizer of S in G
$\dim_k V$	dimension of V over k
$\mathrm{End}_R(V)$	ring of R-endomorphisms of V
$\langle f, g \rangle$	inner product
$|G:H|$	index of H in G
$G°$	p'-elements of G
G'	commutator subgroup of G
G''	commutator subgroup of G'
$GL(n, A)$	general linear group of degree n over A
$\mathrm{Hom}_R(U, V)$	group of homomorphisms from U to V
$i(U, V)$	intertwining number of the pair U, V
iff	if and only if
$\mathrm{Im}\, f$	image of the mapping f
$\mathrm{Ker}\, f$	kernel of the homomorphism f
$\mathrm{Ker}\, V$	kernel of the module V
$\mathrm{Ker}\, \zeta$	kernel of the character ζ
$\mathrm{Mat}(d, A)$	ring of $d \times d$ matrices over A
$N_G(S)$	normalizer of S in G
$O_p(G)$	largest normal p-subgroup of G
$O_{p'}(G)$	largest normal p'-subgroup of G
$\mathrm{Orbit}_G(s)$	orbit of s under G
$PSL(n, q)$	projective special linear group
$\mathbf{Q}_p$	field of p-adic numbers
$\mathrm{rad}\, R$	Jacobson radical of R
$\mathrm{rank}_A V$	rank of V over A
$\mathrm{Stab}(s)$	stabilizer of s
$\mathrm{Sym}(\Omega)$	symmetric group on Ω
$\mathrm{Triv}(G)$	trivial module
$U \otimes_A V$	inner tensor product
U^*	dual of U
V_H	restriction of V to H
V^x	conjugate of V associated with x
V^G	induced module
x_p	p-part of x

NOTATION

$x_{p'}$	p'-part of x
$Z(G)$	center of the group G
$Z(kG)$	center of the group algebra kG
$\mathbf{Z}_p$	ring of p-adic integers
$Z^*(G)$	center of $G/O_{2'}(G)$

CHAPTER I

Representation Modules

The object of this chapter is to give a short introduction to material of a general algebraic nature which will be used in the later chapters. The topics include group algebras, reducibility and decomposability of a module over a ring, tensor products, the structure theory of semisimple algebras (the Wedderburn structure theorem), and the relation that holds between composition factors of a module and intertwining numbers. Some standard theorems are stated without proof; proofs are easily available and can be found, for example, in Lang [1].

1.1 Group algebras and modules

By convention all groups which we consider in this book will be finite unless otherwise specified. Let G be a group and S an arbitrary set. An *action* of G on S is a homomorphism θ of G into the symmetric group $\text{Sym}(S)$ of all permutations of S. We shall also say that G *acts* on S (via θ). Thus if G acts on S via θ, then for each $x \in G$, $\theta(x)$ is a bijection of S onto itself satisfying the conditions: $\theta(1) = 1$ (the identity on S), $\theta(x^{-1}) = \theta(x)^{-1}$, and $\theta(xy) = \theta(x)\theta(y)$ for all x, y in G. The action is called *faithful* if the kernel of the action $\text{Ker } \theta$ equals 1, the trivial subgroup; in this case θ is an injection of G into $\text{Sym}(S)$. When the action is understood we shall suppress the symbol θ and write instead $s \mapsto s^x$ or $s \mapsto sx$ for $\theta(x)$ ($x \in G$, $s \in S$).

EXAMPLE 1 Let H be a subgroup of G. Then H acts on the set G by right multiplication; namely $s^x := sx$ (product) for $s, x \in G$ and by left multiplication by $s^x := x^{-1}s$ (product) for $s, x \in G$.

EXAMPLE 2 G acts on the set Σ of all its subgroups by conjugation; namely $H^x := x^{-1}Hx$ (H, $x^{-1}Hx \in \Sigma$ and $x \in G$).

For any action of a group G on a set S we define the *stabilizer* of an element $s \in S$ by $\text{Stab}(s) := \{x \in G \mid s^x = s\}$ and the *orbit* of an element $s \in S$ by $\text{Orbit}(s) := \{s^x \mid x \in G\}$. G acts *transitively* if $\text{Orbit}(s) = S$ for each $s \in S$.

Note It is readily seen that $\text{Stab}(s)$ is always a subgroup of G and that the orbits in S form a partition of S. Since $s^x = s^y$ if and only if $\text{Stab}(s)x = \text{Stab}(s)y$, the elements of $\text{Orbit}(s)$ are in one-to-one correspondence with the right cosets of $\text{Stab}(s)$ in G; hence $|\text{Orbit}(s)| = |G : \text{Stab}(s)|$. This important relation shows in particular that the length $|\text{Orbit}(s)|$ of each orbit divides the order $|G|$ of G.

Let A be a commutative ring. By convention all rings that we shall consider will be associative rings with unity $1 \neq 0$. For any group G we define the *group algebra* AG of G over A to be the set of all functions $f: G \to A$ with the operations $+$ and $\cdot$ defined by

$$(f+g)(x) := f(x) + g(x) \quad \text{and} \quad (fg)(x) := \sum_{y \in G} f(y)g(y^{-1}x)$$

for all $x \in G$. It is routine to verify that AG is a ring under these operations. We can embed G into AG by the mapping $x \mapsto f_x$, where f_x is defined by

$$f_x(y) := \begin{cases} 1 & \text{if } y = x \\ 0 & \text{otherwise.} \end{cases}$$

Then, if we identify x and f_x, each $f \in AG$ can be written uniquely in the form $f = \sum_{x \in G} \alpha_x x$, where $\alpha_x \in A$; indeed $\alpha_x = f(x)$ for each $x \in G$. Addition and multiplication in AG, then, correspond to

$$\left(\sum_{x \in G} \alpha_x x\right) + \left(\sum_{x \in G} \beta_x x\right) = \sum_{x \in G} (\alpha_x + \beta_x)x$$

$$\left(\sum_{x \in G} \alpha_x x\right)\left(\sum_{x \in G} \beta_x x\right) = \sum_{x \in G} \sum_{y \in G} \alpha_x \beta_y xy.$$

We shall use this latter notation almost exclusively.

We next consider modules over a ring R. By convention (unless explicitly stated otherwise), we shall use the term *R-module* to refer to a finitely generated unital right module V over R. Thus as well as general module properties, V is assumed to have a finite subset $\{v_1, v_2, \ldots, v_n\}$ such that $V = v_1 R + v_2 R + \cdots + v_n R$, and for each $v \in V$ we have $v \cdot 1 = v$ (1 is the unity of R). In the case R is commutative, a (right) R-module V is also a left R-module, and it is sometimes convenient to write the elements from R on the left as well as the right; so $rv = vr$ for all $v \in V$, $r \in R$, in this case. We

1.1 GROUP ALGEBRAS AND MODULES

define the *annihilator* of the R-module V to be $\text{Ann}(V) = \{r \in R \mid Vr = 0\}$. Clearly $\text{Ann}(V)$ is an ideal of R and V can be considered as an $R/\text{Ann}(V)$-module. Obviously the annihilator of the $R/\text{Ann}(V)$-module V is (0). An R-module V is *free* (or more precisely R-*free*) if it has a (finite) set $v_1, v_2, \ldots, v_n$ of elements such that each $v \in V$ can be written in exactly one way in the form $v = v_1 r_1 + v_2 r_2 + \cdots + v_n r_n$ with $r_1, r_2, \ldots, r_n \in R$; in this case $v_1, v_2, \ldots, v_n$ form an R-*basis* of V.

EXAMPLE 3 If K is a field, then every K-module is K-free.

EXAMPLE 4 If A is a (commutative) principal ideal domain, then a basic theorem on the structure of A-modules states that each A-module V can be written as a direct sum $V = v_1 A \oplus v_2 A \oplus \cdots \oplus v_n A$ for suitable v_i in V (see Lang [1], Chapter 15, for a proof). It follows that V is A-free if and only if $va \neq 0$ for each $v \neq 0$ in V and $a \neq 0$ in A. (If this latter condition holds then we say that V is *torsion-free* over A). The structure theorem also shows that for a given torsion-free A-module, there is a number r called the *rank* of V such that all A-bases of V have r elements. This example will be important later.

EXAMPLE 5 For each ring R we have the *regular* R-module consisting of the same set of elements as R and with the natural multiplication by elements from the ring R. The regular R-module R is R-free and has $\{1\}$ as a basis.

Let A be a commutative ring. Then an A-*algebra* is a ring R which is at the same time an A-module such that the two kinds of multiplication are compatible in the sense

$$a(xy) = (ax)y = x(ay) \quad \text{for all } a \in A \text{ and } x, y \in R.$$

EXAMPLE 6 If A is a commutative ring and G is a group, then the group algebra AG is an A-algebra. The set of elements of G forms an A-basis for AG and so AG is A-free.

Let A be a commutative ring and AG the group algebra of a group G over A. Then for any AG-module V we have the action of G on V defined by $v^x = vx$ (module product) for all $v \in V$ and $x \in G$. Conversely, suppose that we are given an A-module V and an action of G on V which is "linear" in the sense that

$$\left(\sum_{i=1}^{n} \alpha_i u_i\right)^x = \sum_{i=1}^{n} \alpha_i u_i^x$$

for all α_i in A, u_i in V, and $x \in G$. Then V becomes an AG-module under the operation

$$v\left(\sum_{x \in G} \alpha_x x\right) = \sum_{x \in G} \alpha_x v^x$$

($v \in V$ and all $\alpha_x \in A$). Thus for an A-module V there is a one-to-one correspondence between the possible definitions of an AG-module structure on V and the possible linear actions of G on the A-module V. It is often useful to define the structure of an AG-module in this way.

If U and V are modules over a given ring R, then we shall write $\text{Hom}_R(U, V)$ to denote the set of R-homomorphisms of U into V. Then $\text{Hom}_R(U, V)$ is a module over the ring of rational integers under the addition defined by $u(f + g) := uf + ug$ for all $f, g \in \text{Hom}_R(U, V)$, and $u \in U$. In particular, $\text{End}_R(U) := \text{Hom}_R(U, U)$ is a ring with multiplication defined by $u(fg) = (uf)g$ for all $f, g \in \text{End}_R(U)$ and $u \in U$. Finally, $\text{Aut}_R(U)$ is the group of units of $\text{End}_R(U)$, consisting of all R-automorphisms of U.

If A is a commutative ring, and if U and V are A-modules, then $\text{Hom}_A(U, V)$ is an A-module with module operation given by $u(fa) := (uf)a$ for all $f \in \text{Hom}_A(U, V)$, $u \in U$, and $a \in A$.

Note Suppose AG is a group algebra of a group G over a commutative ring A and R is a subring of AG. Then for any AG-modules U and V we have $\text{Hom}_{AG}(U, V) \subseteq \text{Hom}_R(U, V)$. If $A \subseteq R$ then $g \in \text{Hom}_R(U, V)$ lies in $\text{Hom}_{AG}(U, V)$ if and only if $(ug)x = (ux)g$ for all $u \in U$ and $x \in G$. In particular, $\text{Hom}_{AG}(U, V) \subseteq \text{Hom}_A(U, V)$.

Consider the case where U is a torsion-free module over a commutative ring A. Assume that U has a basis $\{u_1, u_2, \ldots, u_d\}$ over A and that each basis of U over A has d elements. Then the usual argument from linear algebra shows that there is a corresponding algebra isomorphism $\text{End}_A(U) \simeq \text{Mat}(d, A)$, the ring of all d-by-d matrices over A. Under this mapping $\text{Aut}_A(U) \simeq GL(d, A)$, the *general linear group* of degree d over A. Here $GL(d, A)$ is the group of units of $\text{Mat}(d, A)$ and consists of all matrices whose determinants are units in A.

Let A be a commutative ring and G a group. A (linear) *representation* of G on a nonzero A-module V is a homomorphism $T: G \to \text{Aut}_A(V)$; the representation is *faithful* when the *kernel* Ker T equals 1. Sometimes Ker T will also be denoted by Ker V. A representation of G is just a linear action of G on V in the sense referred to above. Thus to each representation T of G on V there corresponds an AG-module structure on V given by

$$v\left(\sum_{x \in G} \alpha_x x\right) = \sum_{x \in G} \alpha_x [vT(x)]$$

for $v \in V$ and all $\alpha_x \in A$, and conversely. Hence there is a one-to-one correspondence between the class of all nonzero AG-modules and the class of all linear representations of G on A-modules. Moreover, if V is A-free, then $\text{Aut}_A(V) \simeq GL(d, A)$, where d is the number of elements in an A-basis of V.

1.2 REDUCIBLE AND IRREDUCIBLE MODULES

(Again we assume here that any two bases of V have the same number of elements.) Thus to each representation of G on the A-free A-module V there is a *matrix representation* $T^*\colon G \to GL(d, A)$. In particular, if $A = K$ is a field, then all K-modules are K-free. In this case the study of KG-modules, linear representations of G on K-modules, and matrix representations of G over K are essentially equivalent. The theory of representations of groups is concerned with the study of these objects.

Every homomorphism T from G into $\operatorname{Aut}_A(V)$ can be extended to a nonzero homomorphism of AG into $\operatorname{End}_A(V)$ by

$$T\left(\sum_{x \in G} \alpha_x x\right) = \sum_{x \in G} \alpha_x T(x) \quad \text{with } \alpha_x \in A.$$

Conversely, every nonzero homomorphism T from AG into $\operatorname{End}_A(V)$ determines a unique representation of G on V. Thus the study of representations of G and the study of nonzero representations of AG are equivalent.

EXERCISES

1. Suppose that a group G acts on a set S and that for each $s \in S$ and each subgroup H of G we put $\operatorname{Orbit}_H(s) := \{s^x \mid x \in H\}$. Let P be a Sylow p-subgroup of G. Show that for each subgroup H of G, $|\operatorname{Orbit}_H(s)|$ divides $|\operatorname{Orbit}(s)|$ for all $s \in S$. In particular, show that $|\operatorname{Orbit}_P(s)|$ is equal to the largest power of p dividing $|\operatorname{Orbit}(s)|$.
2. Let G act transitively on S. For each s in S, show that the normalizer $N_G(\operatorname{Stab}(s))$ of $\operatorname{Stab}(s)$ ($s \in S$) is transitive on the points fixed by $\operatorname{Stab}(s)$. Also show that the normalizer of a Sylow p-subgroup P of G acts transitively on the points fixed by P.
3. Let G be a group of order 108. By considering the action by conjugation of G on the Sylow 3-subgroups of G, show that G has a normal subgroup of order 9 or 27.

1.2 Reducible and irreducible modules

Let R be a ring and V an R-module. Then V is called *Noetherian* (or said to satisfy the "maximal condition on R-submodules") if every nonempty set Ω of R-submodules of V contains a submodule W_0 which is maximal in Ω in the sense that $W \in \Omega$ and $W \supseteq W_0$ implies $W = W_0$. Dually, V is *Artinian* (or satisfies the "minimal condition on R-submodules") if every nonempty set Ω of submodules of V contains a submodule W_0 which is minimal in Ω in the sense that $W \in \Omega$ and $W_0 \supseteq W$ implies $W = W_0$. The ring R itself is called Noetherian or Artinian, respectively, if the regular R-module R has these properties; in this case "submodule" is equivalent to "right ideal."

EXAMPLE 1 If R is an algebra over a field K, then R is a finitely generated K-module by definition. More generally, since (by convention) each R-module V is finitely generated over R, V is also finitely generated as a K-module. In other words, every R-module V is finite dimensional over K. Since every R-submodule of V is a K-submodule, it follows (by reasons of dimension) that V is both Noetherian and Artinian. In particular, R itself is both Noetherian and Artinian.

Lemma 1.2 (*Fitting's lemma*) Let U be an R-module and let $f \in \mathrm{End}_R(U)$.

(i) If U is Artinian and f is injective, then f is an R-automorphism.
(ii) If U is Noetherian and f is surjective, then f is an R-automorphism.
(iii) If U is both Artinian and Noetherian, then we can write $U = Uf^n \oplus \mathrm{Ker}\, f^n$ as a direct sum of submodules for some integer $n \geq 1$.

Proof (i) Since U is Artinian there exists an integer n such that $U \supset Uf \supset \cdots \supset Uf^n = Uf^{n+1}$. Then for each $u \in U$ there exists $v \in U$ such that $uf^n = vf^{n+1}$. Since f is injective, $u = vf \in Uf$. Hence f is surjective, and so an R-automorphism.

(ii) Since U is Noetherian there exists an integer n such that $0 \subset \mathrm{Ker}\, f \subset \cdots \subset \mathrm{Ker}\, f^n = \mathrm{Ker}\, f^{n+1}$. Since f (and hence f^n) is surjective, each $u \in \mathrm{Ker}\, f$ may be written in the form $u = vf^n$ for some $v \in U$. Since $uf = 0$, this shows that $v \in \mathrm{Ker}\, f^{n+1} = \mathrm{Ker}\, f^n$, and so $u = 0$. Thus $\mathrm{Ker}\, f = 0$, and so f is injective and hence an R-automorphism.

(iii) Since U is Artinian, $Uf^n = Uf^{n+1}$ for some integer $n \geq 1$. Then the restriction of f to the submodule Uf^n is surjective, and so by (ii) this restriction is also injective. Hence $Uf^n \cap \mathrm{Ker}\, f^n = 0$. On the other hand, since $Uf^n = Uf^{2n}$, for each $u \in U$ there exists $v \in U$ such that $(u - vf^n)f^n = 0$. But the latter shows that $u \in vf^n + \mathrm{Ker}\, f^n \subseteq Uf^n + \mathrm{Ker}\, f^n$. Hence $U = Uf^n \oplus \mathrm{Ker}\, f^n$ as asserted.

Let $V \neq 0$ be an R-module. Then V is *reducible* if it has a proper R-submodule $U \neq 0$; otherwise V is *irreducible*. (The zero module 0 is neither reducible nor irreducible.)

A *composition series* for an R-module V is a series of submodules of the form

(1) $$V = V_0 \supset V_1 \supset V_2 \supset \cdots \supset V_l = 0$$

such that for each $i \geq 1$, the factor module V_{i-1}/V_i is irreducible. The *Jordan–Hölder theorem* states that if a module V has a composition series (1), then every series

(2) $$V = U_0 \supset U_1 \supset \cdots \supset U_m = 0$$

1.2 REDUCIBLE AND IRREDUCIBLE MODULES

of submodules can be refined by inserting additional submodules where necessary to make a composition series and, moreover, if (2) is also a composition series, then $m = l$ and there is a permutation $i \mapsto n_i$ of $\{1, 2, \ldots, l\}$ such that $V_{i-1}/V_i \simeq U_{n_i-1}/U_{n_i}$ for $i = 1, 2, \ldots, l-1, l$. An irreducible R-module U is called a *composition factor* or an *irreducible constituent* of V if $U \cong V_{i-1}/V_i$ for some i; and it has *multiplicity j* if $U \cong V_{i-1}/V_i$ for exactly j values of i.

EXAMPLE 2   Not all R-modules have composition series. For example, the ring $\mathbf{Z}$ of integers as a $\mathbf{Z}$-module does not. In the case when R is an algebra over a field K, however, every R-module V does have a composition series. Indeed we saw in Example 1 that V is both Noetherian and Artinian. By the Noetherian condition we can find a sequence $V = V_0, V_1, V_2, \ldots$ of R-submodules of V such that V_i is maximal in V_{i-1}. By the Artinian condition the sequence terminates after a finite number of steps; hence for some $l \geq 0$, $V_l = 0$. The resulting series (1) is clearly a composition series.

Note   Suppose I is a maximal right ideal of the ring R. Then R/I is an irreducible R-module. Conversely, suppose V is an irreducible R-module, and choose any $v \neq 0$ in V. Since $v = v \cdot 1 \in vR$, and vR is an R-submodule of V, the irreducibility of V implies that $vR = V$. The mapping $r \mapsto vr$ is an R-homomorphism of R onto V with kernel I, say. Then I is the annihilator of the element v, and $R/I \simeq V$. Since V is irreducible, I is a maximal right ideal.

Finally, a basic property of irreducible modules is given by *Schur's lemma*. If U and V are irreducible modules over a ring R, then $\text{Hom}_R(U, V) = 0$ if U is not isomorphic to V. If U is isomorphic to V, then each nonzero mapping in $\text{Hom}_R(U, V)$ is bijective; in particular, $\text{End}_R(U)$ is a division ring.

[*Proof*   If $f \neq 0$ in $\text{Hom}_R(U, V)$, then $\text{Ker } f \neq U$. Since U is irreducible, $\text{Ker } f = 0$, and so the image $\text{Im } f \neq 0$. Hence $\text{Im } f = V$ because V is irreducible.]

EXERCISES

1. Let V be a submodule of an R-module U. Then prove that U is Noetherian (Artinian) iff V and U/V are Noetherian (Artinian).
2. Prove that a finite direct sum of Artinian (Noetherian) R-modules is Artinian (Noetherian).
3. Let G be the cyclic group of order 3 generated by a, $\mathbf{Q}$ the rational field, and V the vector space of 2-tuples of $\mathbf{Q}$. If V is a $\mathbf{Q}G$-module such that $(\alpha, \beta)a = (-\beta, \alpha - \beta)$, show that V is irreducible.

4. If in Exercise 3 the field **Q** is replaced by the complex field, show that V is reducible.

Note This example shows that the property of irreducibility depends on the underlying field.

5. Let G be the group generated by a, b where $a^2 = b^2 = (ab)^3 = 1$, **Q** the rational field, and V a **Q**G-module of dimension 3 over **Q**. If $\{v_1, v_2, v_3\}$ is a basis of V such that $(\alpha_1 v_1 + \alpha_2 v_2 + \alpha_3 v_3)a = \alpha_2 v_1 + \alpha_1 v_2 + \alpha_3 v_3$ and $(\alpha_1 v_1 + \alpha_2 v_2 + \alpha_3 v_3)b = \alpha_1 v_1 + \alpha_3 v_2 + \alpha_2 v_3$, show that V is reducible. Write out a composition series of V. Determine the irreducible constituents of V.

6. Let R be a ring and let U and V be Artinian and Noetherian R-modules. If $\text{Hom}_R(U, V) \neq 0$, prove that U and V have at least one common composition factor.

1.3 Semisimple rings and the Wedderburn structure theorem

Let R be a ring and Ω the set of all maximal right ideals of R. For any $I \in \Omega$, R/I is an irreducible R-module and $\text{Ann}(R/I)$ is an ideal of R. The *Jacobson radical* of R is defined to be rad $R := \bigcap_{I \in \Omega} \text{Ann}(R/I)$. It is readily verified that rad $R = \bigcap_{I \in \Omega} I$ and $\text{rad}(R/\text{rad } R) = 0$. Since rad $R \subseteq$ Ann (V) for every irreducible R-module, V can be considered as an irreducible $R/\text{rad } R$-module, and conversely. Thus R and $R/\text{rad } R$ have the "same" irreducible modules and irreducible representations.

Remark Because R is assumed to have a unity, a simple application of Zorn's lemma shows that $\Omega \neq \emptyset$. Note also that whenever U and V are isomorphic R-modules, then $\text{Ann}(U) = \text{Ann}(V)$. In particular, since each irreducible R-module V is isomorphic to R/I for some maximal right ideal I, rad R is the intersection of $\text{Ann}(V)$ taken over all irreducible R-modules V.

Lemma 1.3 Let R be an algebra over a field K, and let V be an irreducible R-module. Put $D := \text{End}_R(V)$ (so D is a division ring by Schur's lemma). Consider the representation $T: R \to \text{End}_K(V)$ afforded by V, and put $S := \text{Im } T$. The ring $\text{End}_K(V)$ has an R-module structure given by $ua := uT(a)$ for all $u \in \text{End}_K(V)$ and $a \in R$, and S is both a subring and submodule of $\text{End}_K(V)$.

(i) All R-composition factors of S are isomorphic to V.
(ii) S is a simple ring.
(iii) $S = \text{End}_D(V)$.

1.3 SEMISIMPLE RINGS AND THE WEDDERBURN STRUCTURE THEOREM

Proof (i) Put $d := \dim_K(V)$. Then $\text{End}_K(V) \simeq \text{Mat}(d, K)$ and it is readily seen that $\text{End}_K(V) = V_1 \oplus \cdots \oplus V_d$, where the V_j are R-submodules that are isomorphic to V (and correspond to the rows of the matrices). Thus all composition factors of $\text{End}_K(V)$ are isomorphic to V, and hence from the Jordan–Hölder theorem the same is true of the submodule S.

(ii) Let J be a proper ideal of R containing $\text{Ann}(V)$. Choose I as a maximal right ideal of R containing J. Since $R/\text{Ann}(V) \simeq S$, (i) shows that $R/I \simeq V$ as R-modules, and so $\text{Ann}(R/I) = \text{Ann}(V)$. But $RJ \subseteq J \subseteq I$, and so $J \subseteq \text{Ann}(R/I)$. Hence $J = \text{Ann}(V)$. This shows that R has no ideal lying properly between R and $\text{Ann}(V)$; and so S, which is isomorphic to $R/\text{Ann}(V)$, is simple.

(iii) Let U be a minimal right ideal in S. Then U is an R-composition factor of S, and so $U \simeq V$ as R-modules by (i). Thus, putting $D_0 := \text{End}_R(U)$ and $E := \text{End}_{D_0}(U)$, we have $D_0 \simeq D$ and $E \simeq \text{End}_D(V)$ as R-modules. We shall first prove that $S \simeq E$.

Consider U as an S-module, and let T_0 denote the representation of S afforded by U. By the definition of D_0, $\text{Im } T_0 \subseteq E$. We claim that $T_0(U)$ is a right ideal in E. To see this, consider the mappings $l_u \in \text{End}_K(U)$ for $u \in U$ defined by $vl_u := uv$ (left multiplication). Clearly $l_u \in D_0$ and so $l_u f = f l_u$ for all $f \in E$ and $u \in U$. Thus for all $u, v \in U$ and $f \in E$ we have

$$(uT_0(v))f = (uv)f = (vl_u)f = (vf)l_u = u(vf) = uT_0(vf).$$

Thus $T_0(v)f = T_0(vf) \in T_0(U)$ for all $v \in U$ and $f \in E$. Since $T_0(U)$ is closed under subtraction, this shows that $T_0(U)$ is a right ideal of E as claimed. Now $SU = S$ by (ii) because SU is a nonzero ideal in S. Therefore $\text{Im } T_0 = T_0(S)T_0(U)$, and so $\text{Im } T_0$ is also a right ideal in E. Since $\text{Im } T_0$ contains 1, $\text{Im } T_0 = E$. Finally, since S is simple, $\text{Ker } T_0 = 0$, and so $S \simeq \text{Im } T_0 = E$ as asserted.

Finally, it is clear that $S \subseteq \text{End}_D(V)$ by the definition of D. Since $S \simeq E \simeq \text{End}_D(V)$ as R-modules, comparison of the R-composition lengths shows that $S = \text{End}_D(V)$. This completes the proof of the lemma.

We shall now prove the basic structure theorem for an algebra over a field. Recall that a right ideal is *nil* if each of its elements is nilpotent.

Theorem 1.3A (*Wedderburn structure theorem*) Let R be an algebra over a field K. Put $\tilde{R} := R/\text{rad } R$. Then

(i) rad R is nilpotent and contains every nil right ideal of R.
(ii) $\tilde{R} = I_1 \oplus \cdots \oplus I_s$, where $I_1, \ldots, I_s$ are the minimal ideals of $\tilde{R}$.

(iii) For each j there is an irreducible $\tilde{R}$-module V_j such that all $\tilde{R}$-composition factors of I_j are isomorphic to V_j, $D_j := \mathrm{End}_{\tilde{R}}(V_j)$ is a division ring containing K in its center, and $I_j \simeq \mathrm{End}_{D_j}(V_j)$ as K-algebras.

Proof (i) Since R is an algebra it has an R-composition series, say, $R = R_0 \supset R_1 \supset \cdots \supset R_t = 0$. For each i, R_{i-1}/R_i is an irreducible R-module, and so $\mathrm{Ann}(R_{i-1}/R_i) \supset \mathrm{rad}\ R$. Thus $R_{i-1}(\mathrm{rad}\ R) \subseteq R_i$ for each i, and hence $(\mathrm{rad}\ R)^t \subseteq R(\mathrm{rad}\ R)^t = 0$. This shows that rad R is nilpotent.

Conversely, suppose that x is an element of some nil right ideal J of R; we have to show that $x \in \mathrm{rad}\ R$. Let V be any irreducible R-module. We shall show that $Vx = 0$. Suppose on the contrary that $Vx \neq 0$ and choose $v \in V$ so that $vx \neq 0$. Since vxR is a nonzero submodule of V, and V is irreducible, $vxR = V$. In particular, for some $a \in R$, $vxa = v$; and so by induction $v(xa)^n = v$ for all integers $n \geq 1$. Since $xa \in J$ is nil, the latter implies that $v = 0$ contrary to the choice of v. Thus $Vx = 0$ for each irreducible R-module V, and so $x \in \mathrm{rad}\ R$ as required.

(ii) As we saw in the last section, each irreducible R-module is isomorphic to an R-composition factor of R. Therefore, by the Jordan–Hölder theorem, there is a finite set $V_1, \ldots, V_s$ of nonisomorphic irreducible R-modules such that each irreducible R-module is isomorphic to one of them; and by the remark above, rad $R = \bigcap_{i=1}^{s} \mathrm{Ann}(V_i)$. Let T_i be the representation of R afforded by the module V_i, and define the K-algebra homomorphism

$$T: R \to \bigoplus_{i=1}^{s} \mathrm{End}_K(V_i) \quad \text{by} \quad T(a) := (T_1(a), \ldots, T_s(a)).$$

Since Ker $T_i = \mathrm{Ann}(V_i)$, Ker $T = \mathrm{rad}\ R$, and so $R/\mathrm{rad}\ R \simeq \mathrm{Im}\ T \subseteq S_1 \oplus \cdots \oplus S_s$, where $S_i := \mathrm{Im}\ T_i$. Since $S_i \simeq R/\mathrm{Ann}(V_i)$, it follows from Lemma 1.3(i) that the multiplicity of V_i as a composition factor in $S_1 \oplus \cdots \oplus S_s$ is no greater than its multiplicity as a composition factor in $R/\mathrm{rad}\ R$. Thus comparing R-composition lengths we conclude that Im $T = S_1 \oplus \cdots \oplus S_s$, and so $\tilde{R} \simeq S_1 \oplus \cdots \oplus S_s$ as K-algebras. Define I_i as the inverse image of S_i under this latter isomorphism. From the direct decomposition it is clear that each I_i is an ideal in $\tilde{R}$, and since S_i is simple by Lemma 1.3(ii), I_i is a minimal ideal. To complete the proof of (ii) it remains to show that $\tilde{R}$ has no other minimal ideals. However, if I is a minimal ideal of $\tilde{R}$, then $I = I\tilde{R} \subseteq II_1 + \cdots + II_s$, and so $II_j \neq 0$ for some j. But II_j is an ideal contained in both I and I_j; therefore the minimality of the latter ideals shows that $I = II_j = I_j$. Thus (ii) is proved.

(iii) As we noted at the beginning of this section, the structure of the V_j as R-modules and as $\tilde{R}$-modules is the same. Thus (iii) follows at once from the proof above together with Lemma 1.7. This completes the proof of the theorem.

1.3 SEMISIMPLE RINGS AND THE WEDDERBURN STRUCTURE THEOREM

Two R-modules U and V are said to be *distinct* if $U \not\simeq V$. A set $\{V_1, V_2, \ldots, V_s\}$ of irreducible R-modules is called a *full set* of irreducible modules if V_i and V_j are distinct for $i \neq j$ and if for any irreducible R-module V there is a j $(1 \leq j \leq s)$ such that $V \simeq V_j$. If U and V are distinct, the associated representations are said to be *inequivalent*; if $U \simeq V$, then the associated representations are *equivalent*.

Note 1 Under the hypothesis of Theorem 1.3A each irreducible $\tilde{R}$-module V (considered as an R-module) is isomorphic to one and only one V_j $(j = 1, 2, \ldots, s)$. [See the proof of part (ii).]

Definition Let K be a field and R an algebra over K. Then K is called a *splitting field* for R if for each irreducible R-module V, $\mathrm{End}_R(V) = K \cdot 1$ [the reason for the term "splitting" field will become clearer in §2.7]. In particular, a splitting field K of the group algebra KG is called a splitting field for G.

Note 2 If V is an irreducible R-module, then by Schur's lemma the K-algebra $\mathrm{End}_R(V)$ is a division ring.

Note 3 If, in Theorem 1.3A, K is a splitting field for R, then $D_j = K$, and so $I_j \simeq \mathrm{End}_K(V_j)$. Conversely, since each irreducible R-module is isomorphic to some V_j (Note 1), the conditions $D_j = K$ for all j imply that K is a splitting field.

Note 4 If K is algebraically closed, then it is a splitting field for every K-algebra R. For suppose $c \in \mathrm{End}_R(V)$ for some irreducible R-module V. Since K is algebraically closed, the K-linear mapping c has an eigenvalue $\gamma \in K$. Then $c - \gamma \cdot 1$ is a singular K-linear mapping in $\mathrm{End}_R(V)$. Since the latter is a division ring by Schur's lemma, $c - \gamma \cdot 1 = 0$; hence $c = \gamma \cdot 1$. This proves that $\mathrm{End}_R(V) = K \cdot 1$.

Corollary (*Burnside*) Under the hypothesis of Theorem 1.3A suppose that K is a splitting field and V is an irreducible R-module. Then the image of R under the representation of R associated with V is all of $\mathrm{End}_K(V)$.

Proof Immediate from Lemma 1.3(iii) since $D = K$ in this case.

A ring R is called *semisimple* if rad $R = 0$. If R is an algebra over a field K, it follows from Theorem 1.3A that R is semisimple iff R is a (finite) direct sum of simple rings.

An R-module V is called *completely reducible* if it can be written as a direct sum of irreducible submodules.

Note 5 If an R-module V can be written as a sum $V = V_1 + V_2 + \cdots + V_m$ of irreducible R-submodules, then it can be written as a direct sum of some subset of $\{V_1, V_2, \ldots, V_m\}$ and so is completely reducible. In fact, after suitable reordering we may suppose that $U = V_1 \oplus V_2 \oplus \cdots \oplus V_n$ is a direct sum for some $n \leq m$, but $U + V_j$ is not a direct sum for any $j > n$. This means that $U \cap V_j \neq 0$ for each j, and hence, since V_j is irreducible, $U \supseteq V_j$ for each j. Thus $U = V$ and we have $V = V_1 \oplus V_2 \oplus \cdots \oplus V_n$.

Note 6 If R is a semisimple algebra over a field K, then R is completely reducible; in other words, $R = \bigoplus_{i=1}^n I_i$ is a direct sum of minimal (that is, irreducible) right ideals I_i. By Theorem 1.3A, $R = \bigoplus_{i=1}^n S_i$, where S_i is a simple ideal of R. Since R is Artinian, S_i contains a minimal right ideal I, say, of R. For any r in R, rI is either 0 or a nonzero right ideal of R which is isomorphic to I and hence irreducible. Since S_i is an ideal, $rI \subseteq S_i$. Moreover, because R is Noetherian there exist $r_1, \ldots, r_m \in R$ such that $J := r_1 I + \cdots + r_m I$ contains the right ideal rI for all $r \in R$. Clearly J is a nonzero two-sided ideal of R contained in S_i; hence $J = S_i$ because S_i is simple. So S_i is a sum of irreducible R-submodules and thus is completely reducible by Note 5. Therefore R is completely reducible as an R-module.

Theorem 1.3B Let R be an algebra over a field K. Then

(i) R is semisimple iff every R-module is completely reducible; and

(ii) (*Maschke's theorem*) the group algebra $R = KG$ of a group G over K is semisimple iff the characteristic of K does not divide $|G|$.

Proof (i) Let R be semisimple and V an R-module. Then $R = \bigoplus_{i=1}^m I_i$, where each I_i is a minimal right ideal and $V = v_1 R + v_2 R + \cdots + v_n R$ for some $v_1, v_2, \ldots, v_n$ in V. Then

$$V = \sum_{i=1}^n v_i R = \sum_{i=1}^n \sum_{j=1}^m v_i I_j.$$

But either $v_i I_j = 0$ or $v_i I_j$ is isomorphic to I_j. Hence V is a sum of irreducible R-modules. By Note 5, V is completely reducible. Conversely, if every R-module is completely reducible, then in particular R is completely reducible; let $R = \bigoplus_{i=1}^n I_i$, where each I_i is irreducible. If J_i is the sum of all I_j except I_i, then J_i is a maximal right ideal of R and $\bigcap_{i=1}^n J_i = 0$. Hence rad $R = 0$, and so R is semisimple.

(ii) Suppose that $R = KG$ is semisimple. Then the element $r := \sum_{x \in G} x$ belongs to the center of KG and $r^2 = r|G|$. Thus if the characteristic of K divided $|G|$, then r would be a nilpotent element in the center of KG, and hence rR would be a nonzero nilpotent ideal. Since rad $R = 0$ this is impossible by Theorem 1.3A. Hence the characteristic of K does not divide $|G|$.

Conversely, suppose KG is not semisimple and choose $\sum_{x \in G} \alpha_x x \neq 0$ in rad R. Multiplying by a suitable element of KG, we may assume that $c = 1 + \sum_{x \neq 1} \gamma_x x$ belongs to rad R. Consider the K-linear transformation $t \in \text{End}_K(KG)$ defined by $vt := vc$ (right multiplication by c). Since

1.4 TENSOR PRODUCTS

$c \in \operatorname{rad}(KG)$, $c^n = 0$ for some integer $n \geq 1$ by part (i) of Theorem 1.3A. This implies that $t^n = 0$, and so all eigenvalues of t are 0. Therefore $\operatorname{tr} t = 0$. But the matrix of t relative to the basis x ($x \in G$) clearly has all its diagonal entries equal to 1, so $\operatorname{tr} t = |G| \cdot 1$. Thus $|G| \cdot 1 = 0$ in K, which implies that the characteristic of K is a factor of $|G|$. This completes the proof of the theorem.

EXERCISES

1. Let U be a completely reducible R-module and let $U = U_1 \oplus U_2 \oplus \cdots \oplus U_n$, where each U_i is a sum of isomorphic irreducible R-modules and U_i and U_j have no isomorphic irreducible constituent in common for $i \neq j$. Prove that $\operatorname{End}_R(U) \simeq \operatorname{End}_R(U_1) \oplus \operatorname{End}_R(U_2) \oplus \cdots \oplus \operatorname{End}_R(U_n)$ as rings.

2. Let G be a group generated by an element a of order 3, K a field of characteristic 3, and V a KG-module of dimension 2 over K. If $\{v_1, v_2\}$ is a basis of V such that $(\alpha_1 v_1 + \alpha_2 v_2)a = -\alpha_2 v_1 + (\alpha_1 - \alpha_2)v_2$, show that V is reducible, but not completely reducible.

1.4 Tensor products

We shall assume a basic knowledge of tensor products of modules and algebras and their most elementary properties (references to this material are given in §1.10). In all cases in which we are interested, the modules and algebras with which we shall deal will be free over some underlying commutative ring or field, and in this case the tensor product construction is especially simple. The following examples are essentially the cases that we shall need.

EXAMPLE 1 Let AG be the group algebra of a group G over a commutative ring A, and let U and V be AG-modules. Then V may also be considered as a left A-module, and so we have the tensor product $U \otimes_A V$ as an A-module. If U and V are free A-modules with A-bases $u_1, u_2, \ldots, u_m$ and $v_1, v_2, \ldots, v_n$, respectively, then $U \otimes_A V$ is a free A-module with an A-basis $u_i \otimes v_j$ ($i = 1, 2, \ldots, m$, $j = 1, 2, \ldots, n$) and an action of G on $U \otimes_A V$ is defined in terms of its action on U and V by

$$(u_i \otimes v_j)x := u_i x \otimes v_j x$$

for all i, j and $x \in G$, and then extended to $U \otimes_A V$ and KG by linearity. The AG-module $U \otimes_A V$ is called the (inner) tensor product of U and V.

EXAMPLE 2 Let K be a subfield of the field L, and let V be a K-module (that is, a finite dimensional vector space over K). Then $V \otimes_K L$ is an L-module. If $v_1, v_2, \ldots, v_n$ form a K-basis of V, then $V \otimes_K L$ has $v_1 \otimes 1, v_2 \otimes 1, \ldots, v_n \otimes 1$ as an L-basis; in particular, $\dim_K V = \dim_L(V \otimes_K L)$. If KG is the group algebra of a group G over K, and V is a KG-module, then $V \otimes_K L$ is an LG-module. The action of G on $V \otimes_K L$ is given by $(v_i \otimes 1)x := v_i x \otimes 1$ for all i and all x in G.

EXAMPLE 3 Let K be a field and A a subring of K such that K is the field of quotients of A. Let AG be the group algebra of a group G over A, and let V be an AG-module. Then $V \otimes_A K$ is a KG-module. If V is a free A-module with an A-basis $v_1, v_2, \ldots, v_n$, then $V \otimes_A K$ has the K-basis $v_1 \otimes 1$, $v_2 \otimes 1$, $\ldots, v_n \otimes 1$, and the action of G on $V \otimes_A K$ is defined by $(v_i \otimes 1)x := v_i x \otimes 1$ for all i and all $x \in G$. In particular, $\mathrm{rank}_A V = \dim_K(V \otimes_A K)$ in the case V is A-free.

Note In these last two examples L need not be finite dimensional over K, nor K finitely generated over A, so they need not be modules by our usual convention.

EXAMPLE 4 Let A be a commutative ring and G a group with subgroup H. Let $\{x_1, x_2, \ldots, x_m\}$ be a right transversal for H in G, so $Hx_1, Hx_2, \ldots, Hx_m$ are the distinct right cosets of H in G. Since we may consider AH as a subring of AG in a natural way, AG can be considered as a (free) left AH-module. Therefore for any AH-module V we can define $V \otimes_{AH} AG$, and the latter is an AG-module. If V is free as an A-module with an A-basis $v_1, v_2, \ldots, v_n$, then $V \otimes_{AH} AG$ is free as an A-module with A-basis $v_i \otimes x_j$ ($i = 1, 2, \ldots, n; j = 1, 2, \ldots, m$). In this case the action of G on $V \otimes_{AH} AG$ is given by $(v_i \otimes x_j)x = v_i z \otimes x_l$, where x_l is the coset representative defined by $x_j x \in Hx_l$ and $z \in H$ is defined by $zx_l := x_j x$ ($i = 1, \ldots n; j = 1, 2, \ldots, m$; and all $x \in G$). The AG-module $V \otimes_{AH} AG$ is called the *induced module*. This example will be very important later when we deal with induced modules.

EXAMPLE 5 Let A be a commutative ring and I an ideal of A. Then A/I is a two-sided A-module. Let G be a group and V an AG-module. Then $V \otimes_A (A/I)$ is defined and is an AG-module. The annihilator of this module contains the ideal IG consisting of all linear combinations of elements of G with coefficients in I. Since $AG/IG \simeq (A/I)G$, the group ring over A/I, we can consider $V \otimes_A (A/I)$ as an $(A/I)G$-module. In particular, if V is A-free with an A-basis $v_1, v_2, \ldots, v_n$, then $V \otimes_A (A/I)$ is (A/I)-free with (A/I)-basis $v_1 \otimes 1, v_2 \otimes 1, \ldots, v_n \otimes 1$.

EXAMPLE 6 Let G be a group and K a field. Let U and V be KG-modules. The *dual* U^* of U is defined as follows. As a K-module $U^* := \mathrm{Hom}_K(U, K)$, the K-space of linear functionals on U; in particular $\dim_K U^* = \dim_K U$. The action of G on U^* is defined by $u(fx) := (ux^{-1})f$ for all $u \in U, f \in U^*$, and $x \in G$. This is clearly a linear action and so defines a KG-module structure on U^*, called the dual of U. Next we describe a K-isomorphism of $U^* \otimes_K V$ onto $\mathrm{Hom}_K(U, V)$ as follows. If $f_1, f_2, \ldots, f_m$ is a basis of U^* and v_1,

$v_2, \ldots, v_n$ is a basis of V over K, then $f_i \otimes v_j$ ($i = 1, 2, \ldots, m; j = 1, 2, \ldots, n$) is a basis of $U^* \otimes V$ over K. Define $h_{ij}: U \to V$ by $uh_{ij} := (uf_i)v_j$ for all i, j and $u \in U$ (note that $uf_i \in K$). Then $h_{ij} \in \mathrm{Hom}_K(U, V)$, and the h_{ij} are linearly independent since if $\alpha_{ij} \in K$ then $\sum_{i,j} \alpha_{ij} h_{ij} = 0$ implies

$$\sum_j \left(\sum_{i=1}^m \alpha_{ij}(uf_i) \right) v_j = 0$$

for all $u \in U$, and so

$$0 = \sum_{i=1}^m \alpha_{ij} uf_i = u \sum_{i=1}^m \alpha_{ij} f_i$$

for all $u \in U$ and all j. But the latter implies

$$\sum_{i=1}^m \alpha_{ij} f_j = 0$$

for all j, and so all $\alpha_{ij} = 0$. This shows that the h_{ij} are linearly independent and hence form a K-basis of $\mathrm{Hom}_K(U, V)$ since $\dim_K \mathrm{Hom}_K(U, V) = mn$. Thus the mapping $f_i \otimes v_j \mapsto h_{ij}$ ($i = 1, 2, \ldots, m; j = 1, 2, \ldots, n$) defines a K-linear mapping which is a K-isomorphism of $U^* \otimes V$ onto $\mathrm{Hom}_K(U, V)$. Finally, if we define the action of G on $\mathrm{Hom}_K(U, V)$ by $h_{ij}^x := x^{-1} h_{ij} x$, then it can be seen that the mapping above gives a KG-isomorphism of $U^* \otimes V$ onto $\mathrm{Hom}_K(U, V)$ with this KG-module structure.

EXERCISES

1. Let H be a subgroup of a group G, K a field, and V a KH-module. If $\mathrm{Ker}\, V = 1$, prove that $\mathrm{Ker}(V \otimes_{KH} KG) = 1$.
2. Let G_1, G_2 be finite groups and let K be a field. Then prove that $K(G_1 \times G_2) \simeq KG_1 \otimes_K KG_2$ as K-algebras.
3. Let R be an algebra over a field K, and let $\mathrm{Mat}(n, R)$ be the algebra of $n \times n$ matrices over K. Prove that $\mathrm{Mat}(n, R) \simeq \mathrm{Mat}(n, K) \otimes_K R$ as K-algebras.

1.5 The number of irreducible KG-modules

In one sense the theory of representations of a group over a field K is the classification of KG-modules to within isomorphism. In the case that KG is semisimple, every KG-module is a direct sum of irreducible KG-modules, and then the problem reduces to a classification of the irreducible KG-modules (see §1.3). If KG is not semisimple (and this is really the case in which we shall be interested) the situation is more complicated, but it is still of basic interest to classify the irreducible KG-modules even though this is only the beginning of a solution of the general problem. In the present section we shall consider the irreducible KG-modules when K is a splitting field for KG.

Lemma 1.5A *(Binomial theorem)* Let K be a field of characteristic $p > 0$ and let R be an algebra over K. Define S as the K-subspace of R spanned by the set of all ring commutators $ab - ba$ $(a, b \in R)$ and write $u \equiv v \pmod{S}$ for $u, v \in R$ when $u - v \in S$. Then

(i) $(x + y)^p \equiv x^p + y^p \pmod{S}$ for all $x, y \in R$; and
(ii) $(x + y)^p = x^p + y^p$ whenever $x, y \in R$ and $xy = yx$.

Proof Suppose $1 \le i \le p - 1$. Then any two products consisting of i factors equal to x and $p - i$ factors equal to y are congruent (mod S). There are $\binom{p}{i} := p!/i!(p-i)!$ of these products, and since $\binom{p}{i} \equiv 0 \pmod{p}$, we conclude that the sum of these products (for fixed i) is in S. Since this is true for each i, $1 \le i \le p - 1$, it follows that $(x + y)^p \equiv x^p + y^p \pmod{S}$. This proves (i), and (ii) then follows if we replace R by the commutative subalgebra generated by 1, x, and y, since then S becomes 0.

Lemma 1.5B Let G be a group and K a splitting field for KG. Let S be the K-subspace of KG spanned by all $ab - ba$ $(a, b \in KG)$.

(i) If K has characteristic $p > 0$, then $T := \{c \in KG \mid c^{p^m} \in S$ for some integer $m \ge 1\}$ is a K-subspace of KG, and the number of minimal ideals of $KG/\text{rad } KG$ [see Theorem 1.3A(ii)] equals $\dim_K(KG/T)$.

(ii) If K has characteristic 0, then the number of minimal ideals of KG equals $\dim_K(KG/S)$.

Proof (i) First note that $c \in S$ implies $c^{p^m} \in S$ for all $m \ge 1$. Indeed by Lemma 1.5A, $(ab - ba)^p \equiv (ab)^p - (ba)^p = a\{(ba)^{p-1}b\} - \{(ba)^{p-1}b\}a \equiv 0 \pmod{S}$. Hence an inductive argument using Lemma 1.5A shows that $c^{p^m} \in S$ for all $c \in S$. In particular, $S \subseteq T$.

Next we see from Lemma 1.5A that $(a + b)^{p^m} \equiv (a^p + b^p)^{p^{m-1}} \equiv \cdots \equiv a^{p^m} + b^{p^m} \pmod{S}$ for all $m \ge 1$. Thus if $a, b \in T$, then for sufficiently large m, $(a + b)^{p^m} \equiv a^{p^m} + b^{p^m} \in S$; hence $a + b \in T$. Clearly $\alpha a \in T$ for all $\alpha \in K$ and $a \in T$; therefore we conclude that T is a K-subspace of KG. Since $\text{rad}(KG)$ is nilpotent [Theorem 1.3A(i)], it follows from the definition of T that $\text{rad}(KG) \subseteq T$.

Now write $R := KG$ and put $\tilde{R} := R/\text{rad } R$ and $\tilde{T} := T/\text{rad } R$; note that $\dim_K(KG/T) = \dim_K(\tilde{R}/\tilde{T})$. In accordance with Theorem 1.3A we write $\tilde{R} = I_1 \oplus I_2 \oplus \cdots \oplus I_r$, where I_j are minimal ideals of $\tilde{R}$. Since K is splitting for R, $I_j \simeq \text{End}_K(V_j)$, where V_j is a minimal right ideal of I_j (see Note 3 of §1.3). Hence $I_j \simeq \text{Mat}(d_j, K)$ for some integer $d_j := \dim_K V_j \ge 1$.

1.5 THE NUMBER OF IRREDUCIBLE KG-MODULES

Put $\tilde{S} = (S + \text{rad } R)/\text{rad } R$. Then clearly $\tilde{S}$ is the K-subspace of $\tilde{R}$ spanned by all $ab - ba$ $(a, b \in \tilde{R})$ and $\tilde{T} = \{c \in \tilde{R} \mid c^{p^m} \in \tilde{S}$ for some integer $m \geq 1\}$. Let S_i, T_i be the analogous K-subspaces defined for the K-algebra I_i. Since $\tilde{R}$ is a direct sum of the I_i, it is easily seen that $S_i = I_i \cap \tilde{S}$ and $T_i = I_i \cap \tilde{T}$ for each i, and as we saw above, $S_i \subseteq T_i$.

We claim that $\dim_K(I_i/T_i) = 1$. Indeed, since $I_i \simeq \text{Mat}(d_i, K)$ we can carry out the calculations in the latter ring. Let $e_{st} \in I_i$ correspond to the $d_i \times d_i$ matrix with its (s, t)th entry equal to 1 and all the other entries 0. Then for any $s \neq t$, $e_{st} = e_{sj}e_{jt} - e_{jt}e_{sj} \in S_i$ and $e_{ss} - e_{tt} = e_{st}e_{ts} - e_{ts}e_{st} \in S_i$. From this it follows that S_i contains the $d_i^2 - 1$ linearly independent elements e_{st} (s, $t = 1, 2, \ldots, d_i$ with $s \neq t$) and $e_{ss} - e_{11}$ ($s = 2, \ldots, d_i$). Since $S_i \subseteq T_i$, this shows that $\dim_K(I_i/T_i) = \dim_K(I_i) - \dim_K(T_i) \leq d_i^2 - (d_i^2 - 1) = 1$. On the other hand, the elements of S_i are linear combinations of elements of the form $ab - ba$ and so correspond in $\text{Mat}(d_i, K)$ to elements of trace 0. Since e_{11} does not correspond to an element of trace 0, $e_{11} \notin S_i$. However $e_{11}^{p^m} = e_{11}$ for all $m \geq 1$, so $e_{11} \notin T_i$. Therefore $T_i \neq I_i$, and so from $\dim_K(I_i/T_i) \leq 1$ we conclude $\dim_K(I_i/T_i) = 1$ as asserted.

Finally, $\dim_K(KG/T) = \dim_K(\tilde{R}/\tilde{T}) = \sum_{i=1}^{r} \dim_K(I_i/T_i) = r$, and part (i) of the lemma is proved.

(ii) The proof is analogous to the last part of (i) taking S_i in place of T_i.

Let x be an element of a group G and let p be a prime. Then x is called a *p-element* (or "*p-singular*") if its order is a power of p, and x is called a *p'-element* (or "*p-regular*") if its order is relatively prime to p (so 1 is the unique element which is both a p-element and a p'-element). In general, suppose that x has order $p^m n$ ($p \nmid n$). Since p^m and n are relatively prime there exist integers r and s such that $rp^m + sn = 1$. If we put $x_p := x^{sn}$ and $x_{p'} := x^{rp^m}$, then we have

(1) $$x = x_p x_{p'} = x_{p'} x_p,$$

where x_p is a p-element and $x_{p'}$ is a p'-element. We call x_p and $x_{p'}$ the *p-part* and *p'-part* of x, respectively.

Note The elements x_p and $x_{p'}$ are completely determined by the conditions given in (1). Indeed, if y is a p-element and z is a p'-element such that $x = yz = zy$, then y and z commute with x and so commute with x_p and $x_{p'}$. Therefore $x_p^{-1}y$ is a p-element and $x_{p'}z^{-1}$ is a p'-element. However from (1) these are equal and so must both equal 1. Hence $y = x_p$ and $z = x_{p'}$.

Theorem 1.5 Let G be a group and K a splitting field for KG. Let r be the number of distinct irreducible KG-modules (up to isomorphism). Then

(i) r equals the number of conjugate classes of G if K has characteristic 0;

(ii) r equals the number of conjugate classes of p'-elements of G if K has characteristic $p > 0$.

Proof Let S be the K-subspace of KG spanned by all $ab - ba$ ($a, b \in KG$). Then it is readily verified that S is also spanned by all $xy - yx$ ($x, y \in G$), and hence that an element $a = \sum_{x \in G} \alpha_x x$ ($\alpha_x \in K$) of KG lies in S if and only if $\sum_{x \in \mathscr{C}} \alpha_x = 0$ for each conjugate class $\mathscr{C}$ of G.

It follows from Note 1 of §1.3 that the number of irreducible KG-modules (up to isomorphism) is equal to the number of minimal ideals of $KG/\text{rad } KG$. Therefore with the notation of Lemma 1.5B, $r = \dim_K(KG/S)$ when K has characteristic 0, and $r = \dim_K(KG/T)$ when K has characteristic $p > 0$.

(i) (K *has characteristic* 0). Let $\mathscr{C}_1, \mathscr{C}_2, \ldots, \mathscr{C}_t$ be the conjugate classes of G and let K^t denote the vector space of all t-tuples over K. Then there is a K-linear mapping ϕ from KG into K^t defined by

$$\sum_{x \in G} \alpha_x x \mapsto (\beta_1, \beta_2, \ldots, \beta_t) \quad \text{where} \quad \beta_i := \sum_{x \in \mathscr{C}_i} \alpha_x.$$

Clearly ϕ is surjective and its kernel is S by the observation at the beginning of this proof. Hence $r = \dim_K(KG/S) = \dim_K K^t = t$ as asserted.

(ii) (K *has characteristic* $p > 0$). Let $a_1, a_2, \ldots, a_l$ be a K-basis of T. Choose the integer m so large that $a_i^{p^m} \in S$ for each i and p^m is at least as large as the order of a Sylow p-subgroup of G. By Lemma 1.5A we have

$$\left(\sum_{i=1}^{l} \gamma_i a_i\right)^{p^m} \equiv \sum_{i=1}^{l} \gamma_i^{p^m} a_i^{p^m} \equiv 0 \pmod{S}$$

for all $\gamma_i \in K$. Therefore $a^{p^m} \in S$ for all $a \in T$. Now let $\sum_{x \in G} \alpha_x x \in KG$. Then by Lemma 1.5A we have

$$(2) \qquad \left(\sum_{x \in G} \alpha_x x\right)^{p^m} \equiv \sum_{x \in G} \alpha_x^{p^m} x^{p^m} \pmod{S}.$$

Let $\mathscr{C}_1, \mathscr{C}_2, \ldots, \mathscr{C}_s$ denote the conjugate classes of p'-elements of G, and put $\mathscr{C}_i^* := \{x \in G \mid x^{p^m} \in \mathscr{C}_i\}$ ($i = 1, 2, \ldots, s$). Then $\mathscr{C}_1^*, \mathscr{C}_2^*, \ldots, \mathscr{C}_s^*$ is a partition of G and it follows from (2) that

$$\sum_{x \in G} \alpha_x x \in T \quad \text{iff} \quad \left(\sum_{x \in \mathscr{C}_i^*} \alpha_x\right)^{p^m} = \sum_{x \in \mathscr{C}_i^*} \alpha_x^{p^m} = 0$$

for $i = 1, 2, \ldots, s$ (using Lemma 1.5A). Thus $\sum_{x \in G} \alpha_x x \in T$ iff $\sum_{x \in \mathscr{C}_i^*} \alpha_x = 0$ for each i. Finally, define ψ from KG into K^s as the K-linear mapping given by $\sum_{x \in G} \alpha_x x \mapsto (\beta_1, \beta_2, \ldots, \beta_s)$, where $\beta_i := \sum_{x \in \mathscr{C}_i^*} \alpha_x$. Then

1.5 THE NUMBER OF IRREDUCIBLE KG-MODULES

ψ is clearly surjective and T is the kernel of ψ. Thus

$$r = \dim_K(KG/T) = \dim_K K^s$$

and so $s = r$ as asserted.

Corollary Under the hypothesis of the theorem,

(i) if G is a p-group and K has characteristic p, then the only irreducible KG-module is the trivial module;

(ii) if G is abelian, then each irreducible KG-module has dimension 1;

(iii) if the characteristic of K does not divide $|G|$, then there are (up to isomorphism) exactly $|G:G'|$ irreducible KG-modules of dimension 1 (G' denotes the commutator group of G).

Proof (i) By the theorem $r = 1$.

(ii) Let V be an irreducible KG-module. The image of G under the representation associated with V maps G into $\text{End}_{KG}(V)$ because G is commutative. Since K is splitting for KG, $\text{End}_{KG}(V) = K \cdot 1$, and so each $x \in G$ acts as a scalar on V. Thus $\text{End}_K(V) \simeq \text{End}_{KG}(V) \simeq K \cdot 1$, and so $\dim_K V = 1$.

(iii) In the case where G is abelian ($G' = 1$), G has $|G|$ conjugate classes. Hence by the theorem there are $|G|$ irreducible KG-modules, and then each has dimension 1 by (ii). In general, if V is a KG-module of dimension 1, then each $x \in G$ acts as a scalar on V. Thus $G/\text{Ker } V$ is an abelian group, and hence $\text{Ker } V \supseteq G'$. Conversely, since G/G' is an abelian group we have just seen that there are $|G/G'|$ irreducible $K(G/G')$-modules of dimension 1. On each of these modules V we can define the action of G by $vx := v(G'x)$ ($v \in V$, $x \in G$, and $G'x \in G/G'$). This gives exactly $|G:G'|$ KG-modules of dimension 1. Since each KG-module of dimension 1 has G' in its kernel, it must be isomorphic to one of these.

EXAMPLE Let K be a field of characteristic $p > 0$, and G a group. Let V be a KG-module such that each composition factor of V is trivial. Then $G/\text{Ker } V$ is a p-group. Indeed, let $V = V_0 \supset V_1 \supset \cdots \supset V_m = 0$ be a composition series of V. By hypothesis V_{i-1}/V_i is trivial for $i = 1, 2, \ldots, m$; that is, $vx + V_i = v + V_i$ for all $v \in V_{i-1}$ and $x \in G$. Hence $V_{i-1}(x-1) \subseteq V_i$ for each i, and so $V(x-1)^m \subseteq V_1(x-1)^{m-1} \subseteq \cdots \subseteq V_m = 0$ for each $x \in G$.

Choose n so that $p^n \geq m$. Since K has characteristic p, the binomial theorem (Lemma 1.5A) shows that $(x-1)^{p^n} = x^{p^n} - 1$ for all $x \in G$, and so $V(x^{p^n} - 1) = V(x-1)^{p^n} = 0$. Hence for all $v \in V$ and all $x \in G$, $v(x^{p^n} - 1) = 0$; that is $x^{p^n} \in \text{Ker } V$. Thus $G/\text{Ker } V$ is a p-group as asserted.

EXERCISES

1. Let G be the group of order 10 generated by a, b satisfying the relations $a^2 = b^5 = (ab)^2 = 1$. Determine the number of distinct irreducible KG-

modules over a splitting field K of (i) characteristic 0, (ii) characteristic 2, (iii) characteristic 3, and (iv) characteristic 5. Also determine the one-dimensional irreducible KG-modules, where K is a splitting field of characteristic 0.

2. Let G be the symmetric group on $\{1, 2, 3, 4\}$ and let $\mathbf{Q}$ be the rational field. Set $a = (12)$, $b = (23)$, and $c = (34)$. Let U and V be $\mathbf{Q}G$-modules which afford the irreducible matrix representations T and S, respectively, with respect to some bases, where

$$T(a) = \begin{bmatrix} 1 & 0 & 0 \\ 0 & 1 & 3 \\ 0 & 0 & -1 \end{bmatrix}, \quad T(b) = \begin{bmatrix} 1 & 4 & 4 \\ 0 & -2 & -3 \\ 0 & 1 & 2 \end{bmatrix},$$

$$T(c) = \begin{bmatrix} -3 & -8 & -12 \\ 1 & 3 & 3 \\ 0 & 0 & 1 \end{bmatrix}$$

and

$$S(a) = \begin{bmatrix} 1 & 3 & 8 \\ 0 & -1 & 0 \\ 0 & 0 & -1 \end{bmatrix}, \quad S(b) = \begin{bmatrix} -2 & -3 & -8 \\ 1 & 2 & 8 \\ 0 & 0 & -1 \end{bmatrix},$$

$$S(c) = \begin{bmatrix} -1 & -1 & -4 \\ 0 & -3 & -8 \\ 0 & 1 & 3 \end{bmatrix}.$$

Show that T and S are inequivalent. Determine other distinct irreducible matrix representations of G over $\mathbf{Q}$. (It can be shown that $\mathbf{Q}$ is a splitting field for G.)

1.6 Indecomposable modules

Let R be any ring. Then an R-module $U \neq 0$ is called *decomposable* if it can be written as a direct sum $U = V \oplus W$ of nonzero R-submodules V and W; otherwise U is called *indecomposable*. (The zero module is not classified as being either decomposable or indecomposable.)

EXAMPLE Any irreducible R-module is indecomposable, but as we shall see later, the converse is not usually true. However, if R is a semisimple algebra over a field, then every R-module is completely reducible (Theorem 1.3B); therefore in this case an R-module is irreducible iff it is indecomposable.

A direct sum decomposition of an R-module U into a sum $U = U_1 \oplus U_2 \oplus \cdots \oplus U_n$ of R-submodules may be described by n projection homomorphisms f_i from U to U_i and n injection homomorphisms g_i from U_i to U

1.6 INDECOMPOSABLE MODULES

by requiring $g_i f_j = 0$ for $i \neq j$, $g_i f_i = 1$ on U_i for all i, and $\sum_{i=1}^{n} f_i g_i = 1$ on U. We observe that $f_i^2 = f_i$ for all i and $f_i \in \mathrm{End}_R(U)$.

Lemma 1.6A Let U_1, U_2, V_1, and V_2 be R-modules, α be an isomorphism from U_1 to V_1, and β be an isomorphism from $U_1 \oplus U_2$ to $V_1 \oplus V_2$ such that $u\beta \in u\alpha + V_2$ for all $u \in U_1$. Then $U_2 \simeq V_2$.

Proof Let f_1 and f_2 be the canonical projections from $V_1 \oplus V_2$ onto V_1 and V_2, respectively. Define the R-homomorphism $\mu : U_1 \oplus U_2 \to V_1 \oplus V_2$ by $\mu := \beta(1 - f_1 \alpha^{-1} \beta f_2)$. We first show that μ is an isomorphism. Indeed, if $u \in U_1$ then $u\beta f_1 = u\alpha$ by the condition on β. Therefore $u\mu = u\beta - u\beta f_2 = u\beta f_1$, and hence $U_1 \mu = V_1$. On the other hand, since $V_2 f_1 = 0$, $(V_2 \beta^{-1})\mu = V_2(1 - f_1 \alpha^{-1} \beta f_2) = V_2$. Thus $V_1 \oplus V_2 \subseteq \mathrm{Im}\ \mu$ and so μ is surjective. On the other hand, if $w \in \mathrm{Ker}\ \mu$ then $w\beta = (w\beta) f_1 \alpha^{-1} \beta f_2$. This shows first that $w\beta \in V_2$, and then that $w\beta = 0$ because $V_2 f_1 = 0$. Since β is invertible, $\mathrm{Ker}\ \mu = 0$. This proves that μ is also injective, and hence is an isomorphism as asserted. Finally, since $U_1 \mu = V_1$ from above, we have

$$U_2 \simeq (U_1 \oplus U_2)/U_1 \simeq (U_1 \oplus U_2)\mu/U_1\mu \simeq (V_1 \oplus V_2)/V_1 \simeq V_2.$$

This proves the lemma.

An element e in a ring R is called an *idempotent* if $e \neq 0$ and $e^2 = e$. A ring R is called *local* if the sum of any two nonunits in R is a nonunit.

Lemma 1.6B Let R be a ring.

(i) R is local iff rad R is the unique maximal right ideal of R.
(ii) If R is local, then 1 is the unique idempotent in R.

Proof (i) First suppose that R is local and let I be a maximal right ideal in R. Let J be any proper right ideal in R. Since I and J consist of nonunits of R, $I + J \neq R$; and since I is maximal, this shows that $J \subseteq I$. Thus I is the only maximal right ideal of R. Since the intersection of the maximal right ideals of any ring is equal to the radical, rad $R = I$ is the unique maximal right ideal.

Conversely, suppose that rad R is the unique maximal right ideal of R. Since every nonunit of R generates a proper right ideal, rad R contains all the nonunits of R and therefore is precisely the set of nonunits. Thus R is local.

(ii) It follows immediately from the definition that (in any ring) 1 is the only idempotent that is also a unit. Now suppose that e is an idempotent in a local ring R. From the relation $e + (1 - e) = 1$ we deduce that either e or $1 - e$ is a unit. Since both e and $1 - e$ are idempotents, the relation $e(1 - e) = 0$ shows that either $e = 0$ or $1 - e = 0$. Since e is an idempotent, $e \neq 0$, and so $e = 1$. This completes the proof of the lemma.

We now have the following basic theorem on direct sum decomposition of R-modules.

Theorem 1.6A (*Krull–Schmidt–Azumaya*) Let R be a ring and V be a nonzero R-module such that

$$V = \bigoplus_{i=1}^{m} U_i = \bigoplus_{j=1}^{n} V_j$$

for some submodules U_i and V_j. Suppose that for each i and j the rings $\operatorname{End}_R(U_i)$ and $\operatorname{End}_R(V_j)$ are local. Then $m = n$ and for some permutation $i \mapsto n_i$ of $\{1, 2, \ldots, n\}$ we have $U_i \simeq V_{n_i}$ for each i.

Proof We may assume that $m \geq n$, and use induction on m. For $m = 1$ the result is trivial, so suppose $m > 1$. Let f_i be the canonical projection from V onto U_i, f'_i the projection from V onto V_i, g_i the injection of U_i into V, and g'_i the injection of V_i into V. Then $g_i f_j = 0$ for $i \neq j$, $g_i f_i = 1$ on U_i, $\sum_{i=1}^{m} f_i g_i = 1$ on V, $g'_i f'_j = 0$ for $i \neq j$, $g'_i f'_i = 1$ on V_i, and $\sum_{i=1}^{n} f'_i g'_i = 1$ on V.

Put $\alpha_i := g_i f'_1 \in \operatorname{Hom}_R(U_i, V_1)$ and $\beta_i := g'_1 f_i \in \operatorname{Hom}_R(V_1, U_i)$. Then each $\beta_i \alpha_i \in \operatorname{End}_R(V_1)$ and

$$\sum_{i=1}^{m} \beta_i \alpha_i = g'_1 \left(\sum_{i=1}^{m} f_i g_i \right) f'_1 = 1.$$

Since $\operatorname{End}_R(V_1)$ is local, at least one of the $\beta_i \alpha_i$ is a unit, and hence an automorphism of V_1. Without loss in generality we may suppose that $\beta_1 \alpha_1$ is an automorphism; put $f := (\beta_1 \alpha_1)^{-1} \beta_1$ and $g := \alpha_1$. Then $fg = 1$ on V_1 and gf is an idempotent of $\operatorname{End}_R(U_1)$ since $g(fg)f = gf \neq 0$. Hence by Lemma 1.6B, $gf = 1$ on U_1, and so g is an isomorphism from U_1 onto V_1. Now it follows from Lemma 1.6A (taking β as the identity mapping on V and g for α) that

$$U_2 \oplus \cdots \oplus U_m \cong V_2 \oplus \cdots \oplus V_n.$$

Finally, by the induction hypothesis, $m - 1 = n - 1$, and there is a permutation $i \mapsto n_i$ of $\{2, 3, \ldots, n\}$ such that $U_i \cong V_{n_i}$ for $i = 2, \ldots, n$. Since $U_1 \cong V_1$, the theorem is proved.

Corollary (*Krull–Schmidt theorem*) Let R be an algebra over a field K, and let V be a nonzero R-module. Then V can be written as a direct sum $V = \bigoplus_{i=1}^{m} U_i$, where the U_i are indecomposable submodules. Moreover, if $V = \bigoplus_{i=1}^{n} V_j$ is another decomposition of this kind, then $n = m$ and (after possibly reordering the V_j) we have $V_i \cong U_i$ for each i.

Proof Since V is a finite dimensional vector space, it is clear that at least one such decomposition exists. Thus the corollary will follow from the

1.6 INDECOMPOSABLE MODULES

theorem once we have shown that $\text{End}_R(U)$ is a local ring for each indecomposable R-module U.

Put $E := \text{End}_R(U)$. Let I be a maximal right ideal of E and let $f \in I$. Since f is a nonunit, $\text{Ker } f \neq 0$ by Lemma 1.2, and so $\text{Ker } f^n \neq 0$ for all $n \geq 1$. Since U is indecomposable, Lemma 1.2 now shows that $Uf^n = 0$ for some $n \geq 1$, and so f is nilpotent. On the other hand, suppose that $g \in E$ and $g \notin I$. Then, since I is maximal, $gE + I = E$, and so $gh + f = 1$ for some $h \in E$. Since $f^n = 0$ for some $n \geq 1$,

$$gh\{1 + f + \cdots + f^{n-1}\} = (1-f)\{1 + f + \cdots + f^{n-1}\}$$
$$= 1 - f^n = 1,$$

and so g is a unit. Thus I is precisely the set of all nonunits of E, and so E is local. This proves the corollary.

A set $\{e_1, e_2, \ldots, e_n\}$ of idempotents in a ring R is called *orthogonal* if $e_i e_j = 0$ whenever $i \neq j$. An orthogonal set of idempotents such that $1 = e_1 + e_2 + \cdots + e_n$ is called a *complete orthogonal* set. In this case it is easily seen that $R = e_1 R \oplus e_2 R \oplus \cdots \oplus e_n R$. Conversely, suppose $R = R_1 \oplus R_2 \oplus \cdots \oplus R_n$, where each R_i is a nonzero right ideal. There are unique elements $e_1, e_2, \ldots, e_n$ such that $1 = e_1 + e_2 + \cdots + e_n$ and $e_i \in R_i$. Since

$$e_i - e_i^2 = e_1 e_i + \cdots + e_{i-1} e_i + e_{i+1} e_i + \cdots + e_n e_i$$
$$\in R_i \cap (R_1 \oplus \cdots \oplus R_{i-1} \oplus R_{i+1} \oplus \cdots \oplus R_n) = (0),$$

it follows that $e_i^2 = e_i$ and $e_i e_j = 0$ for $i \neq j$. Since $R_i \neq 0$, $e_i \neq 0$ and $\{e_1, e_2, \ldots, e_n\}$ is a set of orthogonal idempotents whose sum is 1. Thus there is a one-to-one correspondence between the decompositions of R as a direct sum of nonzero right ideals and the decompositions of 1 as a sum of orthogonal idempotents. If $R = R_1 \oplus R_2 \oplus \cdots \oplus R_n$, then $R_i = e_i R$ for some idempotent e_i. A summand R_i is indecomposable if and only if e_i cannot be written as a sum of orthogonal idempotents. An idempotent e is called *primitive* if it cannot be expressed as a sum of two orthogonal idempotents.

Note If R is any ring and U is an R-module, then U is decomposable iff $\text{End}_R(U)$ possesses an idempotent $e \neq 1$. Indeed, if such an e exists then $U = Ue \oplus U(1-e)$ is a decomposition of U into a direct sum of nonzero R-submodules. Conversely, if $U = U_1 \oplus U_2$, where U_1 and U_2 are nonzero R-submodules, then we can define e to be the canonical projection of U onto U_1. It is readily seen that e is an R-endomorphism of U such that $e^2 = e \neq 0$ or 1.

We shall find the following result on idempotents useful in the following section.

Theorem 1.6B Let K be a field of characteristic $p > 0$ and let R be an algebra over K. Let J be a nilpotent ideal in R. If $e + J$ is an idempotent in R/J, then for some integer $m > 0$, $e_0 := e^{p^m}$ is an idempotent in R such that $e + J = e_0 + J$.

Proof By hypothesis $e^n + J = (e + J)^n = e + J \neq 0$ for all integers $n \geq 1$. Therefore $e^2 - e \in J$ and $e^n \notin J$ for any $n \geq 1$. Since K has characteristic p, the binomial theorem shows that $(e^{p^m})^2 - e^{p^m} = (e^2 - e)^{p^m} = 0$ for sufficiently large m because J is nilpotent. This shows that $e_0 := e^{p^m}$ is an idempotent in R, and from above, $e_0 + J = e + J$.

EXERCISES

1. Let G be a group generated by an element a of order 2, K a field of characteristic 2, and V a KG-module such that $(\alpha_1 v_1 + \alpha_2 v_2)a = \alpha_1 v_1 + (\alpha_1 + \alpha_2)v_2$ for a fixed basis $\{v_1, v_2\}$ of V. Prove that V is indecomposable but not irreducible.

2. Let R be a ring and U an Artinian and Noetherian R-module. Suppose $U = U_1 \oplus U_2 \oplus \cdots \oplus U_n$, where each U_i is an indecomposable submodule. Then prove that any direct summand $V \neq 0$ of U is isomorphic to a direct sum of a subset of the modules $U_1, U_2, \ldots, U_n$.

3. Prove that an Artinian module U over a ring is a finite direct sum of indecomposable submodules. Is the converse true?

1.7 Absolutely indecomposable and absolutely irreducible modules

Let R be an algebra over a field K. An R-module V is *absolutely indecomposable* if for each field extension L of K, $V \otimes_K L$ is indecomposable. Any absolutely indecomposable module is indecomposable, but the converse is not true. However we shall see below that if K is algebraically closed, then each indecomposable R-module is absolutely indecomposable.

Theorem 1.7A Let R be an algebra over a field K, and let V be an R-module. Put $E := \text{End}_R(V)$. Then we have the following criteria.

(i) V is indecomposable iff $E/\text{rad } E$ is a division ring.
(ii) V is absolutely indecomposable iff $E/\text{rad } E = K$.
(iii) If K is algebraically closed and V is indecomposable, then V is absolutely indecomposable.

Remark It is easily seen that a ring is a division ring iff 0 is its only proper right ideal. Thus $E/\text{rad } E$ is a division ring iff rad E is a maximal right ideal of E. Since rad E is the intersection of all maximal right ideals, Lemma 1.6B shows that $E/\text{rad } E$ is a division ring iff E is local.

1.7 ABSOLUTELY INDECOMPOSABLE AND IRREDUCIBLE MODULES

Proof (i) If V is indecomposable, then the proof of the corollary to Theorem 1.6A shows that E is local, and so $E/\text{rad } E$ is a division ring by the remark above. Conversely, if $E/\text{rad } E$ is a division ring, then E is local, and so by Lemma 1.6B, 1 is the unique idempotent of E. In particular, 1 is a primitive idempotent, and so by §1.6, V is indecomposable.

(ii) For each field extension L of K we define $E_L := \text{End}_{R_L}(V \otimes_K L)$, where $R_L := R \otimes_K L$. Elementary properties of the tensor product show that $E_L = E \otimes_K L$, $\text{rad } E_L = (\text{rad } E) \otimes_K L$, and $E_L/\text{rad } E_L \simeq (E/\text{rad } E) \otimes_K L$ (for a proof, work with fixed bases for the K-spaces). Thus, if $E/\text{rad } E = K$, then $E_L/\text{rad } E_L \simeq K \otimes_K L \simeq L$ for each extension field L of K. Hence V is absolutely indecomposable by (i).

Conversely, suppose that V is absolutely indecomposable; then (i) shows that $D = E/\text{rad } E$ is a division ring with the property that $D \otimes_K L$ is a division ring for each field extension L of K. We have to prove $D = K$. Suppose $D \neq K$, and choose $a \in D \setminus K$. Since D is a K-algebra, it is finite dimensional over K, and the powers $1, a, a^2, \ldots$ are not linearly independent over K. Thus there exists a nonzero polynomial $f(x)$ in $K[x]$ with leading coefficient 1 such that $f(a) = 0$. Let L be a splitting field of $f(x)$ over K and write $f(x) = (x - \alpha_1)(x - \alpha_2) \cdots (x - \alpha_m)$. Then in $D \otimes_K L$ we have $(a \otimes 1 - 1 \otimes \alpha_1)(a \otimes 1 - 1 \otimes \alpha_2) \cdots (a \otimes 1 - 1 \otimes \alpha_m) = 0$ with

$$a \otimes 1 \neq 1 \otimes \alpha_i$$

for any i because $a \notin K$. But this shows that $D \otimes_K L$ contains nonzero divisors of 0 and so is not a division ring. Since this is contrary to the hypothesis on V we conclude that $D = K$, and (ii) is proved.

(iii) Suppose V is not absolutely indecomposable. Then it follows from what we proved in (ii) that there exists an algebraic extension L of K for which $V \otimes_K L$ is decomposable. But K is algebraically closed, so $L = K$ and V itself is decomposable.

Let R be an algebra over a field K. An R-module V is called *absolutely irreducible* if for each field extension L of K, $V \otimes_K L$ is irreducible. Any absolutely irreducible R-module is obviously irreducible, but the converse is not true. The following theorem clarifies the concept of splitting field introduced in §1.3.

Theorem 1.7B *Let G be a group and K a field. Then K is a splitting field for KG iff each irreducible KG-module is absolutely irreducible.*

Proof First suppose that each irreducible KG-module is absolutely irreducible. Then for each irreducible KG-module V and each extension L of K, $V \otimes_K L$ is irreducible. In particular, taking L as an algebraic closure of K,

L is a splitting field for LG, so $\text{End}_{LG}(V \otimes_K L) = L \cdot 1$. On the other hand, if $c \in \text{End}_{KG}(V)$, then c commutes with the action of G, and so $c \otimes 1 \in \text{End}_{LG}(V \otimes_K L) = L \cdot 1$. This shows that $c \in K \cdot 1$. Since $K \cdot 1 \subseteq \text{End}_{KG}(V)$, we conclude therefore that $\text{End}_{KG}(V) = K \cdot 1$. Hence K is a splitting field for G.

Conversely, suppose that K is a splitting field for KG, and let V be an irreducible KG-module of dimension n. Let T be the representation of KG associated with V. Then by the corollary of Theorem 1.3A we know that $T(KG) = \text{End}_K(V)$ as a K-space. For any extension field L of K, let T_L denote the representation of LG associated with $V \otimes_K L$. Then $T_L(LG) = T(KG) \otimes_K L$, and so $\dim_L T_L(KG) = \dim_K T(KG) = n^2$; hence $T_L(LG)$ equals all of $\text{End}_L(V \otimes_K L)$. Since $\text{Aut}_L(V \otimes_K L)$ acts transitively on $V \otimes_K L$, this shows that $V \otimes_K L$ is an irreducible LG-module. Hence V is absolutely irreducible.

EXERCISES

1. Let G be the cyclic group of order 3, x a generator of G, **Q** the rational field, and V a **Q**G-module of dimension 2. If $\{v_1, v_2\}$ is a basis of V over **Q** such that $(\alpha_1 v_1 + \alpha_2 v_2)a = -(\alpha_1 + \alpha_2)v_1 + \alpha_1 v_2$, show that V is irreducible but not absolutely irreducible.

2. Let G be the group generated by a, b that satisfy $a^3 = b^3 = (ab)^2 = 1$, let K be a field of 3 elements, and let V be a KG-module of dimension 3. If $\{v_1, v_2, v_3\}$ is a basis of V over K such that

$$(\alpha_1 v_1 + \alpha_2 v_2 + \alpha_3 v_3)a = \alpha_1 v_1 + (\alpha_1 + \alpha_2)v_2 + (\alpha_1 + \alpha_3)v_3$$

and

$$(\alpha_1 v_1 + \alpha_2 v_2 + \alpha_3 v_3)b = (\alpha_1 + \alpha_2)v_1 + v_2 + (\alpha_1 - \alpha_3)v_3,$$

show that V is absolutely indecomposable.

1.8 Principal indecomposable modules

Let R be an algebra over a field K. Then the corollary to Theorem 1.6A shows that we can write

(1) $$R = U_1 \oplus \cdots \oplus U_n$$

as a direct sum of indecomposable R-submodules U_i where the U_i are uniquely determined up to isomorphism and the order in which they appear. We shall call the summands in (1) the *principal indecomposable R-modules*. From §1.6 we know that there exists a complete set $\{e_1, \ldots, e_n\}$ of primitive

1.8 PRINCIPAL INDECOMPOSABLE MODULES

orthogonal idempotents such that $U_i = e_i R$ for $i = 1, \ldots, n$. Except when R is semisimple, the principal indecomposable R-modules form only a small subclass of the class of all indecomposable R-modules. However, it is the class which plays a very important role in the theory of modular representations.

Theorem 1.8 Let R be an algebra over a field K. Then

   (i) each principal indecomposable R-module $U = eR$ (e an idempotent) has a unique maximal submodule $U' = e\,\mathrm{rad}\,R = U \cap \mathrm{rad}\,R$, and

   (ii) if U and V are two principal indecomposable R-modules with maximal submodules U' and V', respectively, then $U \simeq V$ iff $U/U' \simeq V/V'$.

Proof (i) Since R is a K-algebra, U certainly contains at least one maximal submodule. Moreover it is easily verified that the submodules $e\,\mathrm{rad}\,R$ and $U \cap \mathrm{rad}\,R$ are equal. Thus to prove (i) it is enough to show that each proper submodule of U is contained in $e\,\mathrm{rad}\,R$.

Suppose that V is a submodule of U such that $V \nsubseteq e\,\mathrm{rad}\,R$. Then we shall show that $U = V$, from which it follows that $e\,\mathrm{rad}\,R$ is the unique maximal submodule of U. Assume that $V \neq U$. Then there is a maximal submodule W of U with $V \subseteq W$. Since $V \subseteq W \nsubseteq e\,\mathrm{rad}\,R$ and W is maximal, $U = W + e\,\mathrm{rad}\,R$, and so $e \in Ue = We + e(\mathrm{rad}\,R)e$. Thus $e = w + h$ for some $w \in We$ and $h \in e(\mathrm{rad}\,R)e$. Note that $eh = he = h$ and that h is nilpotent by Theorem 1.3A. Then for some $n \geq 1$,

$$w(e + h + \cdots + h^{n-1}) = (e - h)(e + h + \cdots + h^{n-1}) = e - h^n = e.$$

This implies $e \in wR \subseteq W$ and so $eR = W$, a contradiction. This proves the first part.

   (ii) Clearly $U \simeq V$ implies $U/U' \simeq V/V'$. On the other hand, suppose g is an R-isomorphism of U/U' onto V/V'. Then $u \mapsto (u + U')g$ is an R-homomorphism of U onto V/V'. Let e be the idempotent of R such that $U = eR$, and choose $v_0 \in V$ such that $(e + U')g = v_0 + V'$. Since $v_0 R \subseteq V$, we can define an R-homomorphism $f: U \to V$ by $uf := v_0 u$. Because g is an R-homomorphism, $v_0 e + V' = v_0 + V'$, and since g is surjective, the latter is not 0. Thus the image of f contains $v_0 e$ and the latter is not contained in V'. Hence by (i), $\mathrm{Im}\,f = V$. In particular, this shows that $\dim_K U \geq \dim_K V$. Now interchanging the roles of U and V we can show similarly that $\dim_K V \geq \dim_K U$. Hence U and V have the same dimension, and the surjectivity of f then implies that f is bijective. Thus f is an R-isomorphism from U onto V. This completes the proof of the theorem.

Corollary Suppose that the decomposition (1) holds for R. Then there is a unique integer r such that (possibly after reordering the U_i) $\{U_1, U_2, \ldots, U_r\}$ is a full set of nonisomorphic principal indecomposable R-modules and $\{U_1/U'_1, U_2/U'_2, \ldots, U_r/U'_r\}$ is a full set of irreducible R-modules.

Proof From part (ii) of the theorem it is enough to show that each irreducible R-module W is isomorphic to U/U' for some principal indecomposable R-module U. Since $W = WR$, (1) shows that $WU_i \neq 0$ for some i. Then choose $w \in W$ such that $wU_i \neq 0$; the submodule wU_i of W equals W because W is irreducible. Thus the mapping $u \mapsto wu$ is an R-homomorphism of U_i onto W, and since W is irreducible the kernel must be a maximal submodule of U_i. Since U_i' is the unique maximal submodule of U_i, this shows that $U_i/U_i' \cong W$.

EXERCISES

1. Let G be the group of order 4 generated by a and b that satisfy $a^2 = b^2 = (ab)^2 = 1$, K a field of characteristic 2, and V a KG-module of dimension 2. If $\{v_1, v_2\}$ is a basis of V over K such that

$$(\alpha_1 v_1 + \alpha_2 v_2)a = (\alpha_1 + \alpha_2)v_1 + \alpha_2 v_2$$

and

$$(\alpha_1 v_1 + \alpha_2 v_2)b = (\alpha_1 + z\alpha_2)v_1 + \alpha_2 v_2,$$

where z is a fixed element of K, then show that V is an indecomposable KG-module, which is not isomorphic to a principal indecomposable KG-module.

2. With the notation of Problem 1, show that $(1 + a)KG$ is an indecomposable but not an irreducible KG-module.

3. An R-module U is called *projective* if given R-modules V and W and a surjective homomorphism $g \in \text{Hom}_R(V, W)$, then for any $h \in \text{Hom}_R(U, W)$ there is an $f \in \text{Hom}_R(U, V)$ such that $h = fg$. Assume that R is Artinian. Prove that a projective R-module is isomorphic to a direct sum of principal indecomposable R-modules. Hence a projective indecomposable R-module is isomorphic to a principal idecomposable R-module.

4. Let $R = KG$ be the group algebra of a p-group G over a field K of characteristic $p > 0$. Prove that the regular R-module R is indecomposable and hence that R is the unique principal indecomposable R-module.

5. If $R = KG$ is the group algebra of a group G over a field K, prove that every direct summand of R is projective and hence is a direct sum of principal indecomposable R-modules.

6. Let U be a principal indecomposable KG-module and U^* the dual of U (see Example 6 of §1.4). Show that U^* is isomorphic to a principal indecomposable KG-module. If $U' \neq 0$, show that the set of all elements f of U^* such that $f(U') = 0$ is the unique minimal submodule of U^*. Deduce that U has a unique minimal submodule isomorphic to U/U'.

7. If U is a principal indecomposable KG-module, prove that $U \simeq U^*$ iff $U/U' \simeq (U/U')^*$.

1.9 Composition factors and intertwining numbers

Let KG be the group algebra of a group G over a field K. Let V be a KG-module and suppose that

(1) $$V = V_0 \supset V_1 \supset \cdots \supset V_l = 0$$

is a composition series of V. Then by the Jordan–Hölder theorem we know that the number of the irreducible constituents V_{i-1}/V_i ($i = 1, 2, \ldots, l$) that are isomorphic to a given irreducible KG-module W is independent of the series (1). The theorems of the present section show how to calculate the multiplicities of the composition factors.

Theorem 1.9A Let K be a splitting field for the group G, and let V be a KG-module. Let U be a principal indecomposable KG-module with maximal submodule U'. Then the multiplicity of U/U' as an irreducible constituent of V is equal to $\dim_K \mathrm{Hom}_{KG}(U, V)$.

Proof By the corollary of Theorem 1.8 each irreducible constituent of V is isomorphic to U/U' for some principal indecomposable KG-module U. Suppose that (1) is a composition series for V; we shall proceed by induction on the length l. Since the result is trivial when $l = 0$, suppose that $l \geq 1$. To simplify notation we write Hom for Hom_{KG}.

Let $h: V \to V/V_1$ denote the canonical homomorphism and define $\phi: \mathrm{Hom}(U, V) \to \mathrm{Hom}(U, V/V_1)$ by $\phi(f) := fh$. Then ϕ is a K-linear mapping such that $f \in \mathrm{Ker}\,\phi$ iff $Uf \subseteq V_1$; hence $\mathrm{Ker}\,\phi \simeq \mathrm{Hom}(U, V_1)$ as vector spaces. On the other hand, ϕ is surjective. Indeed, since U is a principal indecomposable module we can write $U = e(KG)$ for some idempotent e. Suppose $f_0 \in \mathrm{Hom}(U, V/V_1)$ and $ef_0 = v + V_1$, say. Then since f_0 is a KG-homomorphism, $af_0 = (ea)f_0 = (ef_0)a = va + V_1$ for all $a \in e(KG) = U$. If we define $f \in \mathrm{Hom}(U, V)$ by $af := va$; then $\phi(f) = f_0$. This proves that ϕ is surjective.

We next show that $\mathrm{Hom}(U, V/V_1) \simeq \mathrm{Hom}(U/U', V/V_1)$ (as vector spaces). Since V/V_1 is irreducible, each $g \neq 0$ in $\mathrm{Hom}(U, V/V_1)$ has $\mathrm{Im}\,g = V/V_1$ and hence $\mathrm{Ker}\,g = U'$, since U' is the unique maximal submodule of U. Thus for all $g \in \mathrm{Hom}(U, V/V_1)$, $\mathrm{Ker}\,g \supseteq U'$, and so there is a natural isomorphism $\mathrm{Hom}(U, V/V_1) \simeq \mathrm{Hom}(U/U', V/V_1)$.

Finally, since K is a splitting field for G we have, using Schur's lemma,

(2) $$\dim_K \mathrm{Hom}(U/U', V/V_1) = \begin{cases} 1 & \text{if } U/U' \simeq V/V_1 \text{ as } KG\text{-modules} \\ 0 & \text{otherwise.} \end{cases}$$

Therefore, counting, we obtain

$$\dim_K \text{Hom}(U, V) = \dim_K \text{Hom}(U, V_1) + \dim_K \text{Hom}(U, V/V_1)$$

$$= \begin{cases} \dim_K \text{Hom}(U, V_1) + 1 & \text{if } U/U' \simeq V/V_1 \\ \dim_K \text{Hom}(U, V_1) & \text{otherwise.} \end{cases}$$

Induction now completes the proof of the theorem.

Definition Let U and V be any KG-modules. Then the *intertwining number* for U and V is defined to be $i(U, V) := \dim_K \text{Hom}_{KG}(U, V)$.

Note 1 Since $\text{Hom}_{KG}(U \oplus V, W) \simeq \text{Hom}_{KG}(U, W) \oplus \text{Hom}_{KG}(V, W)$, we have $i(U \oplus V, W) = i(U, W) + i(V, W)$ for any KG-modules U, V, and W. Similarly $i(U, V \oplus W) = i(U, V) + i(U, W)$.

Note 2 Theorem 1.9A shows that when K is a splitting field and U is a principal indecomposable KG-module, then $i(U, V)$ equals the multiplicity of U/U' as an irreducible constituent of V.

Theorem 1.9B Let K be a splitting field for the group G and let

(3) $$KG = U_1 \oplus U_2 \oplus \cdots \oplus U_n$$

be a decomposition of KG into a direct sum of principal indecomposable KG-modules. Let U be any principal indecomposable KG-module. Then the number of U_i isomorphic to U [in (3)] equals $\dim_K(U/U')$.

Proof Suppose that U is isomorphic to m of the U_i. Since $\text{Hom}_{KG}(KG, V) \simeq V$, $\dim_K(U/U') = i(KG, U/U') = i(U_1 \oplus \cdots \oplus U_n, U/U') = i(U_1, U/U') + i(U_2, U/U') + \cdots + i(U_n, U/U')$. By Theorem 1.8(ii) $U_j/U'_j \simeq U/U'$ iff $U_j \simeq U$. Therefore by Theorem 1.9A,

$$i(U_j, U/U') = \begin{cases} 1 & \text{if } U_j \simeq U \\ 0 & \text{otherwise.} \end{cases}$$

Thus $\dim_K(U/U') = m$ as required.

Corollary Let $U_1, \ldots, U_r$ be a full set of nonisomorphic principal indecomposable KG-modules and let K be a splitting field for G. Put $d_j := \dim_K(U_j/U'_j)$. Then

$$|G| = \sum_{i=1}^{r} d_i \dim_K U_i.$$

In particular, if KG is semisimple (see Theorem 1.3B), then each U_i is irreducible. Hence in this case, $U'_i = 0$ and

$$|G| = \sum_{i=1}^{r} d_i^2.$$

1.9 COMPOSITION FACTORS AND INTERTWINING NUMBERS

Later on we shall need two other facts about the intertwining numbers. We record these in the following result.

Theorem 1.9C Let K be a splitting field for G.

(i) If KG is semisimple, then $i(U, V) = i(V, U)$ for any KG-modules U and V.

(ii) For any idempotent e of KG, and any KG-module V, $i(e(KG), V) = \dim_K Ve$.

Remark In general Ve need not be a KG-module.

Proof (i) Since KG is semisimple, by Theorem 1.3B every KG-module is completely reducible, and so we can write $U = U_1 \oplus U_2 \oplus \cdots \oplus U_m$ and $V = V_1 \oplus V_2 \oplus \cdots \oplus V_n$ as sums of irreducible modules. By Note 1 above

$$i(U, V) = \sum_{l=1}^{m} \sum_{j=1}^{n} i(U_l, V_j)$$

and there is a similar expression for $i(V, U)$. However, since K is a splitting field for G, Schur's lemma shows that $i(U_l, V_j) = 0$ if $U_l \not\simeq V_j$ and $i(U_l, V_j) = 1$ if $U_l \simeq V_j$. Therefore $i(U, V) = i(V, U)$.

(ii) Define $\phi: \mathrm{Hom}_{KG}(e(KG), V) \to V$ by $\phi(f) := ef$. Then clearly ϕ is a K-linear mapping; and $\mathrm{Ker}\, \phi = 0$ since $ef = 0$ implies $(ef)a = (ea)f = 0$ for all $a \in KG$. On the other hand, $\mathrm{Im}\, \phi = Ve$. Indeed, $\mathrm{Im}\, \phi \subseteq Ve$ because for all f, $\phi(f) = e^2 f = (ef)e \in Ve$, whilst $\mathrm{Im}\, \phi \supseteq Ve$ because for all $v \in V$ we can define $f \in \mathrm{Hom}_{KG}(e(KG), V)$ by $af := va$ and then $\phi(f) = ef = ve$. Hence ϕ is injective with image Ve, and so

$$i(e(KG), V) = \dim_K \mathrm{Hom}_{KG}(e(KG), V) = \dim_K Ve.$$

EXAMPLE Suppose that G is a p-group and that K is a splitting field of characteristic p for G; then the regular KG-module KG is indecomposable. (It follows from Theorem 2.7B that any field of characteristic p is a splitting field for every p-group.) Indeed, by the corollary of Theorem 1.5 the trivial KG-module is the only irreducible KG-module; so the corollary of Theorem 1.8 shows that (up to isomorphism) there is a unique principal indecomposable KG-module, say U, and U/U' is the trivial KG-module of dimension 1. Hence Theorem 1.9B shows that $KG = U$, and so KG is indecomposable.

EXERCISES

1. Let U be a KG-module. An invariant of U is an element $u \in U$ such that $ux = u$ for all $x \in G$. Show that the set of all invariants of U is a KG-submodule.

2. Let U and V be KG-modules. Then show that $i(U, V)$ is equal to the dimension of the space of invariants of $U^* \otimes_K V$, where U^* denotes the dual of U.

3. Let G be a group and K a splitting field for G. Show that a completely reducible KG-module is irreducible iff $i(U, U) = 1$.

4. Let G be a group, K a field, and $U = e(KG)$ a principal indecomposable KG-module generated by the idempotent e. Show that a KG-module V has a composition factor isomorphic to U/U' iff $Ve \neq 0$.

5. Let K be a field, $U_1, U_2, \ldots, U_r$ a full set of principal indecomposable KG-modules, and U'_i the unique maximal submodule of U_i. If

$$|G| = \sum_{i=1}^{r} [\dim_K(U_i/U'_i)]^2,$$

prove that KG is semisimple.

1.10 Notes and comments

The concept of a group acting on a set is a generalization of the concept of permutation, and so a permutation group G on a set Ω acts naturally on Ω. The definition of a group algebra AG given in §1.1 can be extended to the group algebra RG of any (not necessarily finite) group G and any ring R. For ring theoretic properties of RG see Passman [1]. Other proofs of Wedderburn structure theorem (Theorem 1.3A) may be found in Lang [1], pp. 443–445, or Curtis and Reiner [1], §§25–26. These two books also give detailed treatments of the tensor product construction. The proof of Theorem 1.5 is due to Brauer [8]. Theorem 1.5 has been generalized to the case where K is not necessarily a splitting field; the case where the characteristic of the field does not divide the order of the group is dealt with in Berman [1] and the case of modular representations over an arbitrary field in Reiner [1].

The principal indecomposable KG-modules form only a small subclass of the class of indecomposable KG-modules (when the characteristic of K divides $|G|$). The determination of all indecomposable KG-modules is a difficult step, except when the Sylow p-subgroups of G are cyclic. For the proof of this result see §8.3, and for further theory of indecomposable modules see Green [1, 2, 3]. Most of the results of §1.9 hold for a finite dimensional algebra over a field. For further results on intertwining numbers, see Curtis and Reiner [1].

CHAPTER II

Induced Modules and Characters

This chapter describes the connection between the representations of a group and the representations of its subgroups. In the first two sections we shall prove two crucial results which have important roles in our work. The first of these results, due to Mackey, describes the behavior of the restriction of an induced module, while the second result, due to Clifford [1], describes the behavior of an irreducible module on restriction to a normal subgroup. As an immediate application of Clifford's result, we shall prove a theorem of Blichfeldt on nilpotent groups. The third section is an introduction to the theory of group characters and the fourth describes the orthogonal property of ordinary characters. The next two sections deal with induced characters and Brauer's induction theorem. In the seventh section we use Brauer's induction theorem to show that every group has "small" splitting fields.

2.1 Induced modules

Let A be a commutative ring. Let G be a group and H a subgroup. Then we can identify the group ring AH as a subring of AG. If V is an AG-module then we shall denote by V_H the AH-module obtained by the restriction of the ring; thus as an A-module V_H equals V, but only action of H (and not all of G) is defined on V_H. This process will be called *restriction* and it permits us to go from any AG-module V to a (uniquely determined) AH-module V_H.

There is a dual process of *induction*. Let W be any AH-module. Then we saw in Example 4 of §1.4 that we can define an AG-module structure on the tensor product $W \otimes_{AH} AG$. This is the *induced module* and we denote it by W^G.

Note 1 If V and W are A-free modules (in particular, if A is a field), then $\text{rank}_A V_H = \text{rank}_A V$, but $\text{rank}_A W^G = |G:H|\text{rank}_A W$ (see §1.4).

Note 2 If $V = V_1 \oplus V_2$ and $W = W_1 \oplus W_2$, then it is easily seen that $V_H = (V_1)_H \oplus (V_2)_H$ and $W^G = (W_1)^G \oplus (W_2)^G$.

Note 3 If two modules are isomorphic then restriction (respectively, induction) gives modules which are also isomorphic. Moreover induction and restriction are transitive in the following sense. If L is a subgroup of G with $H \subseteq L \subseteq G$, then

$$(V_L)_H = V_H \quad \text{and} \quad (W^L)^G \simeq W^G.$$

The latter relation follows from

$$(W^L)^G = (W \otimes_{AH} AL) \otimes_{AL} AG \simeq W \otimes_{AH} (AL \otimes_{AL} AG) \simeq W \otimes_{AH} AG = W^G.$$

The process of induction is related to the concept of imprimitivity of a module.

Definition Let A be a commutative ring and G a group. Then an AG-module V is called *imprimitive* if V can be written as a direct sum $V = V_1 \oplus V_2 \oplus \cdots \oplus V_m$ of A-submodules V_i with $m > 1$ such that G acts on the set $\{V_1, V_2, \ldots, V_m\}$ by right multiplication. The set $\{V_1, V_2, \ldots, V_m\}$ is called a *system of imprimitivity* for V. If V cannot be written in this form, then V is *primitive*.

Note 4 If $\Omega = \{V_1, V_2, \ldots, V_m\}$ is a system of imprimitivity for V, then the fact that G acts on this set means that for each $x \in G$, $V_i \mapsto V_i x$ is a permutation of Ω. If V is irreducible, it is easily seen that Ω forms a single orbit under G; so G acts transitively on Ω.

EXAMPLE Let H be a proper subgroup of G and let W be an AH-module for some commutative ring A. Then W^G cannot be primitive. For suppose that $y_1, y_2, \ldots, y_h$ is a right transversal of H in G. Then $W^G = (W \otimes y_1) \oplus (W \otimes y_2) \oplus \cdots \oplus (W \otimes y_h)$ (as A-modules). Moreover, it is clear that $W \otimes y_1, \ldots, W \otimes y_h$ is a system of imprimitivity for W^G and the action is transitive. Indeed $h = |G:H| > 1$, and for each i and each $x \in G$ we have $(W \otimes y_i)x = W \otimes y_j$, where y_j is defined by $Hy_j = Hy_i x$.

This example gives a method of constructing imprimitive AG-modules. The following lemma shows that essentially all imprimitive AG-modules can be constructed in this way.

2.1 INDUCED MODULES

Lemma 2.1 Let A be a commutative ring and G a group. Let V be an imprimitive AG-module with $\{V_1, V_2, \ldots, V_m\}$ as a system of imprimitivity and suppose that G acts transitively on $\{V_1, V_2, \ldots, V_m\}$. Define $H := \{x \in G \mid V_1 x = V_1\}$ as the stabilizer of V_1 under the action of G. Then V_1 is an AH-module and $V \simeq V_1^G$.

Proof Since G acts transitively on $\{V_1, V_2, \ldots, V_m\}$ and $H = \text{Stab}(V_1)$, it follows from §1.1 that $m = |G:H|$. Thus we can choose a right transversal of H in G, say $y_1 = 1, y_2, \ldots, y_m$ such that $V_i = V_1 y_i$ ($i = 1, 2, \ldots, m$). Since $V_1^G = (V_1 \otimes y_1) \oplus (V_1 \otimes y_2) \oplus \cdots \oplus (V_1 \otimes y_m)$ and

$$V = V_1 y_1 \oplus V_1 y_2 \oplus \cdots \oplus V_1 y_m$$

as A-modules, we have an A-isomorphism θ of V_1^G onto V given by

$$\sum_{i=1}^m v_i \otimes y_i \mapsto \sum_{i=1}^m v_i y_i,$$

where $v_1, v_2, \ldots, v_m \in V_1$. However, for each $x \in G$ and each i there exist uniquely determined $z \in H$ and j such that $y_i x = z y_j$; and $(v_i \otimes y_i) x = v_i z \otimes y_j$ and $(v_i y_i) x = (v_i z) y_j$ for $v_i \in V_1$. Thus θ is an AG-isomorphism and the lemma is proved.

A basic relation between the process of induction and restriction is given by the following theorem.

Theorem 2.1A (*Mackey*) Let H and L be subgroups of the group G and let $\{x_1, x_2, \ldots, x_n\}$ be a set of double coset representatives for (H, L) in G. Let A be a commutative ring and let V be an AH-module. For each $x \in G$, define V_x to be the $A(x^{-1}Hx \cap L)$-module $V \otimes x$ with the action $(v \otimes x)y := vxyx^{-1} \otimes x$ for all $v \in V$ and $y \in x^{-1}Hx \cap L$. Then $(V^G)_L \simeq \bigoplus_{i=1}^n (V_{x_i})^L$.

Proof Let $y_1, y_2, \ldots, y_m$ be a right transversal of H in G. Then $V^G = (V \otimes y_1) \oplus (V \otimes y_2) \oplus \cdots \oplus (V \otimes y_m)$ as A-modules. Put

$$\Omega := \{V \otimes y_1, \ldots, V \otimes y_m\}.$$

Then G, and in particular L, acts on Ω. Moreover $V \otimes y_i$ and $V \otimes y_j$ lie in the same orbit under L iff y_i and y_j lie in the same double (H, L)-coset. Let W_i denote the sum of the $V \otimes y_j$ for which $y_j \in Hx_i L$. Then each W_i is an AL-module and $(V^G)_L = W_1 \oplus W_2 \oplus \cdots \oplus W_n$. It remains to show that each $W_i \simeq (V_{x_i})^L$. Consider W_1, say, and suppose $W_1 = (V \otimes y_1) \oplus \cdots \oplus (V \otimes y_l)$, where by definition of W_1, $Hy_1 \cup Hy_2 \cup \cdots \cup Hy_l = Hx_1 L$. Without loss of generality we may assume that the y_j are chosen so that one of them, say y_1, equals x_1. Now L acts transitively on the set $\{V \otimes y_1,$

$V \otimes y_2, \ldots, V \otimes y_t\}$ and under this action the stabilizer of $V \otimes y_1 = V \otimes x_1$ is

$$\{z \in L \mid V \otimes x_1 z = V \otimes x_1\} = \{z \in L \mid x_1 z x_1^{-1} \in H\} = x_1^{-1} H x_1 \cap L.$$

Hence by the definition of V_{x_1} and Lemma 2.1 we conclude that $W_1 \simeq (V_{x_1})^L$. Similarly $W_i \simeq (V_{x_i})^L$ for each i, and the theorem is proved.

Our next result also relates the induction and restriction processes.

Theorem 2.1B Let A be a commutative ring, G a group, and H a subgroup of G. Let U be an AH-module and V an AG-module, and suppose that U and V are both free as A-modules (in particular, the latter always holds if A is a field). Then

$$(U \otimes V_H)^G \simeq U^G \otimes V.$$

Proof Let $u_1, u_2, \ldots, u_m$ and $v_1, v_2, \ldots, v_n$ be A-bases of U and V, respectively. Let $x_1, x_2, \ldots, x_h$ be a right transversal of H in G. Then $(u_i \otimes v_j) \otimes x_l$ $(i = 1, 2, \ldots, m; j = 1, 2, \ldots, n;$ and $l = 1, 2, \ldots, h)$ is an A-basis of $(U \otimes V_H)^G$ and $(u_i \otimes x_l) \otimes v_j x_l$ $(i = 1, 2, \ldots, m; j = 1, 2, \ldots, n;$ and $l = 1, 2, \ldots, h)$ is an A-basis of $U^G \otimes V$. Hence

(1) $$(u_i \otimes v_j) \otimes x_l \mapsto (u_i \otimes x_l) \otimes v_j x_l$$

defines an A-isomorphism of $(U \otimes V_H)^G$ onto $U^G \otimes V$; it remains to check that this is an AG-isomorphism. To do this we must show that it respects the action of G. If $y \in G$, then for each l we have uniquely defined $z \in H$ and k such that $x_l y = z x_k$, since $x_1, x_2, \ldots, x_h$ is a right transversal of H. But then

$$((u_i \otimes v_j) \otimes x_l) y = (u_i \otimes v_j) z \otimes x_k = (u_i z \otimes v_j z) \otimes x_k.$$

On the other hand

$$((u_i \otimes x_l) \otimes v_j x_l) y = (u_i \otimes x_l) y \otimes v_j x_l y = (u_i z \otimes x_k) \otimes (v_j z) x_k.$$

This shows that the A-isomorphism defined by (1) is in fact an AG-isomorphism, and so the result is proved.

EXERCISES

1. Prove that in Theorem 2.1A, $(V_x)^L \simeq (V_y)^L$ when x and y lie in the same double (H, L)-coset.
2. Suppose that H and L are subgroups of the group G such that $G = HL$ and $H \cap L = 1$. Let K be a field and let V be a KL-module. Put $n := \dim_K V$. Show that $(V^G)_H$ is isomorphic to the direct sum of n copies of KH (as a KH-module).
3. Let G be a group and K a splitting field for G. If K has characteristic 0, deduce from Theorem 2.1A that the number of distinct irreducible representations of G is equal to the number of conjugate classes of G.

2.2 CLIFFORD'S THEOREM

4. Let A be a commutative ring, H a subgroup of the group G, and V an AH-module. Define an action of G on $\text{Hom}_{AH}(AG, V)$ by $af^y := (ay^{-1})f$ for all $a \in AG$, $f \in \text{Hom}_{AH}(AG, V)$, and $y \in G$. Show that under this action the A-module $\text{Hom}_{AH}(AG, V)$ becomes an AG-module which is isomorphic to V^G.

2.2 Clifford's theorem

Let K be a field and let G be a group with H as a normal subgroup. In this situation we can say more about the restriction–induction process. The key result is Clifford's theorem.

In the case where $H \triangleleft G$ we say that two KH-modules V_1 and V_2 are *conjugate* (under G) with respect to some $y \in G$ if there is a K-isomorphism $v \mapsto v^y$ of V_1 onto V_2 such that for all $x \in H$ and $v \in V$ we have $(vx)^y = v^y y^{-1} xy$ (note that $y^{-1}xy \in H$ because $H \triangleleft G$). Clearly if V_1 and V_2 are KH-modules that are conjugate under G, then V_1 is irreducible (respectively, indecomposable) iff V_2 has this property.

EXAMPLE Let H be a normal subgroup of G and suppose that V is a KG-module. Let U be a KH-submodule of V_H. Then for each $y \in G$, Uy is also a KH-module and U is conjugate to Uy with respect to y.

Theorem 2.2A (*Clifford*) Let H be a normal subgroup of the group G and let K be a field. Let V be an irreducible KG-module. Then

(i) V_H is completely reducible and its irreducible components are all conjugate (under G);

(ii) there exists a subgroup S of G such that $H \subseteq S \subseteq G$ and for some irreducible KS-module W we have $V \simeq W^G$;

(iii) each irreducible component of V_H occurs with the same multiplicity, and there are exactly $|G:S|$ nonisomorphic irreducible constituents.

Proof (i) Let U be an irreducible submodule of V_H. Put $\Omega := \{Uy \mid y \in G\}$. Then by the example above each element of Ω is a KH-module which is conjugate under G to U and hence also irreducible. The sum of the modules in Ω is clearly a KG-module and so equals V by the irreducibility of V. Thus V_H is a sum of irreducible KH-modules Uy ($y \in G$); hence by Note 5 of §1.3, V_H is completely reducible and can be written

(1) $$V_H = Uy_1 \oplus Uy_2 \oplus \cdots \oplus Uy_n$$

for some $y_i \in G$ with $y_1 = 1$. This proves (i).

(ii) Let $U_1, U_2, \ldots, U_m$ denote the distinct (up to isomorphism) irreducible constituents of V_H, and define W_i ($i = 1, 2, \ldots, m$) as the sum of all Uy

$(y \in G)$ such that $Uy \simeq U_i$. Then W_i is a KH-submodule of V_H which is a direct sum of certain Uy; in particular, every irreducible constituent of W_i is isomorphic to U_i. Thus by the Jordan–Hölder theorem, $(\sum_{i \neq j} W_i) \cap W_j = 0$ for each j since $\sum_{i \neq j} W_i$ and W_j have no isomorphic irreducible constituent in common. Therefore from (1) we conclude that

(2) $$V_H = W_1 \oplus W_2 \oplus \cdots \oplus W_m.$$

On the other hand it is clear that G acts on the set $\{W_1, W_2, \ldots, W_m\}$. This action is transitive because V is irreducible.

If we define S to be the stabilizer of W_1 under the latter action, then $|G:S| = m$, and it follows from Lemma 2.1 that W_1 is a KS-module and $V \simeq (W_1)^G$. Finally W_1 is irreducible because V is (see Note 2 of §2.1). Hence taking $W = W_1$, (ii) follows.

(iii) Let $d = \dim_K V$. Since G acts transitively on $\{W_1, W_2, \ldots, W_m\}$, each W_i has the same dimension, and so by (ii) we have $\dim_K W_i = d/m$. On the other hand, by (i) each irreducible constituent of V has dimension equal to $\dim_K U$. Therefore by (ii) the multiplicity of U_i as an irreducible constituent in V_H is $\dim W_i / \dim U_i = d/m \dim U$, which is independent of i. As we noted above, $m = |G:S|$, and so (iii) is proved.

Remark The KH-modules W_i defined in part (ii) of the proof above are called the *homogeneous components* of V_H.

Corollary Under the hypothesis of the theorem

(i) if V_H has at least two nonisomorphic irreducible components, then V is imprimitive;

(ii) if V_H has at least one trivial irreducible component, then $H \subseteq \text{Ker } V$.

Proof (i) It follows from (iii) that $S \neq G$. Hence G is imprimitive by the example of §2.1.

(ii) If U is a trivial KH-submodule of V_H, then its conjugates under G are all trivial. Hence by part (i) of the theorem, all irreducible components of the completely reducible KH-module V_H are trivial. This means H acts trivially on V, and so $H \subseteq \text{Ker } V$.

If a KG-module V has dimension 1, then its structure is very simple (see for example, the corollary of Theorem 1.5). The situation is only a little more complicated when V can be written as an induced module $V \simeq W^G$, where W is a KH-module of dimension 1 for some subgroup H (see Exercise 3). In the latter case we say that V is *monomial*; in particular, every KG-module of dimension 1 is monomial. The following classical theorem of Blichfeldt shows that for a nilpotent group and a splitting field, the irreducible KG-modules are all monomial.

2.3 GROUP CHARACTERS

Theorem 2.2B (*Blichfeldt*) Let G be a nilpotent group and K a field such that K is a splitting field for each subgroup of G. Then every irreducible KG-module is monomial.

Proof We shall proceed by induction on $|G|$; the result is certainly true if $G = 1$. Let V be an irreducible KG-module. If $\dim_K V = 1$, then V is monomial, so we may assume that $\dim_K V > 1$. Put $\tilde{G} = G/\text{Ker } V$. Then V is also an irreducible $K\tilde{G}$-module, and so by the corollary of Theorem 1.5, $\tilde{G}$ is not abelian because $\dim'_K V > 1$. This means that the center $\tilde{Z}$ of $\tilde{G}$ is a proper subgroup; choose $\tilde{y} \in \tilde{G} \setminus \tilde{Z}$ so that $\tilde{Z}\tilde{y}$ lies in the center of $\tilde{G}/\tilde{Z}$. (Note that the center of $\tilde{G}/\tilde{Z}$ is nontrivial because G is nilpotent.) Let $\tilde{H}$ be the subgroup of $\tilde{G}$ generated by $\tilde{Z}$ and $\tilde{y}$; then $\tilde{H}/\tilde{Z} \triangleleft \tilde{G}/\tilde{Z}$ and $\tilde{H}$ is abelian because $\tilde{Z}$ centralizes $\tilde{y}$. Take H as the full inverse image of $\tilde{H}$ under the canonical homomorphism $G \to \tilde{G}$. The subgroup H is normal in G since $\tilde{H}/\tilde{Z} \triangleleft \tilde{G}/\tilde{Z}$ and $H/\text{Ker } V \simeq \tilde{H}$ is abelian. Choose $y \in G$ mapping onto $\tilde{y}$. Since $\tilde{y} \notin \tilde{Z}$, the action of y on V is not scalar. Now by the corollary of Theorem 1.5, the irreducible constituents of V_H all have dimension 1. Since the action of y is not scalar, and $y \in H$, therefore not all irreducible constituents of V_H are isomorphic. Hence by Theorem 2.2A, there exists a proper subgroup S of G and an irreducible KS-module W such that $V \simeq W^G$. By the induction assumption, W is monomial (since $|S| < |G|$). Hence V is monomial by Note 3 of §2.1.

EXERCISES

1. Let G be the dihedral group of order 8 and let $\mathbf{C}$ be the complex field. Determine a full set of irreducible $\mathbf{C}G$-modules (use Theorem 2.2B).
2. Let G be a group with a faithful KG-module V of dimension n over any field K. Suppose that V is monomial. Show that G has a normal abelian subgroup H such that G/H is isomorphic to a subgroup of the symmetric group Sym(n) on n symbols. Moreover, if V is irreducible, then show that the latter group is transitive.
3. Let K be a field, G a group, and V a completely reducible KG-module. Use Theorem 2.2A to show that V_H is a completely reducible KH-module whenever H is a subnormal subgroup of G. (Recall that H is *subnormal* in G if it appears in some composition series for G.)

2.3 Group characters

Let G be a group and K a field. Then to each KG-module V we have the associated representation $T: G \to \text{Aut}_K(V)$ defined by $vT(x) := vx$ for all $v \in V$ and $x \in G$. Since $T(x)$ is a linear transformation of the vector space V, we can define its trace tr $T(x)$. The function $\zeta: G \to K$ given by $\zeta(x) := $ tr $T(x)$ is called the (Frobenius) *character* of G afforded by V.

EXAMPLE The trivial KG-module V is a one-dimensional vector space such that $vx = v$ for all $v \in V$ and $x \in G$. The character afforded by V is denoted by 1_G and is called the *principal* character; we have $1_G(x) = 1$ for all $x \in G$.

Note 1 A character ζ is a *class function* on G; that is, its value is constant on any conjugate class $\mathscr{C}$ of G. Indeed, for all $x, y \in G$ we have $\zeta(y^{-1}xy) = $ tr $T(y^{-1}xy) = \text{tr}\{T(y^{-1})T(x)T(y)\} = $ tr $T(x) = \zeta(x)$ by standard properties of the trace.

Note 2 If U and V are isomorphic KG-modules and afford the characters ζ_U and ζ_V, respectively, then $\zeta_U = \zeta_V$. For suppose that $f: U \to V$ is a KG-isomorphism, and let T_U and T_V be the representations of G associated with U and V, respectively. Then for all $u \in U, x \in G$ we have $(uf)x = (ux)f$, which gives $(uf)T_V(x) = (uT_U(x))f$. Hence $fT_V(x)f^{-1} = T_U(x)$ for all $x \in G$. This shows that $\zeta_U(x) = $ tr $T_U(x) = \text{tr}\{fT_V(x)f^{-1}\} = $ tr $T_V(x) = \zeta_V(x)$ for all $x \in G$; hence $\zeta_U = \zeta_V$.

Note 3 Let V be a KG-module and U a proper nonzero submodule; so $V \supset U \supset 0$. If ζ_1 and ζ_2 are characters afforded by U and V/U, respectively, then V affords the character $\zeta_1 + \zeta_2$. [*Proof* Take a K-basis $v_1, v_2, \ldots, v_n$ of V such that $v_1, v_2, \ldots, v_m$ is a basis for U and $v_{m+1} + U, \ldots, v_n + U$ is a basis for V/U. Calculate the traces relative to these bases.]

Note 4 If U and V are KG-modules which afford the characters ζ_U and ζ_V, respectively; then $U \otimes V$ (see §1.4) affords the character $\zeta_U \zeta_V$ (the product). [*Proof* Let $u_1, u_2, \ldots, u_m$ and $v_1, v_2, \ldots, v_n$ be K-bases of U and V, respectively; then $u_i \otimes v_j$ ($i = 1, 2, \ldots, m; j = 1, 2, \ldots, n$) is a K-basis of $U \otimes V$. Calculate the traces relative to these bases.]

Let G be a group and K a field. Then every representation $T: G \to \text{Aut}_K(V)$ of G can be extended to a unique representation $T: KG \to \text{End}_K(V)$ of the K-algebra KG (we denote the latter representation also by T). The annihilator Ann V of V is then equal to $\{a \in KG \mid T(a) = 0\}$.

Given any representation T of G (or KG) associated with V (also said to be *afforded* by V) and any fixed K-basis $v_1, v_2, \ldots, v_n$ of V, there is a corresponding matrix representation $T^*: G \to GL(n, K)$, where $n := \dim_K V$. The relation between T and T^* is given by $T^*(x) := [\alpha_{ij}(x)]$, where $v_i T(x) = v_i x = \sum_{j=1}^n \alpha_{ij}(x)v_j$ for $i = 1, 2, \ldots, n$. The following classical theorem is due to Burnside, Frobenius, and Schur.

Theorem 2.3 Let G be a group and K a splitting field for KG. Let $V_1, V_2, \ldots, V_s$ be a full set of nonisomorphic irreducible KG-modules, and let $T_1^*, T_2^*, \ldots, T_s^*$ be matrix representations of G afforded by these modules. Suppose $T_l^*(x) := [\alpha_{ij}^l(x)]$ for each $x \in G$ and put $d_l := \dim_K V_l$. Then the set of

2.3 GROUP CHARACTERS

functions $\alpha_{ij}^l: G \to K$ ($i, j = 1, \ldots, d_l$; $l = 1, 2, \ldots, s$) is linearly independent over K.

Remark The functions α_{ij}^l ($i, j = 1, 2, \ldots, d_l$) are called the *coordinate functions* of the matrix representation T_l^*.

Proof We shall use the notation of Theorem 1.3A with $R = KG$. The note following that theorem shows that we may take our V_j to be the irreducible KG-modules referred to in the theorem. Furthermore, it was also shown there that Ann V_l contains each $a \in KG$ such that $a + \text{rad } KG \in I_i$ for some $i \neq l$; in other words, $KG/\text{rad } KG = (I_l + \text{Ann } V_l)/\text{rad } KG$. On the other hand it follows from the corollary of Theorem 1.3A that the image of KG under the representation afforded by V_l is the whole of $\text{End}_K(V_l)$ because K is a splitting field. Interpreting this in terms of the matrix representation shows that the image of KG under T_l^* is $\text{Mat}(d_l, K)$. Thus for any $m, n = 1, 2, \ldots, d_l$ we can choose $a \in KG$ such that $\alpha_{mn}^l(a) = 1$ and $\alpha_{ij}^l(a) = 0$ when $(i, j) \neq (m, n)$. Moreover, we may suppose that $a + \text{rad } KG \in I_l$ since as we saw above, $KG/\text{rad } KG = (I_l + \text{Ann } V_l)/\text{rad } KG$. But when $a + \text{rad } KG \in I_l$, then from above $a \in \text{Ann } V_t$ for each $t \neq l$, and so $T_t^*(a) = [\alpha_{ij}^t(a)] = 0$. Thus we have shown that for each function α_{mn}^l there exists $a \in KG$ such that $\alpha_{mn}^l(a) = 1$ but $\alpha_{ij}^t(a) = 0$ for all $(t, i, j) \neq (l, m, n)$. This clearly implies that α_{mn}^l ($m, n = 1, 2, \ldots, d_l$; $l = 1, 2, \ldots, s$) are linearly independent, and the theorem is proved.

Corollary 1 With the hypothesis and notation of the theorem, let $\zeta_1, \zeta_2, \ldots, \zeta_s$ be the characters afforded by the irreducible KG-modules $V_1, V_2, \ldots, V_s$. Then these characters are linearly independent (as functions from G into K). In particular, two irreducible KG-modules are isomorphic iff they afford the same character.

Proof Since

$$\zeta_l := \sum_{i=1}^{d_l} \alpha_{ii}^l \quad (l = 1, 2, \ldots, s),$$

the first assertion follows at once from the theorem. Since every irreducible KG-module is isomorphic to one (and only one) of the V_j, the second assertion now follows using Note 2.

Corollary 2 Let G be a group and K a field of characteristic 0 such that K is a splitting field for G. Then two KG-modules are isomorphic iff they afford the same character.

Proof Let U and V be two KG-modules and suppose they afford the characters ζ_U and ζ_V, respectively. Then using the notation of the theorem, let m_j (respectively, n_j) be the multiplicity of V_j as an irreducible constituent of U (respectively, V) for $i = 1, 2, \ldots, s$. Since K has characteristic 0, Maschke's theorem (Theorem 1.3B) shows that both U and V are completely reducible. Therefore $U \simeq V$ iff $m_j = n_j$ for $j = 1, 2, \ldots, s$. On the other hand, if ζ_j is the character afforded by V_j, then using Note 3 it follows that

$$\zeta_U = \sum_{j=1}^{s} m_j \zeta_j \quad \text{and} \quad \zeta_V = \sum_{j=1}^{s} n_j \zeta_j.$$

Thus by Corollary 1, $\zeta_U = \zeta_V$ iff $m_j = n_j$ for $j = 1, 2, \ldots, s$ (here we are using the fact that K has characteristic 0). Hence we conclude $U \simeq V$ iff $\zeta_U = \zeta_V$.

Let G be a group and K a field. A *class function* from G into K is defined to be any function $\theta: G \to K$ with the property that θ is constant on each conjugate class $\mathscr{C}$ of G. [Equivalently, $\theta(y^{-1}xy) = \theta(x)$ for all $x, y \in G$.] We shall denote the set of all class functions of G into K by $\text{Class}_K(G)$. It is readily verified that $\text{Class}_K(G)$ is a K-algebra under the operations defined by $(\theta_1 + \theta_2)(x) := \theta_1(x) + \theta_2(x)$, $(\theta_1 \theta_2)(x) := \theta_1(x)\theta_2(x)$ and $(\alpha\theta_1)(x) := \alpha(\theta_1(x))$ for all $x \in G$, $\alpha \in K$, and $\theta_1, \theta_2 \in \text{Class}_K(G)$. In particular, if G has t conjugate classes, then it is evident that $\dim_K \text{Class}_K(G) = t$; for example, we could choose as a K-basis the t functions $\theta_1, \theta_2, \ldots, \theta_t$, where θ_i is 1 on the ith conjugate class and 0 on all other conjugate classes.

The conjugate classes of G also come up in what at first appears to be a different context. Consider the group algebra KG, and let $Z(KG)$ denote the center of KG. Then $a = \sum_{x \in G} \alpha_x x$ ($\alpha_x \in K$) lies in $Z(KG)$ iff it commutes with all elements of KG; and since $\{y \mid y \in G\}$ is a K-basis of KG, this is equivalent to $ay = ya$ for all $y \in G$. Comparing coefficients, this gives $\sum_{x \in G} \alpha_x x \in Z(KG)$ iff $\alpha_x = \alpha_{y^{-1}xy}$ for all $x, y \in G$. Let $\mathscr{C}_1, \mathscr{C}_2, \ldots, \mathscr{C}_t$ be the conjugate classes of G and define the ith *class sum* in KG as $c_i := \sum_{x \in \mathscr{C}_i} x$ for $i = 1, 2, \ldots, t$. Then $c_1, c_2, \ldots, c_t$ are linearly independent over K (since the elements of G are); and from what we have just shown, $c_1, c_2, \ldots, c_t$ is a K-basis of $Z(KG)$. It is now easy to see that $\sum_{x \in G} \alpha_x x$ lies in $Z(KG)$ iff the function $x \mapsto \alpha_x$ lies in $\text{Class}_K(G)$.

Since $c_1, c_2, \ldots, c_t$ is a K-basis for $Z(KG)$, there exists γ_{ijl} ($i, j, l = 1, 2, \ldots, t$) in K such that

(1) $$c_i c_j = \sum_{l=1}^{t} \gamma_{ijl} c_l \quad \text{for} \quad i, j = 1, 2, \ldots, t.$$

Rewriting these equations in terms of the basis $\{x \mid x \in G\}$ of KG, we see that the γ_{ijl} are nonnegative integer multiples of the unity 1 of K. Moreover,

(2) $$\gamma_{ij1} = \begin{cases} |\mathscr{C}_i| & \text{if } \mathscr{C}_j = \{x^{-1} \mid x \in \mathscr{C}_i\} \\ 0 & \text{otherwise.} \end{cases}$$

EXERCISES

1. Let G be a finite group, K an arbitrary field of characteristic 0, T^1, T^2, ..., T^n a set of pairwise inequivalent irreducible matrix representations of G over K, d_i the degree of T^i, and α_{uv}^i the coordinate functions of T^i. Prove that $\{\alpha_{uv}^i \mid u, v = 1, 2, \ldots, d_i;\ i = 1, 2, \ldots, n\}$ is a linearly independent set.
2. Let G be a group and K an arbitrary field of characteristic 0. Prove that the irreducible characters of G over K are linearly independent.

2.4 The theory of ordinary characters

Throughout this section we shall consider a group G of order g and a field K where K is a splitting field of characteristic 0 for KG. Under these hypotheses we shall call the characters and representations of G over K the *ordinary* characters and representations. There is an extensive theory of ordinary characters, but here we shall just touch on a few basic results which we shall be using later.

Since K has characteristic 0, all KG-modules are completely reducible (Theorem 1.3B); and since K is also a splitting field, the number of nonisomorphic irreducible KG-modules is equal to the number of conjugate classes of G (Theorem 1.5). By Corollary 1 of Theorem 2.3 this latter number is also the number of distinct characters afforded by irreducible KG-modules; we shall denote this common number by s.

It follows from Corollary 2 of Theorem 2.3 that two KG-modules are isomorphic iff they have the same character. Therefore (in this situation of ordinary characters) we can refer unambiguously to a character ζ of G as being *reducible* or *irreducible* depending whether it is afforded by a reducible or irreducible KG-module, and define the *irreducible constituents* of ζ as the characters of the irreducible constituents of a KG-module that affords ζ. Similarly, the *kernel* of ζ (written Ker ζ) is defined as the kernel of any KG-module that affords ζ, and the *degree* of ζ (written deg ζ) is the dimension of this module.

Let V be a KG-module that affords ζ and let T be the representation of G afforded by V. Each element $x \in G$ has order dividing $g := |G|$, and so $T(x)^g = 1$. Let ω be a primitive gth root of unity over K. Then $X^g - 1$ splits into linear factors in the field extension $K(\omega)$ of K, and its roots are precisely $1, \omega, \omega^2, \ldots, \omega^{g-1}$. Each eigenvalue of $T(x)$ is then one of the roots of unity. Without loss of generality we may assume that the field $\mathbf{Q}$ of rationals is contained in K (since K has characteristic 0) and identify $\mathbf{Q}(\omega)$ with the corresponding subfield of the complex field. In particular, the values of ζ all lie in $\mathbf{Q}(\omega)$. Moreover, if $T(x)$ has the eigenvalues $\omega^{i_1}, \omega^{i_2}, \ldots, \omega^{i_d}$ (where $d = \deg \zeta := \dim_K V$), then we have

(1) $\quad |\zeta(x)| = |\omega^{i_1} + \omega^{i_2} + \cdots + \omega^{i_d}| \leq |\omega^{i_1}| + |\omega^{i_2}| + \cdots + |\omega^{i_d}| = d$

with equality iff all ω^{i_s} are equal. Since K has characteristic 0, it follows from $T(x)^g = 1$ that the minimal polynomial for $T(x)$ has distinct roots. Thus if all the eigenvalues ω^{i_t} of $T(x)$ are equal, then the minimal polynomial has the single root ω^{i_t} and $T(x)$ is scalar. Thus from (1) we conclude

(2) $\quad |\zeta(x)| \leq \zeta(1) \quad$ for all $x \in G \quad$ and $\quad |\zeta(x)| = \zeta(1) \quad$ iff $T(x)$ is a scalar.

In particular, $x \in \text{Ker } \zeta$ iff $T(x) = 1$, so from (2) we have

(3) $\qquad \text{Ker } \zeta = \{x \in G \mid \zeta(x) = \zeta(1)\}.$

Next we define an *inner product* $\langle \, , \, \rangle$ on the K-algebra $\text{Class}_K(G)$ of class functions by

$$\langle \theta, \phi \rangle := \frac{1}{g} \sum_{x \in G} \theta(x)\phi(x^{-1}).$$

Clearly $\langle \, , \, \rangle$ is bilinear and symmetric. (Note that this definition does not require K to be a splitting field for KG.) A set of functions $\theta_1, \theta_2, \ldots, \theta_n$ in $\text{Class}_K(G)$ is called *orthogonal* if $\langle \theta_i, \theta_j \rangle = 0$ for all $i \neq j$, and *orthonormal* if it is orthogonal and $\langle \theta_i, \theta_i \rangle = 1$ for each i. Since G has s conjugate classes, $\dim_K \text{Class}_K(G) = s$ (see §2.3). On the other hand, we also know that G has s distinct irreducible characters and each of these lies in $\text{Class}_K(G)$. Since these characters are linearly independent by Corollary 1 of Theorem 2.3, they must form a K-basis for $\text{Class}_K(G)$. In the next theorem we shall prove an even stronger result.

To standardize the notation let $\mathscr{C}_1 = \{1\}, \mathscr{C}_2, \ldots, \mathscr{C}_s$ be the conjugate classes of G and let $c_i := \sum_{x \in \mathscr{C}_i} x$ be the corresponding class sums in KG. For each class $\mathscr{C}_i$, the set $\{x^{-1} \mid x \in \mathscr{C}_i\}$ is also a conjugate class which we denote by $\mathscr{C}_{i^*}$. Put $h := |\mathscr{C}_i|$ and for each character ζ of G define ζ_i as the value ζ takes on the class $\mathscr{C}_i$ ($i = 1, 2, \ldots, s$). We shall denote the s irreducible characters of G over K by $\zeta^1, \zeta^2, \ldots, \zeta^s$ (using upper indices). As usual δ_{ij} denotes the Kronecker delta (equal to 1 if $i = j$ and 0 otherwise).

EXAMPLE Consider the KG-module KG and let ρ be the character that it affords (ρ is called the *regular* character of G). Since KG is semisimple by Maschke's theorem (Theorem 1.3B), it follows from Theorem 1.3A that $KG = I_1 \oplus I_2 \oplus \cdots \oplus I_s$, where the irreducible constituents of I_j are all isomorphic to a single irreducible KG-module V_j. By Note 3 of §1.3, $I_j \simeq \text{End}_K(V_j)$ as K-algebras because K is a splitting field. Hence, counting dimensions, we see that $\dim_K I_j = d_j^2$, where $d_j := \dim_K V_j$, and so V_j occurs with multiplicity d_j as an irreducible constituent of I_j. By the note following

2.4 THE THEORY OF ORDINARY CHARACTERS

Theorem 1.3A, the characters afforded by the V_j are precisely the distinct irreducible characters of G, so we may suppose that V_j affords ζ^j. Thus we conclude

$$\rho = \sum_{j=1}^{s} d_j \zeta^j = \sum_{j=1}^{s} \zeta^j(1)\zeta^j.$$

On the other hand, using the basis $\{y \mid y \in G\}$ for KG to calculate traces, we find at once that $\rho(1) = g$ and $\rho(x) = 0$ for all $x \neq 1$ in G.

Theorem 2.4A (*Orthogonality relations*) With the above notation

(1) $$\sum_{l=1}^{s} h_l \zeta_i^l \zeta_{j*}^l = g\, \delta_{ij}$$

and

(2) $$\sum_{l=1}^{s} h_l \zeta_l^i \zeta_{l*}^j = g\, \delta_{ij}$$

for $i, j = 1, 2, \ldots, s$. In particular, the characters $\zeta^1, \zeta^2, \ldots, \zeta^s$ form an orthonormal basis of $\text{Class}_K(G)$.

Proof For each character ζ^i choose a KG-module U_i which affords ζ^i and let T_i be the representation afforded by U_i. Since c_j lies in the center $Z(KG)$ of KG, $T_i(c_j) \in \text{End}_{KG}(U_i)$. But K is a splitting field and so the latter equals $K \cdot 1$. Hence for each i and j there exist $\omega_j^i \in K$ such that $T_i(c_j) = \omega_j^i \cdot 1$. Taking traces then shows that $h_j \zeta_j^i = \sum_{x \in \mathscr{C}_j} \zeta^i(x) = \text{tr } T_i(c_j) = \omega_j^i \cdot \zeta^i(1)$, and so

(3) $$\omega_j^i = h_j \zeta_j^i / \zeta^i(1) \qquad \text{for all} \quad i, j.$$

On the other hand it follows from (1) of §2.3 that for each i and j and each l we have

$$T_l(c_i) T_l(c_j) = \sum_{m=1}^{s} \gamma_{ijm} T_l(c_m).$$

Therefore using (3) we obtain

(4) $$h_i h_j \zeta_i^l \zeta_j^l = \zeta^l(1) \sum_{m=1}^{s} \gamma_{ijm} h_m \zeta_m^l \qquad (i, j, l = 1, 2, \ldots, s).$$

Now consider the regular character ρ of G (see the example above). Since $\rho = \sum_{l=1}^{s} \zeta^l(1)\zeta^l$, (4) shows that

$$h_i h_j \sum_{l=1}^{s} \zeta_i^l \zeta_j^l = \sum_{m=1}^{s} \gamma_{ijm} h_m \sum_{l=1}^{s} \zeta^l(1) \zeta_m^l = \sum_{m=1}^{s} \gamma_{ijm} h_m \rho_m = g \gamma_{ij1}$$

because $\rho_1 = g$ and $\rho_i = 0$ for $i \neq 1$. Finally from (2) of §2.3 we conclude $\gamma_{ij1} = h_i \delta_{ij*}$. Replacing j by j^* (and j^* by j) this proves (1).

Now the equations in (1) can be written as a matrix identity in the form $YZ = I$, where Y and Z are s-by-s matrices over K whose (i, j)th entries are $h_i \zeta_i^j$ and ζ_{j*}^i/g, respectively. Hence $Y = Z^{-1}$, and so $ZY = I$. Using the fact that $(l^*)^* = l$ and $h_l = h_{l*}$, the latter matrix identity is equivalent to Eqs. (2).

Finally (2) can be written in the form

$$\langle \zeta^i, \zeta^j \rangle = \frac{1}{g} \sum_{x \in G} \zeta^i(x) \zeta^j(x^{-1}) = \frac{1}{g} \sum_{l=1}^{s} h_l \zeta_l^i \zeta_{l*}^j = \delta_{ij}$$

and this shows that $\zeta^1, \zeta^2, \ldots, \zeta^s$ are orthonormal. Since $\text{Class}_K(G)$ has dimension s, they form an orthonormal basis for this space.

Corollary If ζ is any character of G, then we can write $\zeta = \sum_{l=1}^{s} m_l \zeta^l$, where the nonnegative integer m_l is the multiplicity of ζ^l in ζ. Since $\zeta^1, \zeta^2, \ldots, \zeta^s$ are orthonormal, this shows that $m_l = \langle \zeta, \zeta^l \rangle$ for each l, and $\langle \zeta, \zeta \rangle = \sum_{i=1}^{s} m_i^2$. In particular, ζ is irreducible iff $\langle \zeta, \zeta \rangle = 1$.

Since K has characteristic 0 its prime subfield is isomorphic to $\mathbf{Q}$, so without loss of generality we may suppose that K contains the field $\mathbf{Q}$ of rational numbers and hence the ring $\mathbf{Z}$ of integers. Consider the ring $\mathbf{Z}[X]$ of polynomials over $\mathbf{Z}$; a polynomial $f(X) \in \mathbf{Z}[X]$ is called *monic* if its leading coefficient is 1. An element $\alpha \in K$ is called an *algebraic integer* if it is a root of some monic polynomial $f(X) \in \mathbf{Z}[X]$.

Note 1 It is well known (and easily proved), that $\alpha \in K$ is an algebraic integer iff the subring $\mathbf{Z}[\alpha]$ is a (finitely generated) $\mathbf{Z}$-module. It follows immediately that if α and β are algebraic integers of K, then $\alpha \pm \beta$ and $\alpha\beta$ are also algebraic integers of K. Thus the set $\text{Int}(K)$ of all algebraic integers of K is a subring of K containing $\mathbf{Z}$. Moreover, if m and n are nonzero relatively prime integers, $f(X) := X^d + a_1 X^{d-1} + \cdots + a_d \in \mathbf{Z}[X]$ and $f(m/n) = 0$, then $m^d/n = -a_1 m^{d-1} - a_2 m^{d-2} n - \cdots - a_d n^{d-1} \in \mathbf{Z}$ and so $n = \pm 1$. This shows that $\mathbf{Q} \cap \text{Int}(K) = \mathbf{Z}$. [Compare with §3.2.]

Theorem 2.4B With the notation above

(i) ζ_i^l and $h_i \zeta_i^l/\zeta^l(1)$ are algebraic integers in K for all $i, l = 1, 2, \ldots, s$; and
(ii) $\zeta^l(1)$ divides g.

Proof (i) As we saw earlier, ζ_i^l is a sum of gth roots of unity. Since each gth root of unity is clearly an algebraic integer, this shows that $\zeta_i^l \in \text{Int}(K)$. Now fix i and let C be the $s \times s$ matrix whose (j, m)th entry is γ_{ijm}. As we

2.4 THE THEORY OF ORDINARY CHARACTERS

noted in §2.3 the γ_{ijm} are nonnegative integers in K. The equation (4) above can be written

$$C \begin{bmatrix} h_1 \zeta_1^l \\ h_2 \zeta_2^l \\ \vdots \\ h_s \zeta_s^l \end{bmatrix} = \frac{h_i \zeta_i^l}{\zeta^l(1)} \begin{bmatrix} h_1 \zeta_1^l \\ h_2 \zeta_2^l \\ \vdots \\ h_s \zeta_s^l \end{bmatrix},$$

which shows that $h_i \zeta_i^l / \zeta^l(1)$ is an eigenvalue of the matrix C. Since C has integer entries, the characteristic polynomial $\det(X \cdot 1 - C)$ is a monic polynomial in $\mathbf{Z}[X]$, and so (i) is proved.

(ii) Theorem 2.4A shows that for any l we have $g/\zeta_1^l = \sum_{i=1}^{s} \zeta_{i*}^l (h_i \zeta_i^l / \zeta^l(1))$. But ζ_{i*}^l and $h_i \zeta_i^l / \zeta^l(1)$ both lie in $\text{Int}(K)$ by (i), and so $g/\zeta_1^l \in \mathbf{Q} \cap \text{Int}(K) = \mathbf{Z}$. Hence $\zeta^l(1)$ divides g.

EXERCISES

1. Let G and H be two groups and let K be a splitting field of characteristic 0 for both KG and KH. Suppose ζ is a character of G over K and let χ be a character of H over K. Show that $\psi \in \text{Class}_K(G \times H)$ defined by $\psi((x,y)) := \zeta(x)\chi(y)$ for all $(x,y) \in G \times H$ is a character of $G \times H$ over K and prove that it is irreducible iff ζ and χ are both irreducible.

2. Let G be a group and K a field of characteristic 0 which is splitting for G. If $Z(G)$ is the center of G, prove that the degree of an irreducible representation of G over K divides $|G: Z(G)|$. [*Hint*: Define the relation $\sim$ on the set of conjugate classes of G by $\mathscr{C}_i \sim \mathscr{C}_j$ iff $\mathscr{C}_i = \mathscr{C}_j z$ for some $z \in Z(G)$. Show that if $\chi_i \neq 0$, then the $\sim$-class containing $\mathscr{C}_i$ has $|Z(G)|$ members.]

3. Let G be a finite group, K a splitting field for G of characteristic 0, ζ a character of G, n the number of conjugate classes of G, and ζ_i the value of ζ at the ith conjugate class. Prove that $\zeta_1, \zeta_2, \ldots, \zeta_n$ are the eigenvalues of a matrix with nonnegative integer coefficients. [*Hint*: Consider $\zeta^i \zeta = \sum_{j=1}^{n} a_{ij} \zeta^j$, where $\zeta^1, \zeta^2, \ldots, \zeta^n$ are the irreducible characters of G.]

4. In Exercise 3, prove that $\zeta_1, \zeta_2, \ldots, \zeta_n$ are the eigenvalues of a permutation matrix iff $\zeta_1 = 1$.

5. Let θ be an ordinary character of G over a subfield of the complex numbers. Show that $\theta(x^{-1})$ equals the complex conjugate of $\theta(x)$. If θ is irreducible and x is an involution, show that $\theta(x)$ is an integer and $\theta(x) \equiv \theta(1) \pmod{2}$.

6. Let χ be a faithful ordinary character of a group G. If χ takes only rational values show the following: (i) the values of χ are all rational integers with $-n \leq \chi(x) \leq n$ ($n := \chi(1)$); and (ii) $g := |G|$ divides $(2n)!$. [*Hint*: Put $\theta = \prod_{j=-n}^{n-1} (\chi - j)$ and calculate $\langle \theta, \theta \rangle$.]

2.5 Induced characters

Let G be a group and let K be any field. Suppose that H is a subgroup of G and V is a KH-module. Consider the induced KG-module V^G. If $v_1, v_2, \ldots, v_d$ is a K-basis of V, and $x_1, x_2, \ldots, x_h$ is a right transversal of H in G, then $v_l \otimes x_j$ ($l = 1, 2, \ldots, d; j = 1, 2, \ldots, h$) is a K-basis for V^G. By definition of V^G we have $(v_l \otimes x_i)y := v_l(x_i y x_j^{-1}) \otimes x_j$ for each $y \in G$, where x_j is the coset representative such that $x_i y x_j^{-1} \in H$. Suppose that V and V^G, respectively, afford the matrix representations S and T with respect to the given bases. Then for each $y \in G$, $T(y)$ can be written as the "block" matrix

$$T(y) = \begin{bmatrix} \check{S}(x_1 y x_1^{-1}) & \check{S}(x_1 y x_2^{-1}) & \cdots & \check{S}(x_1 y x_h^{-1}) \\ \check{S}(x_2 y x_1^{-1}) & \check{S}(x_2 y x_2^{-1}) & \cdots & \check{S}(x_2 y x_h^{-1}) \\ \vdots & \vdots & \vdots & \vdots \\ \check{S}(x_h y x_1^{-1}) & \check{S}(x_h y x_2^{-1}) & \cdots & \check{S}(x_h y x_h^{-1}) \end{bmatrix},$$

where for all $z \in G$ we define the $d \times d$ matrix $\check{S}(z)$ by

$$\check{S}(z) = \begin{cases} S(z) & \text{if } z \in H \\ 0 & \text{otherwise.} \end{cases}$$

Thus we find that the characters χ and ζ afforded by V and V^G, respectively, are related by

(1) $$\zeta(y) = \sum_{i=1}^{h} \check{\chi}(x_i y x_i^{-1})$$

where

$$\check{\chi}(z) := \begin{cases} \chi(z) & \text{if } z \in H \\ 0 & \text{if } z \in G \setminus H. \end{cases}$$

The character ζ afforded by V^G and given in terms of χ by (1) is called the character *induced* from χ, and we shall use the notation χ^G for ζ. It is clear from the definition of ζ that the value of ζ given by (1) is independent of the choice of the right transversal $x_1, x_2, \ldots, x_h$ of H in G.

We now specialize to the case where K has characteristic 0. In this case, since G can be written as a disjoint union of $|H|$ right transversals of H in G, we have from (1) that

$$\chi^G(y) = \frac{1}{|H|} \sum_{x \in G} \check{\chi}(x^{-1} y x) \quad \text{for all } y \in G.$$

2.5 INDUCED CHARACTERS

More generally, for any class function $\theta \in \text{Class}_K(H)$ we shall define $\theta^G: G \to K$ by

$$\theta^G(y) := \frac{1}{|H|} \sum_{x \in G} \breve{\theta}(x^{-1}yx), \tag{2}$$

where $\breve{\theta}(z) := \theta(z)$ if $z \in H$ and 0 otherwise. Clearly $\theta^G \in \text{Class}_K(G)$, and from above θ^G is a character of G whenever θ is a character of H. The induction mapping $\theta \mapsto \theta^G$ is a function from $\text{Class}_K(H)$ into $\text{Class}_K(G)$. There is also the restriction mapping $\phi \mapsto \phi_H$ from $\text{Class}_K(G)$ into $\text{Class}_K(H)$ (where ϕ_H denotes the restriction of the function ϕ to H). The basic properties of these mappings are given in the following theorem.

Theorem 2.5A Let K be a field of characteristic 0. Let G be a group and H and S subgroups such that $H \subseteq S \subseteq G$. Then for all $\theta \in \text{Class}_K(H)$ and $\phi \in \text{Class}_K(G)$ we have

(i) $(\theta^S)^G = \theta^G$
(ii) $\theta^G \phi = (\theta \phi_H)^G$
(iii) (*Frobenius reciprocity*) $\langle \theta, \phi_H \rangle_H = \langle \theta^G, \phi \rangle_G$.

Proof (i) As x ranges over S and z ranges over G, the product xz ranges over G taking each value $|S|$ times. Therefore from (2) we have for all $y \in G$

$$(\theta^S)^G(y) = \frac{1}{|S|} \frac{1}{|H|} \sum_{x \in G} \sum_{x \in S} \breve{\theta}(z^{-1}x^{-1}yxz) = \frac{1}{|H|} \sum_{w \in G} \breve{\theta}(w^{-1}yw) = \theta^G(y).$$

(ii) (*Compare with Theorem 2.1B*) Put $\psi := \theta \phi_H \in \text{Class}_K(H)$. Then for all $x, y \in G$ we have

$$\breve{\theta}(x^{-1}yx)\phi(y) = \breve{\theta}(x^{-1}yx)\phi(x^{-1}yx) = \breve{\psi}(x^{-1}yx)$$

because ϕ is a class function on G. Therefore from (2) we get

$$(\theta^G \phi)(y) = \theta^G(y)\phi(y) = \psi^G(y)$$

and so $\theta^G \phi = \psi^G = (\theta \phi_H)^G$, as asserted.

(iii) Again we have for all $x, y \in G$, $\phi(y^{-1}) = \phi(x^{-1}y^{-1}x)$. On the other hand, $x^{-1}yx$ ranges over G as y ranges over G. Therefore using (2) we have

$$\langle \theta^G, \phi \rangle_G = \frac{1}{|G|} \sum_{y \in G} \left\{ \frac{1}{|H|} \sum_{x \in G} \breve{\theta}(x^{-1}yx)\phi(x^{-1}y^{-1}x) \right\}$$

$$= \frac{1}{|H|} \sum_{y \in G} \breve{\theta}(y)\phi(y^{-1}) = \frac{1}{|H|} \sum_{y \in H} \theta(y)\phi(y^{-1})$$

$$= \langle \theta, \phi_H \rangle_H \,.$$

Corollary Under the hypothesis of the theorem suppose that K is a splitting field for both H and G, and let χ and ζ be irreducible (ordinary) characters of H and G, respectively. Then the multiplicity of χ as an irreducible constituent of ζ_H is $\langle \chi, \zeta_H \rangle$ (corollary of Theorem 2.4A) and the multiplicity of ζ as an irreducible constituent of χ^G is $\langle \chi^G, \zeta \rangle$. These multiplicities are equal by part (iii) of the theorem.

EXAMPLE Let H be a subgroup of the group G and let 1_H denote the principal character of H. Let $x_1, x_2, \ldots, x_h$ be a right transversal of H in G. Then

$$1_H^G(y) = \sum_{i=1}^{h} \check{1}_H(x_i y x_i^{-1}).$$

Since $\check{1}_H(x_i y x_i^{-1}) = 1$ if $H x_i y = H x_i$ and 0 otherwise, this shows that $1_H^G(y)$ equals the number of cosets Hx of H in G such that $Hxy = Hx$. Equivalently, if we consider the action of G on the set $\Omega := \{Hx \mid x \in G\}$ of right cosets of H in G by right multiplication, then $1_H^G(y)$ equals the number of points fixed by y. In particular, $y \in \operatorname{Ker} 1_H^G$ iff $y \in x^{-1} H x$ for all $x \in G$; therefore $\operatorname{Ker} 1_H^G = \bigcap_{x \in G} x^{-1} H x$, which is the largest normal subgroup of G contained in H.

By Theorem 2.5A(iii), $\langle 1_H^G, 1_G \rangle = \langle 1_H, 1_H \rangle = 1$ and so 1_G is an irreducible constituent of 1_H^G with multiplicity 1. In general, ζ is an irreducible constituent of 1_H^G iff 1_H is an irreducible constituent of ζ_H. In the case where $H \triangleleft G$, the corollary of Theorem 2.2B shows that $\langle 1_H, \zeta_H \rangle \neq 0$ iff $H \subseteq \operatorname{Ker} \zeta$. Hence in this case

$$1_H^G = \sum_{i=1}^{t} m_i \zeta^i \quad \text{with} \quad m_i = \zeta^i(1),$$

where the sum is over all irreducible characters ζ^i of G with $H \subseteq \operatorname{Ker} \zeta^i$. (Compare with the example of §2.4, noting that $\rho = 1^G$, where 1 is the principal character of the subgroup 1).

The use of the induction process is an important method of obtaining characters of a group G from the characters of a subgroup H. One problem which arises is that the induced character χ^G is not usually an irreducible character of G even when χ is an irreducible character of H, and it is difficult to separate out the irreducible constituents. The following theorem due to Brauer and Suzuki [1] is therefore of importance. It shows how we can give a partial analysis of the induced characters under suitable restrictions. We shall only need this theorem in Chapter 8. To state the theorem we shall require the concept of a *trivial intersection set*. A subset S of a group G is called a trivial intersection set if

(a) $1 \notin S$;
(b) $S \subseteq N_G(S)$ and $x^{-1} S x \cap S = \emptyset$ for all $x \in G \setminus N_G(S)$.

2.5 INDUCED CHARACTERS

Theorem 2.5B Let G be a group and let S be a trivial intersection set in G; put $H := N_G(S)$. Let K be a field of characteristic 0 such that K is a splitting field for both H and G.

(i) If $\theta, \phi \in \text{Class}_K(H)$ that are 0 on $H \setminus S$, then $\theta^G(y) = \theta(y)$ and $\phi^G(y) = \phi(y)$ for all $y \in S$, $\theta^G(1) = \phi^G(1) = 0$, and $\langle \theta, \phi \rangle_H = \langle \theta^G, \phi^G \rangle_G$.

(ii) Let $\chi^1, \chi^2, \ldots, \chi^n$ ($n \geq 2$) be distinct irreducible characters of H over K such that for each i, $\chi^i(y) = 0$ for all $y \neq 1$ in $H \setminus S$, and $\chi^1(1) = \chi^2(1) = \cdots = \chi^n(1)$. Then there exists $\varepsilon = 1$ or -1 and distinct irreducible characters $\zeta^1, \zeta^2, \ldots, \zeta^n$ of G over K such that $(\chi^i - \chi^j)^G = \varepsilon(\zeta^i - \zeta^j)$ for $i, j = 1, 2, \ldots, n$.

Proof (i) Since S is a trivial intersection set and θ is a class function, we have for all $x \in G$ and $y \in S$

$$\theta(x^{-1}yx) = \begin{cases} \theta(y) & \text{if } x \in H := N_G(S) \\ 0 & \text{if } x \in G \setminus H. \end{cases}$$

Therefore it follows from (2) that $\theta^G(y) = \theta(y)$; and similarly, $\phi^G(y) = \phi(y)$ for all $y \in S$. Again from the hypothesis on S, the number of distinct conjugates of S in G is $|G:H| = |G:N_G(S)|$, and any two distinct conjugates of S are disjoint. It follows from (2) and the hypothesis on θ and ϕ that $\theta^G(y) = \phi^G(y) = 0$ for each $y \in G$ not conjugate to an element of S. Put $S_0 := \{y \in G \mid y \in S \text{ and } y^{-1} \in S\}$. Then using the fact that θ^G and ϕ^G are class functions we obtain

$$\langle \theta^G, \phi^G \rangle_G = \frac{1}{|G|} \sum_{y \in G} \theta^G(y) \phi^G(y^{-1}) = \frac{|G:H|}{|G|} \sum_{y \in S_0} \theta^G(y) \phi^G(y^{-1})$$

$$= \frac{1}{|H|} \sum_{y \in S_0} \theta(y) \phi(y^{-1}) = \frac{1}{|H|} \sum_{y \in H} \theta(y) \phi(y^{-1}) = \langle \theta, \phi \rangle_H.$$

(ii) First consider the case where $n = 2$, and put $\theta := \chi^1 - \chi^2$. Then θ satisfies the hypothesis of (i), and so

$$\langle \theta^G, \theta^G \rangle_G = \langle \theta, \theta \rangle_H = \langle \chi^1 - \chi^2, \chi^1 - \chi^2 \rangle_H = 2$$

by the orthogonality relations (corollary of Theorem 2.4A). By the same corollary it is clear that the relation $\langle \theta^G, \theta^G \rangle_G = 2$ implies that for some distinct irreducible characters ζ^1, ζ^2 of G we have $\theta^G = \pm \zeta^1 \pm \zeta^2$. Since $\theta^G(1) = 0$ by (i), we have $\theta^G = \varepsilon(\zeta^1 - \zeta^2)$ with $\varepsilon = 1$ or -1. This proves (ii) in the case $n = 2$.

In the case $n \geq 3$ put $\theta_i := \chi^1 - \chi^i$ ($i = 3, 4, \ldots, n$). Then θ_i satisfies the hypothesis of (i) in place of ϕ and so we have

$$1 = \langle \chi^1 - \chi^2, \chi^1 - \chi^i \rangle_H = \langle \theta, \theta_i \rangle_H = \langle \theta^G, \theta_i^G \rangle_G$$

and

$$2 = \langle \chi^1 - \chi^i, \chi^1 - \chi^i \rangle_H = \langle \theta_i, \theta_i \rangle_H = \langle \theta_i^G, \theta_i^G \rangle_G.$$

The second relation shows that θ_i^G has exactly two irreducible constituents each with multiplicity ± 1. Then the condition $\theta_i^G(1) = 0$ and the first relation show that either $\theta_i^G = \varepsilon(\zeta^1 - \zeta^2)$ or $\theta_i^G = \varepsilon(\zeta^i - \zeta^2)$ for some irreducible character ζ^i different from ζ^1 or ζ^2. Now consider θ_3^G. If $\theta_3^G = \varepsilon(\zeta^3 - \zeta^2)$, then we interchange ζ^1 and ζ^2 in our notation; this also changes ε to $-\varepsilon$. As a result θ_3^G becomes $\varepsilon(\zeta^1 - \zeta^3)$. Thus we may choose the notation so that $\theta_3^G = \varepsilon(\zeta^1 - \zeta^3)$. Then from the relation

$$1 = \langle \chi^1 - \chi^3, \chi^1 - \chi^i \rangle_H = \langle \theta_3, \theta_i \rangle_H = \langle \theta_3^G, \theta_i^G \rangle_G$$

for all $i \geq 4$, we see that $\theta_i^G = \varepsilon(\zeta^1 - \zeta^i)$ for all θ_i (including $\theta_2 := \theta$ and $\theta_1 := 0$). Moreover, it follows from the relation $1 = \langle \theta_i^G, \theta_j^G \rangle$ for all $i > j > 1$ that $\zeta^i \neq \zeta^j$ when $i \neq j$. Finally, for all $i, j = 1, 2, \ldots, n$ we have

$$(\chi^i - \chi^j)^G = (\theta_j - \theta_i)^G = \varepsilon(\zeta^1 - \zeta^j) - \varepsilon(\zeta^1 - \zeta^i) = \varepsilon(\zeta^i - \zeta^j)$$

and the theorem is proved.

EXERCISES

1. Let H and L be subgroups of the group G. Let V be a **C**G-module (**C** the complex field) which affords a character ζ. Put $V_0 := \{v \in V \mid vx = v \text{ for all } x \in G\}$. Show that V_0 is a **C**G-module with $\dim_{\mathbf{C}} V_0 = \langle \zeta, 1_G \rangle$. Deduce that if H and L together generate G, then

$$\langle \zeta_{H \cap L}, 1_{H \cap L} \rangle + \langle \zeta, 1_G \rangle \geq \langle \zeta_H, 1_H \rangle + \langle \zeta_L, 1_L \rangle.$$

2. Let H be a normal subgroup of G and χ an irreducible ordinary character of G. Prove that there exists an irreducible character ψ of H such that $\chi = \psi^G$ iff χ vanishes outside H and χ_H is a sum of distinct irreducible characters of H.

2.6 Brauer's theorem on induced characters

Throughout the present section we shall be dealing with a group G and a field K of characteristic 0 such that K is a splitting field for KH for each subgroup H of G. Let $\zeta^1, \zeta^2, \ldots, \zeta^s$ denote the set of all irreducible characters of G over K. Then a class function $\theta \in \text{Class}_K(G)$ is called a *generalized character* of G if it can be written in the form $\theta = \sum_{i=1}^{s} m_i \zeta^i$ for some integers m_i. The set of all generalized characters is denoted by $\text{Char}(G)$.

Note 1 Since $\zeta^1, \zeta^2, \ldots, \zeta^s$ are linearly independent, $\text{Char}(G)$ is a free **Z**-module of rank s. Moreover by Note 4 of §2.3 the product of two characters is again a character, and so it follows easily that $\text{Char}(G)$ is a subring of $\text{Class}_K(G)$. It will be shown in Theorem 2.7A that the ordinary characters $\zeta^1, \zeta^2, \ldots, \zeta^s$ are essentially independent of K.

2.6 BRAUER'S THEOREM ON INDUCED CHARACTERS

If H is a subgroup of G, and $\theta \in \text{Char}(H)$, then clearly $\theta^G \in \text{Char}(G)$. The first of the important theorems of this section shows that $\text{Char}(G)$ is generated as a **Z**-module by characters induced from an especially simple class of subgroups (the "elementary" subgroups). The second theorem is a dual; it characterizes the elements of $\text{Char}(G)$ as those class functions whose restrictions to the elementary subgroups are always generalized characters.

Definition A subgroup E of a group G is called *elementary* (or more explicitly *q-elementary*) if for some prime q it can be written as a direct product $E = Q \times C$, where Q is a q-subgroup and C a cyclic q'-group.

We shall call a subgroup H of G *quasi-elementary* if for some prime q it can be written in the form $H = QC$, where Q is a q-subgroup and C a normal cyclic q'-subgroup of H; briefly, H has a normal cyclic q-complement. Clearly subgroups of elementary subgroups are elementary and subgroups of quasi-elementary subgroups are quasi-elementary.

Note 2 If R denotes the **Z**-module spanned by all characters of the form 1_H^G with H quasi-elementary, then R is a subring of $\text{Char}(G)$. Indeed it is clear that R is closed under addition. To see that it is also closed under multiplication, proceed as follows. If H and L are quasi-elementary and $\theta := 1_L^G$, then $1_H^G \theta = (\theta_H)^G$ by Theorem 2.5A. Since θ is a permutation character, $\theta_H = (1_L^G)_H = \sum_{i=1}^m 1_{H_i}^H$, where m is the number of orbits and H_i the stabilizer of a member of the ith orbit (compare with the example of §2.5). Since each $H_i \subseteq H$, we have $1_H^G 1_L^G = (\theta_H)^G = \sum_{i=1}^m 1_{H_i}^G \in R$.

Lemma 2.6 With the notation above the principal character $1_G \in R$.

Proof We first show that for each $x \in G$ and each prime q there exists a quasi-elementary subgroup H of G such that $q \nmid 1_H^G(x)$. Indeed, take C as the q-complement of $\langle x \rangle$ and put $N := N_G(C)$. We choose H/C as a Sylow q-subgroup of N/C; clearly H is quasi-elementary. By the example of §2.5 we know that $1_H^G(x)$ equals the number of right cosets Hy in G such that $Hyx = Hy$. But $Hyx = Hy$ implies $\langle yxy^{-1} \rangle \subseteq H$, which in turn implies $y \in N$ because C is the unique q-complement of H. Therefore $1_H^G(x) = 1_H^N(x)$. Since C is normal in N and contained in H, C lies in the kernel of 1_H^N. Since Cx is a q-element in N/C, the nontrivial orbits in the set of right cosets of H in N all have lengths which are multiples of q. Therefore the number of orbits of length 1 equals $1_H^N(x) \equiv |N:H| \not\equiv 0 \pmod{q}$.

Now consider the set $I_x := \{\phi(x) \mid \phi \in R\}$ for $x \in G$. Clearly I_x is an ideal in **Z**; and from what we have just proved, $I_x \not\subseteq q\mathbf{Z}$ for any prime q. Therefore $I_x = \mathbf{Z}$. This means that for each $x \in G$ we can choose $\phi_x \in R$ such that $\phi_x(x) = 1$. Then the generalized character $\prod_{x \in G} (\phi_x - 1_G)$ is 0. Expanding this product, we can write 1_G as a **Z**-linear combination of products of ϕ_x's. Hence by Note 2 above, $1_G \in R$.

II INDUCED MODULES AND CHARACTERS

Theorem 2.6A (*Brauer's theorem on induced characters*) The ring Char(G) of generalized characters of G is generated as a **Z**-module by characters of the form λ^G, where λ is a character of degree 1 of an elementary subgroup of G. Hence each character of G is a **Z**-linear combination of induced characters of this form.

Proof We first note that it is enough to prove that we can write

(1) $$1_G = \sum_{i=1}^{m} m_i \phi_i^G$$

for suitable integers m_i, where the ϕ_i are characters of some elementary subgroups E_i ($i = 1, 2, \ldots, m$). Indeed, if 1_G is of this form and χ is any character of G, then

$$\chi = \chi \cdot 1_G = \sum_{i=1}^{m} m_i (\chi_{E_i} \phi_i)^G$$

by Theorem 2.5A(ii). Since each elementary subgroup is nilpotent, Theorem 2.2B shows that $\chi_{E_i} \phi_i$ is of the form $\lambda_i^{E_i}$ for a character λ_i of degree 1 of some subgroup of E_i. Since subgroups of elementary subgroups are elementary, $\chi = \sum_{i=1}^{m} m_i \lambda_i^G$ has the required form.

We shall proceed by induction on $|G|$. Since (1) is trivially true if G is elementary, we may suppose that G is not elementary.

Case 1 G *is quasi-elementary.* Then for some prime q, $G = QC$, where C is a normal cyclic q'-subgroup and Q a q-group. Put $H = N_G(Q)$. Note that if M is a normal subgroup of G, then $H \subseteq M$ implies $M = G$. Indeed, M contains all Sylow q-subgroups of G and so by the Sylow theorems, counting the number of Sylow q-subgroups of G gives $|M : H| = |G : H|$; hence $M = G$. Also $H \neq G$ because G is assumed to be not elementary. By the example of §2.5, 1_G is an irreducible constituent of 1_H^G with multiplicity 1; so we can write $1_H^G = 1_G + \sum_i m_i \chi_i$, where χ_i are the irreducible constituents of 1_H^G different from 1_G. If we can show that each χ_i has the form ϕ_i^G where ϕ_i is the character of some proper subgroup of G, then $1_G = 1_H^G - \sum_i m_i \phi_i^G$, and so (1) will follow from Theorem 2.5A(i) and the induction hypothesis.

First note that $C \triangleleft G$. It follows from the definition of 1_H^G that

$$1_H^G(x) = \begin{cases} 0 & \text{if } x \in C \setminus H \\ |G : H| & \text{if } x \in C \cap H. \end{cases}$$

Since $G = CH$, $|G : H| = |C : C \cap H|$, and so $(1_H^G)_C = (1_{H \cap C})^C$. Now C is abelian and $H \cap C \triangleleft C$, and so by the example of §2.5,

$$H \cap C = \text{Ker}(1_{H \cap C})^C \quad \text{and} \quad (1_{H \cap C})^C = \sum_{j=1}^{n} \lambda_j(1) \lambda_j,$$

2.6 BRAUER'S THEOREM ON INDUCED CHARACTERS 55

where $\lambda_1, \lambda_2, \ldots, \lambda_n$ are the irreducible characters of C with $H \cap C \subseteq$ Ker λ_j. Moreover $\lambda_i(1) = 1$ for each i, and $n = |C : H \cap C|$ by the corollary of Theorem 1.5.

Now let χ_i be an irreducible constituent of 1_H^G different from 1_G. The Frobenius reciprocity theorem (Theorem 2.5A) shows that $\langle 1_H, (\chi_i)_H \rangle = \langle 1_H^G, \chi_i \rangle \geq 1$. If $\chi_i(1) = 1$, this would show that $(\chi_i)_H = 1_H$, and so Ker $\chi_i \supseteq H$. As we noted above, G is the only normal subgroup of G containing H; so Ker $\chi_i = G$ and $\chi_i = 1_G$, contrary to the choice of χ_i. Hence $\chi_i(1) > 1$. On the other hand, since $(1_H^G)_C = (1_{H \cap C})^C = \sum_{j=1}^n \lambda_j$, where the λ_i are distinct characters of degree 1, $(\chi_i)_C$ is a sum of distinct characters of degree 1. Therefore by the corollary of Theorem 2.2A, any module affording χ_i is imprimitive. Hence $\chi_i = \phi_i^G$ for some character ϕ_i of a proper subgroup of G. As we saw above, this proves the theorem in the case where G is quasi-elementary.

Case 2 G *is not quasi-elementary.* In this case the theorem follows immediately from Lemma 2.6 and the induction hypothesis. This completes the proof of the theorem.

Theorem 2.6B (*Characterization of characters*) Let $\theta \in \text{Class}(G)$ for the group G. Then

  (i) $\theta \in \text{Char}(G)$ if (and only if) $\theta_E \in \text{Char}(E)$ for every elementary subgroup E of G;

  (ii) θ is an irreducible character of G if (and only if) $\theta \in \text{Char}(G)$, $\theta(1) > 0$, and $\langle \theta, \theta \rangle = 1$.

Proof The proof of the "only if" statements is trivial so we consider only the direct statements.

  (i) Using Theorem 2.6A we write $1_G = \sum_1^n m_i \phi_i^G$ as in (1). Then

$$\theta = \theta 1_G = \sum_1^n m_i(\theta \phi_i^G) = \sum_{i=1}^n m_i(\theta_{E_i} \phi_i)^G$$

by Theorem 2.5A. Since $\text{Char}(E_i)$ is a ring, it follows from the hypothesis that for each i, $\theta_{E_i} \phi_i \in \text{Char}(E_i)$; hence $\theta \in \text{Char}(G)$.

  (ii) Since $\theta \in \text{Char}(G)$, we can write $\theta = \sum_{i=1}^s n_i \chi_i$, where the n_i are integers and the χ_i are the distinct irreducible characters of G. The condition $\langle \theta, \theta \rangle = 1$ shows that $\sum_{i=1}^s n_i^2 = 1$ since $\langle \chi_i, \chi_j \rangle = \delta_{ij}$; hence for some j, $n_j = \pm 1$ and $n_i = 0$ for all $i \neq j$. Thus $\theta = \pm \chi_j$ and the condition $\theta(1) > 0$ shows that $\theta = \chi_j$.

EXERCISES

1. Let $K = \mathbf{Q}(\omega)$, where ω is a primitive $|G|$th root of unity, and let $\text{Int}(K)$ be the ring of integers of K. Assume that K is a splitting field for G. Show

that the set of all elements of $\text{Class}_K(G)$ that are linear combinations of irreducible characters of G with coefficients in $\text{Int}(K)$ is a subring R of $\text{Class}_K(G)$.

2. With the notation of Exercise 1, prove that an element $\theta \in \text{Class}_K(G)$ belongs to R iff $\langle \theta_H, \lambda \rangle \in \text{Int}(K)$ whenever H is an elementary subgroup of G and λ is a character of degree 1 of H.

2.7 Splitting fields

Let G be a group and K a field. Then recall that K is defined to be a splitting field for KG if $\text{End}_{KG}(V) = K \cdot 1$ for every irreducible KG-module V. It was shown in §1.3 that an algebraically closed field is a splitting field for every group, and in §1.7 that K is a splitting field for KG iff each irreducible KG-module is absolutely irreducible. In the present section we give some further criteria for a field to be a splitting field. In particular, we shall show that if G is a group of order g and the polynomial $X^g - 1$ splits into linear factors over K, then K is a splitting field for G.

Lemma 2.7 Let G be a group, K a field, and L an extension field of K such that L is a splitting field for LG. Let U and V be completely reducible KG-modules.

(i) If $\text{Hom}_{LG}(U \otimes_K L, V \otimes_K L) \neq 0$, then U and V have an irreducible KG-constituent in common.

(ii) Suppose that V is irreducible and that each irreducible submodule of $V \otimes_K L$ is isomorphic to an LG-module of the form $U \otimes_K L$ where U is a KG-module. Then $V \otimes_K L$ is irreducible.

Proof (i) Write $U = U_1 \oplus U_2 \oplus \cdots \oplus U_m$ and $V = V_1 \oplus V_2 \oplus \cdots \oplus V_n$ as sums of irreducible KG-modules. Then as vector spaces

$$\text{Hom}_{KG}(U, V) \simeq \bigoplus_{i=1}^{m} \bigoplus_{j=1}^{n} \text{Hom}_{KG}(U_i, V_j).$$

Suppose that no U_i is isomorphic to any V_j. Then by Schur's lemma, $\text{Hom}_{KG}(U_i, V_j) = 0$ for all i, j. Hence $\text{Hom}_{KG}(U, V) = 0$, and so $\text{Hom}_{LG}(U \otimes_K L, V \otimes_K L) \simeq \text{Hom}_{KG}(U, V) \otimes_K L = 0$. This proves (i).

(ii) Choose U as a KG-module such that $U \otimes_K L$ is isomorphic to an irreducible LG-submodule of $V \otimes_K L$. Since $U \otimes_K L$ can be embedded into $V \otimes_K L$, $\text{Hom}_{LG}(U \otimes_K L, V \otimes_K L) \neq 0$. Moreover, U is irreducible because $U \otimes_K L$ is, and so (i) and the irreducibility of V show that $U \simeq V$. Thus $V \otimes_K L$ is isomorphic to $U \otimes_K L$ and hence is irreducible.

Recall that the *exponent* of a group G is the least integer $d > 0$ such that $x^d = 1$ for all $x \in G$. In particular, d always divides $|G|$. Note also that if T is

2.7 SPLITTING FIELDS

a representation of G over some field K, then for each $x \in G$, $T(x)^d = T(x^d) = 1$. Therefore the eigenvalues (in a suitable extension field of K) are all dth roots of unity. It follows that the values of each character of G over K all lie in the field generated by the dth roots of unity. However it should be noted that often the character values may all lie in a much smaller field.

Theorem 2.7A (*Brauer*) Let G be a group of exponent d, and let K be a field of characteristic 0 over which the polynomial $x^d - 1$ splits into linear factors. Then

  (i) every character of G over an extension field L of K is afforded by a KG-module;
  (ii) K is a splitting field for G.

Proof (i) Let ζ be a character of G over L. Without loss in generality we may take L to be algebraically closed and hence assume that L is a splitting field for LG. Moreover, since it is enough to prove the result for each irreducible constituent of ζ, we may assume that ζ is irreducible. By Theorem 2.6A we can write $\zeta = \sum_{i=1}^{m} m_i \lambda_i^G$ as a $\mathbb{Z}$-linear combination of characters λ_i^G induced by characters λ_i of degree 1 of certain subgroups H_i of G. Since K contains the dth roots of unity, each of the characters λ_i is afforded by some KH_i-module V_i, and so the character λ_i^G is afforded by the KG-module V_i^G. Hence separating out the terms with $m_i \geq 0$ from those with $m_i < 0$, we can write $\zeta = \theta - \phi$, where θ and ϕ are both characters of G (possibly $\phi = 0$) that are afforded by KG-modules. Changing notation, we may suppose that θ is a character of smallest possible degree such that $\theta = \zeta + \phi$ and θ and ϕ are characters (possibly $\phi = 0$) afforded by KG-modules. It remains to prove $\phi = 0$.

Suppose $\phi \neq 0$. Then there exist KG-modules U and V which afford ϕ and θ, respectively, such that $U \otimes_K L$ is isomorphic to a submodule of $V \otimes_K L$ (all modules are completely reducible by Maschke's theorem 1.3B). Then $\text{Hom}_{LG}(U \otimes_K L, V \otimes_K L) \neq 0$ and so by Lemma 2.7(i) we conclude that U and V have a common irreducible KG-constituent. But this means that θ and ϕ have a common irreducible constituent (as characters over K) contrary to the minimality of θ. This shows that $\phi = 0$, and (i) is proved.

  (ii) Let V be an irreducible KG-module. It follows from (i) and Lemma 2.7(ii) that $V \otimes_K L$ is irreducible for all extension fields L of K, and hence V is absolutely irreducible. Then (ii) follows by Theorem 1.7B.

The situation for fields of characteristic $p > 0$ is similar but in some ways simpler; the result is stated in Theorem 2.7B. The proof is quite different and uses a classical result of Wedderburn, namely: every finite division ring is a field. (For a proof see Curtis and Reiner [1], §68.)

Theorem 2.7B Let G be a group, K a field of characteristic $p > 0$, L an extension field of K so that L is a splitting field for G, and ζ an irreducible character of G over L. Then

(i) If $\zeta(x) \in K$ for all x in G, there is a KG-module of dimension $\deg \zeta$ which affords ζ.

(ii) If d is the exponent of G and the polynomial $X^d - 1$ splits into linear factors over K, then K is a splitting field for G.

Proof (i) We first observe that K can be taken as a finite field. Since K has characteristic $p > 0$, the prime subfield K_0 of K has p elements. Each value $\zeta(x)$ ($x \in G$) of ζ is algebraic over K_0 since it is a sum of roots of unity. Thus the subfield K_1 of K generated by the values of ζ is algebraic over K_0 and hence finite. If we can prove that ζ is afforded by a $K_1 G$-module U_1, then $U := U_1 \otimes_{K_1} K$ will be a KG-module which affords ζ. Thus it is enough to prove (i) in the case K is finite.

Let V be an LG-module which affords ζ, and let T be the representation afforded by V. By the corollary of Theorem 1.3A, $T(LG) = \text{End}_L(V)$; and so if $n := \dim_L V$, then there exists an L-basis $T(x_1) = 1$, $T(x_2)$, ..., $T(x_{n^2})$ of $\text{End}_L(V)$ with each $x_i \in G$. Since K is a subfield of L, we can identify $R := T(KG)$ as a K-algebra embedded in $\text{End}_L(V)$. We claim that $T(x_1)$, $T(x_2)$, ..., $T(x_{n^2})$ is a K-basis of R. Certainly those vectors are linearly independent over K since they are linearly independent over L. Thus it is only necessary to show that

$$T(x) = \sum_{i=1}^{n^2} \alpha_i T(x_i)$$

with $\alpha_i \in K$ for each $x \in G$. This is equivalent to

$$T(xx_j) = \sum_{i=1}^{n^2} \alpha_i T(x_i x_j)$$

for all $j = 1, 2, \ldots, n^2$, and so taking the traces we have

(1) $\qquad \zeta(xx_j) = \sum_{i=1}^{n^2} \alpha_i \zeta(x_i x_j) \qquad$ for $\quad j = 1, 2, \ldots, n^2$.

We assert that the $n^2 \times n^2$ matrix $[\zeta(x_i x_j)]$ is nonsingular. For otherwise there are $\beta_i \in L$ not all zero such that

$$c := \sum_{i=1}^{n^2} \beta_i x_i \notin \text{Ann } V \qquad \text{and} \qquad \text{tr}\{T(c)T(x_j)\} = 0$$

for $j = 1, 2, \ldots, n^2$; and since $T(x_1)$, $T(x_2)$, ..., $T(x_{n^2})$ form a basis of $T(LG)$, this would mean that $\text{tr}(T(c)T(x)) = 0$ for all $x \in G$. However since $c \notin \text{Ann } V$, this gives a nontrivial linear relation among the coordinate func-

2.7 SPLITTING FIELDS

tions of a matrix representation associated with T, contrary to Theorem 2.3. Therefore the equations (1) have a unique solution for $\alpha_1, \alpha_2, \ldots, \alpha_{n^2}$ which can be obtained using Cramer's rule. Since $\zeta(y) \in K$ for all $y \in G$, this shows that each $\alpha_i \in K$ as required. In particular, $\dim_K R = n^2$.

We next show that R is a simple ring. For if I is a proper ideal of R, then I is a K-subalgebra of dimension m, say, where $m < n^2$. This shows that $I_0 := I \otimes_K L$ is an L-space of dimension m. But clearly I_0 is an ideal of $T(LG) = \text{End}_L(V)$ that is simple. Therefore $I_0 = 0$, $m = 0$, and hence $I = 0$.

Since R is a simple K-algebra, Theorem 1.3A shows that $R \simeq \text{End}_D(U)$ for some division ring D with K in its center and some R-module U. Since K is finite, R is finite, and so D is finite. By the theorem of Wedderburn quoted above, D is a field. Since L is a splitting field, $\text{End}_{LG}(V) = L \cdot 1$. We claim that the center $Z(R)$ of R is $L \cdot 1 \cap R = K \cdot 1$. Indeed,

$$c = \sum_{i=1}^{n^2} \alpha_i T(x_i) \in Z(R)$$

iff $cT(x_i) = T(x_i)c$ for all $i = 1, 2, \ldots, n^2$ and each $\alpha_i \in K$. This is so if and only if $c \in L \cdot 1$ and $\alpha_i \in K$. Thus $c \in Z(R)$ iff $c = \alpha_1 T(x_1) = \alpha_1 \cdot 1 \in K \cdot 1$. But the center of $\text{End}_D(U)$ is $D \cdot 1$ because D is a field, and $K \subseteq D$. So we conclude that $K = D$. Since $\dim_K R = n^2$, it follows that $\dim_K(U) = n$; hence $R \simeq \text{End}_K(U) \simeq \text{Mat}(n, K)$.

Let $e_{ij} \in R$ $(i, j = 1, 2, \ldots, n)$ be elements of R corresponding to the matrices in $\text{Mat}(n, K)$ with an entry 1 in the (i, j)th position and 0 elsewhere. Then the isomorphism $R \simeq \text{Mat}(n, K)$ shows that $|e_{ij} R| = |K|^n$ for each i and j. Since $VR \neq 0$, there exists some $v \in V$ such that $ve_{ij} \neq 0$ for some i, j. Put $W := ve_{ij} R$. Then W is a KG-module with $|W| \leq |e_{ij} R| = |K|^n$, so $\dim_K W \leq n$. Moreover LW is a nonzero LG-submodule of V. By the irreducibility of V, $LW = V$ and $\dim_L(LW) = \dim_L V = n$. Therefore $\dim_K W = n$ and $V \simeq W \otimes_K L$. Hence W is a KG-module of dimension $\deg \zeta$ which affords ζ. This proves (i).

(ii) Since $X^d - 1$ splits into linear factors over K, K contains the values of all characters of G over L. If V is an irreducible KG-module, then (i) and Lemma 2.7(ii) show that $V \otimes_K L$ is irreducible for any field extension L of K, and so V is absolutely irreducible. The result follows from Theorem 1.7B, and (ii) is proved.

EXAMPLE Let G be a group of order g, K a field, and L the extension of the field K generated by all roots of $X^g - 1$. Then by Theorems 2.7A and 2.7B, L is a splitting field for G. Moreover L is a separable normal extension of K. Let V be a LG-module which affords the character χ, and let σ be an automorphism of L over K. Suppose $v_1, v_2, \ldots, v_n$ is a L-basis of V and $v_i x = \sum_{j=1}^{n} \alpha_{ij}(x) v_j$ with $\alpha_{ij}(x) \in L$ for all $x \in G$. Then $\chi(x) = \sum_{i=1}^{n} \alpha_{ii}(x)$ for any $x \in G$. Let V^σ be the LG-module whose underlying vector space is V and whose module action is given by $v_i x := \sum_{j=1}^{n} \alpha_{ij}(x)^\sigma v_j$ for all $x \in G$ and $j = 1, 2, \ldots, n$, where $\alpha_{ij}(x)^\sigma$ is the image of $\alpha_{ij}(x)$ under σ. If χ^σ is the character of G afforded by V^σ, then $\chi^\sigma(x) = \chi(x)^\sigma$ for all $x \in G$. The character χ^σ is called an *algebraic conjugate* of χ. Thus the Galois group of L over K acts on the characters of G over L, and two characters of G over L are algebraically conjugate if and only if they are in the same orbit.

EXERCISES

1. Let G be the group of order 8 generated by a and b with relations $a^4 = 1$, $a^2 = b^2$, and $ab = b^3 a$, $\mathbf{Q}$ the rational field, and $K = \mathbf{Q}(i)$, where $i^2 = -1$. Show that the matrix representation T over K defined by

$$T(a) := \begin{bmatrix} 0 & i \\ i & 0 \end{bmatrix} \quad \text{and} \quad T(b) := \begin{bmatrix} 0 & -1 \\ 1 & 0 \end{bmatrix}$$

is absolutely irreducible, and its character has its values in $\mathbf{Q}$. Also show that T is not equivalent to a representation over $\mathbf{Q}$. Finally verify that K is a splitting field for G, but $\mathbf{Q}$ is not.

2. Give an example to show that Theorem 2.7B(ii) cannot be sharpened in that a smaller field is not a splitting field in general.

3. Let k be a field of p (prime) elements and let $K = k(t)$ be a transcendental extension of k. Let G be the group generated by a and b with relations $a^p = b^p = 1$ and $ab = ba$. Show that the matrix representation T of G over K defined by

$$T(a) := \begin{bmatrix} 1 & 0 \\ t & 1 \end{bmatrix} \quad \text{and} \quad T(b) := \begin{bmatrix} 1 & 0 \\ 1 & 1 \end{bmatrix}$$

is not equivalent to a representation over k. [This example shows that the conclusion of Theorem 2.7B(i) need not hold if the character is reducible.]

4. Let G be a group of order g, and let K be the extension of the rational field $\mathbf{Q}$ generated by the roots of $X^g - 1$. Let χ and ϕ be characters of G over K and let σ be an automorphism of K over $\mathbf{Q}$. Show that $\langle \chi, \phi \rangle = \langle \chi^\sigma, \phi^\sigma \rangle$.

5. Let G, g, K, and $\mathbf{Q}$ be as in Exercise 4, and let χ be any character of G over K. Suppose σ is an automorphism of K over $\mathbf{Q}$ such that for some integer n relatively prime to g, $\omega^\sigma = \omega^n$ for all gth roots ω of 1. Show that $\chi^\sigma(x) = \chi(x^n)$ for all $x \in G$.

2.8 Notes and comments

There is an extensive theory of ordinary representations and characters. For example, see Burnside [1], Huppert [1], and Curtis and Reiner [1]. A comprehensive treatment of applications of ordinary characters is given in

2.8 NOTES AND COMMENTS

Feit [1] and Isaacs [3]. Also see Brauer [6] and Glauberman [2]. Brauer's theorem on induced characters [Theorem 2.6A] was first proved in Brauer [7] and later a simpler ring theoretic proof was given by Brauer and Tate [1]. This result was generalized by Berman [3]. The proof we have given here is due to Goldschmidt and Isaacs [1]. A proof of Wedderburn's theorem [which asserts the commutativity of a finite division ring] may be found in Herstein's book "Topics in Algebra" [1]. Also see Curtis and Reiner [1]. The theorems of §2.7 were first proved by Brauer [4] and later the proofs were simplified by Feit.

CHAPTER **III**

Modular Representations and Characters

When the characteristic of a field K is $p > 0$, the theory of representations of G over K (called "modular representation theory") becomes more involved, since in this case the characters of G over K do not in general determine the representations (even up to equivalence). By defining the (modular) characters in a different way, the property that the characters determine the representations (Corollary 2 of Theorem 2.3) can be partly regained. In order to establish the relations between the ordinary and modular characters, it is simplest to work in a p-adic field. The necessary theory of p-adic fields is developed in the first sections and is then used to establish an important relation between the Cartan invariants and the decomposition numbers. This relation in turn implies the orthogonality relations for modular characters. In conclusion, we consider an application of the theory to prove a result of Fong and Swan on the representations of a p-solvable group.

3.1 The p-adic integers

Let p be a prime and let $\mathbf{Z}$ denote the ring of integers. Consider the sequence of rings

$$\mathbf{Z}/p\mathbf{Z} \xleftarrow{v_2} \mathbf{Z}/p^2\mathbf{Z} \xleftarrow{v_3} \mathbf{Z}/p^3\mathbf{Z} \xleftarrow{v_4} \cdots$$

3.1 THE p-ADIC INTEGERS

together with the canonical homomorphisms. The *ring of p-adic integers* $\mathbf{Z}_p$ is defined as the "inverse limit" of this sequence. In concrete terms it can be described as follows.

Let $S = \prod_{n=1}^{\infty} (\mathbf{Z}/p^n\mathbf{Z})$ be the Cartesian product of the (finite) sets $\mathbf{Z}/p^n\mathbf{Z}$. The elements of S may be written as sequences (x_n), where $x_n \in \mathbf{Z}/p^n\mathbf{Z}$ and S is a ring under componentwise addition and multiplication. Then

$$\mathbf{Z}_p := \{(x_n) \in S \mid x_n v_n = x_{n-1} \text{ for } n = 2, 3, \ldots\}.$$

Clearly $\mathbf{Z}_p$ is a subring of S because the v_n are ring homomorphisms. Moreover $\mathbf{Z}$ is embedded as subring of $\mathbf{Z}_p$ by the mapping

$$x \mapsto (x_n) \quad \text{where} \quad x_n := x + p^n\mathbf{Z} \quad \text{for each } n.$$

Theorem 3.1A *The ring $\mathbf{Z}_p$ has the following properties.*

(i) $(x_n) \in \mathbf{Z}_p$ is a unit in $\mathbf{Z}_p$ iff $x_1 \neq 0$.

(ii) $p\mathbf{Z}_p$ is the unique maximal ideal of $\mathbf{Z}_p$, and so rad $\mathbf{Z}_p = p\mathbf{Z}_p$; moreover $\mathbf{Z}_p/p\mathbf{Z}_p \simeq \mathbf{Z}/p\mathbf{Z}$.

(iii) The proper nonzero ideals of $\mathbf{Z}_p$ are precisely $p^n\mathbf{Z}_p$ ($n = 1, 2, 3, \ldots$); and $\bigcap_{n=1}^{\infty} p^n\mathbf{Z}_p = 0$. In particular, $\mathbf{Z}_p$ is a principal ideal domain.

(iv) $\mathbf{Z}_p$ is an integral domain of characteristic 0.

Remark The field of quotients of $\mathbf{Z}_p$ denoted by $\mathbf{Q}_p$ is the *field of p-adic numbers*.

Proof (i) If (x_n) has an inverse, say (y_n), then in particular, $x_1 y_1 = 1$, so $x_1 \neq 0$. Conversely, suppose $(x_n) \in \mathbf{Z}_p$ and $x_1 \neq 0$. Since v_n is surjective, it follows that $x_n \in p\mathbf{Z}/p^n\mathbf{Z}$ iff $x_n v_n \in p\mathbf{Z}/p^{n-1}\mathbf{Z}$. Since $x_1 \notin p\mathbf{Z}/p\mathbf{Z}$, this shows that for each n, $x_n \notin p\mathbf{Z}/p^n\mathbf{Z}$. Thus x_n is a unit in $\mathbf{Z}/p^n\mathbf{Z}$, and hence has an inverse y_n, say, for $n = 1, 2, 3, \ldots$. Then $1 = x_n y_n$; and $1 = (x_n v_n)(y_n v_n) = x_{n-1}(y_n v_n)$ shows that $y_n v_n = x_{n-1}^{-1} = y_{n-1}$ for all $n \geq 2$. Hence $(y_n) \in \mathbf{Z}_p$ and $(x_n)(y_n) = 1$. Thus (x_n) is a unit in $\mathbf{Z}_p$, and (i) is proved.

(ii) It follows at once from (i) that $\mathbf{Z}_p$ is a local ring and hence has a unique maximal ideal by Lemma 1.6B. (Note that there is no distinction between right and two-sided ideals because $\mathbf{Z}_p$ is commutative.) On the other hand, it is readily verified that there is a ring homomorphism of $\mathbf{Z}_p$ onto $\mathbf{Z}/p\mathbf{Z}$ given by $(x_n) \mapsto x_1$ which has kernel $p\mathbf{Z}_p$. Hence $\mathbf{Z}_p/p\mathbf{Z}_p \simeq \mathbf{Z}/p\mathbf{Z}$ and $p\mathbf{Z}_p$ is the unique maximal ideal. Lemma 1.6B now shows that $p\mathbf{Z}_p = \text{rad } \mathbf{Z}_p$.

(iii) If $(x_n) \in p^m\mathbf{Z}_p$, then $x_i = 0$ for $i = 1, 2, \ldots, m$. Therefore $\bigcap_{m=1}^{\infty} p^m\mathbf{Z}_p = 0$. Now let I be any proper nonzero ideal in $\mathbf{Z}_p$. Then from what we have just shown, there exists $m \geq 1$ such that $p^{m+1}\mathbf{Z}_p \not\supseteq I$; choose m as small as possible, so $p^m\mathbf{Z}_p \supseteq I$. Let $x \in I \backslash p^{m+1}\mathbf{Z}_p$. Then $x = p^m y$ for some $y \in \mathbf{Z}_p \backslash p\mathbf{Z}_p$. By (ii) y is a unit, and so $I \supseteq p^m y \mathbf{Z}_p = p^m \mathbf{Z}_p$. Hence $I = p^m \mathbf{Z}_p$. This proves (iii).

(iv) Since $p^n\mathbf{Z}_p p^m\mathbf{Z}_p = p^{m+n}\mathbf{Z}_p \neq 0$, it follows from (iii) that the product of any two nonzero ideals in $\mathbf{Z}_p$ is nonzero. Hence the product of any two nonzero elements is also nonzero. Thus $\mathbf{Z}_p$ is an integral domain. As we noted at the beginning of this section, $\mathbf{Z}$ is embedded as a subring in $\mathbf{Z}_p$, so $\mathbf{Z}_p$ has characteristic 0. This completes the proof of the theorem.

The construction used for $\mathbf{Z}_p$ can be applied in slightly more general situations. Let R be a free $\mathbf{Z}_p$-algebra and let $u_1, u_2, \ldots, u_s$ be a $\mathbf{Z}_p$-basis for R. Then for each $n \geq 0$ we have

(1) $$p^n R = p^n \mathbf{Z}_p u_1 \oplus p^n \mathbf{Z}_p u_2 \oplus \cdots \oplus p^n \mathbf{Z}_p u_s$$

as $\mathbf{Z}_p$-modules. Since $\bigcap_{n=1}^{\infty} p^n \mathbf{Z}_p = 0$ by Theorem 3.1A, we see that $\bigcap_{n=1}^{\infty} p^n R = 0$. Let $S_R := \prod_{n=1}^{\infty} (R/p^n R)$. Then S_R is a ring under componentwise addition and multiplication. We define a ring homomorphism $\psi: R \to S_R$ by $a\psi = (a_n)$, where $a_n := a + p^n R$ for each n. Since $\bigcap_{n=0}^{\infty} p^n R = 0$, Ker $\psi = 0$, and so ψ is injective. On the other hand it follows from (1) and the definition of $\mathbf{Z}_p$ that $(a_n) \in S_R$ lies in Im ψ iff $a_n \mu_n = a_{n-1}$ for each n, where $\mu_n: R/p^n R \to R/p^{n-1} R$ is the canonical homomorphism.

EXAMPLE Let R be a free $\mathbf{Z}_p$-algebra. Then every element of the form $1 - pb$ ($b \in R$) is a unit in R. Indeed, put $a_n := 1 + pb + \cdots + p^{n-1}b^{n-1} + p^n R$. Then $a_n \mu_n = a_{n-1}$ for each $n \geq 2$, and so $(a_n) \in$ Im ψ; hence there exists $a \in R$ such that $a_n = a + p^n R$ for each n. Since $(1 - pb)a_n = 1 - p^n b^n + p^n R = 1 + p^n R$, this shows that $(1 - pb)a - 1 \in p^n R$ for each n; and so $(1 - pb)a - 1 \in \bigcap_{n=1}^{\infty} p^n R = 0$. Hence $a = (1 - pb)^{-1}$. In particular, this shows that $pR \subseteq \mathrm{rad}\ R$. Indeed if M is a maximal right ideal of R such that $M \not\supseteq pR$, then $R = M + pR$, and so some element $c \in M$ has the form $c = 1 - pb$ with $b \in R$. Since c is a unit this contradicts the choice of M as a proper right ideal of R.

We now turn to a very important property of $\mathbf{Z}_p$-algebras. The following theorem may be compared with Theorem 1.6B. The final version appears as Theorem 3.4A.

Theorem 3.1B (*Idempotent refinement*) Let R be a free $\mathbf{Z}_p$-algebra and suppose J is an ideal of R such that J/pR is a nilpotent ideal of R/pR. Let f be an idempotent in R/J. Then there exists an idempotent e in R such that $e + J = f$.

Remark Since $p\mathbf{Z}_p$ annihilates R/pR, we may consider R/pR as a $\mathbf{Z}_p/p\mathbf{Z}_p$-algebra. By Theorem 3.1A, $\mathbf{Z}_p/p\mathbf{Z}_p$ is a field with p elements, and so we can apply Theorem 1.3A to conclude that $\mathrm{rad}(R/pR)$ is nilpotent. By the example above $\mathrm{rad}\ R \supseteq pR$, and so $\mathrm{rad}(R/pR) = \mathrm{rad}\ R/pR$. Thus the hypothesis of the theorem holds with $J = \mathrm{rad}\ R$.

Proof By Theorem 1.6B there exists an idempotent e_1 of R/pR such that e_1 maps onto f under the canonical mapping $R/pR \to R/J$. We shall now define e_n ($n \geq 2$) recursively such that for each n

(2) $\qquad e_n$ is an idempotent in $R/p^n R \qquad$ and $\qquad e_n \mu_n = e_{n-1}$,

where $\mu_n \colon R/p^n R \to R/p^{n-1} R$ is the canonical homomorphism. Suppose e_{n-1} has been defined and choose any $a \in R/p^n R$ such that $a\mu_n = e_{n-1}$. Then put $b := a^2 - a$; note that $b \in \operatorname{Ker} \mu_n = p^{n-1} R/p^n R$ and so $b^2 = 0$ because $n \geq 2$. Define $e_n := a + (1 - 2a)b$. Clearly $ab = ba$ and so

$$e_n^2 - e_n = a^2 - a + (2a-1)(1-2a)b + (1-2a)^2 b^2$$
$$= \{1 - 1 + 4a - 4a^2\}b$$
$$= -4b^2 = 0.$$

Since $b \in \operatorname{Ker} \mu_n$, we also have $e_n \mu_n = a\mu_n = e_{n-1}$; hence e_n satisfies (2). Thus in the notation above we can define an element $(e_n) \in S_R$ such that $e_n \mu_n = e_{n-1}$ for all $n \geq 1$. Therefore from what we saw there, there is a unique $e \in R$ such that $e_n = e + p^n R$ for each n. Since $(e_n^2) = (e_n)$, therefore $e^2 = e$. Finally, since e_1 maps onto f under the natural mapping $R/pR \to R/J$, and $e_1 = e + pR$, therefore $e + J = f$. In particular, $e \neq 0$, and so e is an idempotent with the required properties.

EXERCISES

1. Show that the rational field $\mathbf{Q}$ is a subfield of the p-adic field $\mathbf{Q}_p$ for any prime p.
2. The following exercise is intended to provide an alternative approach for the study of the ring $\mathbf{Z}_p$.

Let p be a fixed prime. For any integer $n > 0$, let x_n be a primitive (complex) p^nth root of unity, and let U be the group generated by the set $\{x_1, x_2, x_3, \ldots\}$. Show that every sequence of nonnegative integers

(1) $\qquad\qquad\qquad (k_1, k_2, \ldots, k_n, \ldots)$

such that $0 \leq k_n < p^n$ and $k_{n+1} \equiv k_n \pmod{p^n}$ determines a unique endomorphism f of U by $x_n f := x_n^{k_n}$ for $n = 1, 2, \ldots$. Conversely, show that every endomorphism f of U determines a unique sequence of the form (1). The set E of all sequences of the form (1) forms a ring with respect to componentwise addition and multiplication. Prove that $E \simeq \mathbf{Z}_p$.

3.2 *p*-adic algebras

We continue with the notation of §3.1. Recall that $\mathbf{Q}_p$ is the field of quotients of $\mathbf{Z}_p$. Let K be an extension field of finite degree over $\mathbf{Q}_p$, and define A to be the set of all $\alpha \in K$ that are roots of monic polynomials of

$\mathbf{Z}_p[X]$. Elements of A are said to be *integral* over $\mathbf{Z}_p$, and A is called the *integral closure* of $\mathbf{Z}_p$ in K.

Note 1 Suppose that $\alpha \in K$. Then $\alpha \in A$ iff the ring $\mathbf{Z}_p[\alpha]$ is a finitely generated $\mathbf{Z}_p$-module. For suppose $\alpha \in A$. Then by definition there exists an integer $m > 0$ and $a_{m-1}, a_{m-2}, \ldots, a_0 \in \mathbf{Z}_p$ such that $\alpha^m = a_{m-1}\alpha^{m-1} + \cdots + a_0$, and hence $\mathbf{Z}_p[\alpha] = \mathbf{Z}_p \alpha^{m-1} + \mathbf{Z}_p \alpha^{m-2} + \cdots + \mathbf{Z}_p$. Conversely, assume that $\mathbf{Z}_p[\alpha]$ is a finitely generated $\mathbf{Z}_p$-module. Suppose

$$\mathbf{Z}_p[\alpha] = \mathbf{Z}_p\, f_1(\alpha) + \mathbf{Z}_p\, f_2(\alpha) + \cdots + \mathbf{Z}_p\, f_n(\alpha)$$

with each $f_i(X) \in \mathbf{Z}_p[X]$. Choose N greater than the degrees of all the $f_i(X)$. Since $\alpha^N \in \mathbf{Z}_p[\alpha]$, we may write

$$\alpha^N = a_1 f_1(\alpha) + a_2 f_2(\alpha) + \cdots + a_n f_n(\alpha).$$

This shows that α is a root of a monic polynomial of $\mathbf{Z}_p[X]$, and hence $\alpha \in A$.

Note 2 It follows at once from Note 1 that if $\alpha, \alpha' \in A$, then $\alpha - \alpha'$ and $\alpha\alpha'$ also lie in A. From the hypothesis $\mathbf{Z}_p[\alpha]$ and $\mathbf{Z}_p[\alpha']$ are finitely generated $\mathbf{Z}_p$-modules, say $\mathbf{Z}_p[\alpha] = \sum_{i=1}^{m} \mathbf{Z}_p \alpha_i$ and $\mathbf{Z}_p[\alpha'] = \sum_{i=1}^{n} \mathbf{Z}_p \beta_i$. Then

$$\mathbf{Z}_p[\alpha, \alpha'] = (\mathbf{Z}_p[\alpha])[\alpha'] = \sum_{j=1}^{n} \mathbf{Z}_p[\alpha]\beta_j = \sum_{i=1}^{m}\sum_{j=1}^{n} \mathbf{Z}_p \alpha_i \beta_j,$$

and so every submodule of $\mathbf{Z}_p[\alpha, \alpha']$ is a finitely generated $\mathbf{Z}_p$-module (because $\mathbf{Z}_p$ is a principal ideal domain). In particular, $\mathbf{Z}_p[\alpha + \alpha']$ and $\mathbf{Z}_p[\alpha\alpha']$ are finitely generated $\mathbf{Z}_p$-modules. Hence A is a ring contained in K (and it contains $\mathbf{Z}_p$). We shall see below that A is finitely generated as a $\mathbf{Z}_p$-module and hence is a $\mathbf{Z}_p$-algebra.

Theorem 3.2 With the notation above:

(i) K is the field of quotients of A;
(ii) $A \cap \mathbf{Q}_p = \mathbf{Z}_p$;
(iii) A is a $\mathbf{Z}_p$-algebra which is free as a $\mathbf{Z}_p$-module;
(iv) A has a unique maximal ideal M (which equals rad A), and

$$\bigcap_{n=1}^{\infty} M^n = 0;$$

(v) the maximal ideal $M = \pi A$ for some $\pi \in A$, and the proper nonzero ideals of A are precisely $\pi^n A$ ($n = 1, 2, \ldots$) (in particular, A is a principal ideal domain);

(vi) $k := A/\pi A$ is a finite field of characteristic p.

3.2 p-ADIC ALGEBRAS

Proof (i) First note that for each $\alpha \in K$ there exists $c \neq 0$ in $\mathbf{Z}_p$ such that $c\alpha \in A$. Indeed, since α is algebraic over $\mathbf{Q}_p$, there exists a polynomial

$$a_m X^m + a_{m-1} X^{m-1} + \cdots + a_0 \in \mathbf{Z}_p[X] \quad \text{with} \quad a_m \neq 0$$

which has α as a root. Then $a_m \alpha$ is a root of the monic polynomial

$$X^m + a_{m-1} X^{m-1} + a_{m-2} a_m X^{m-2} + \cdots + a_0 a_m^{m-1} \in \mathbf{Z}_p[X]$$

and so lies in A. In particular, K is the field of quotients of A.

(ii) Each $\alpha \neq 0$ in $\mathbf{Q}_p$ can be written $\alpha = bp^t$, where b is a unit in $\mathbf{Z}_p$ and $t \in \mathbf{Z}$. Suppose $\alpha \in A$. We have to show that $t \geq 0$, and then $\alpha \in \mathbf{Z}_p$. Indeed, let

$$X^m + a_{m-1} X^{m-1} + \cdots + a_0 \in \mathbf{Z}_p[X]$$

be a monic polynomial which has α as a root. If $t < 0$, then substituting α into this polynomial and multiplying through by $p^{m|t|}$ gives

$$b^m = -a_{m-1} b^{m-1} p^{|t|} - \cdots - a_0 p^{m|t|} \in p\mathbf{Z}_p.$$

Since b is a unit, this is a contradiction. Hence $t \geq 0$ and $\alpha \in \mathbf{Z}_p$ as asserted.

(iii) Each $\alpha \in K$ acts as a $\mathbf{Q}_p$-endomorphism h_α of K by right multiplication. Temporarily, we shall write $\operatorname{tr} \alpha := \operatorname{tr} h_\alpha$, the trace of this linear transformation h_α over $\mathbf{Q}_p$. (tr α is equal to the usual field trace introduced in algebraic number theory.) By definition $\operatorname{tr} \alpha \in \mathbf{Q}_p$ for all $\alpha \in K$. We also have $\operatorname{tr} \alpha \in \mathbf{Z}_p$ whenever $\alpha \in A$. Indeed, if $\alpha \in A$, then α is a root of some monic polynomial $q(X) \in \mathbf{Z}_p[X]$. Let $m(X) \in \mathbf{Q}_p[X]$ be the minimal polynomial for the linear transformation h_α. Then since $q(h_\alpha) = 0$, $m(X) \mid q(X)$ in $\mathbf{Q}_p[X]$, and so each root of $m(X)$ is a root of $q(X)$ and hence is integral over $\mathbf{Z}_p$ in some extension field of K. But all eigenvalues of h_α are roots of $m(X)$, and the trace is the sum of the eigenvalues (with appropriate multiplicities). Since a sum of elements integral over $\mathbf{Z}_p$ is again integral over $\mathbf{Z}_p$, this shows that $\operatorname{tr} \alpha = \operatorname{tr} h_\alpha$ is integral over $\mathbf{Z}_p$. But $\operatorname{tr} \alpha \in \mathbf{Q}_p$ from above, so by (ii), $\operatorname{tr} \alpha \in \mathbf{Z}_p$ whenever $\alpha \in A$.

Now we show that for any $\mathbf{Q}_p$-basis $\eta_1, \ldots, \eta_n$, the $n \times n$ matrix $[\operatorname{tr}(\eta_i \eta_j)]$ is invertible. Indeed otherwise the rows would be linearly dependent and (using the linearity property of the trace and the linear independence of the η_i) that would mean that there exists $\alpha = \sum_{i=1}^n a_i \eta_i \neq 0$ with $a_i \in \mathbf{Q}_p$, such that $\operatorname{tr}(\alpha \eta_j) = 0$ for all j. Since $\alpha \neq 0$, we can write $\alpha^{-1} = \sum_{j=1}^n b_j \eta_j$ for some $b_j \in \mathbf{Q}_p$, and then $0 = \operatorname{tr} \alpha\alpha^{-1} = \operatorname{tr} 1$, which is impossible because K has characteristic 0.

Finally, using the observation at the beginning of the proof we can choose a $\mathbf{Q}_p$-basis $\eta_1, \ldots, \eta_n$ of K such that each $\eta_i \in A$. Since A is a ring, it then follows from above that $\operatorname{tr}(\alpha \eta_j) \in \mathbf{Z}_p$ for every $\alpha \in A$. On the other hand each

$\alpha \in A$ can be written in the form $\alpha = \sum_{i=1}^{n} a_i \eta_i$ ($a_i \in \mathbf{Q}_p$). Then multiplying by η_j and taking traces we obtain

$$\sum_{i=1}^{n} a_i \, \mathrm{tr}(\eta_i \eta_j) = \mathrm{tr}(\alpha \eta_j) \qquad (j = 1, \ldots, n).$$

Since $d := \det[\mathrm{tr}(\eta_i \eta_j)] \in \mathbf{Z}_p$ is not 0, we can use Cramer's rule to solve this system of equations and write each a_i in the form b_i/d, where $b_i \in \mathbf{Z}_p$ because each $\mathrm{tr}(\alpha \eta_j) \in \mathbf{Z}_p$. Thus

$$\alpha = \sum_{i=1}^{n} a_i \eta_i \in \mathbf{Z}_p(\eta_1/d) + \cdots + \mathbf{Z}_p(\eta_n/d)$$

for each $\alpha \in A$. Thus the ring A is contained in a $\mathbf{Z}_p$-module. Since $\mathbf{Z}_p$ is a principal ideal domain, the structure theorem for (finitely generated) modules over principal ideal domains shows that A is also finitely generated as a $\mathbf{Z}_p$-module; and A is $\mathbf{Z}_p$-free because it is $\mathbf{Z}_p$-torsion-free (see §1.1). Because $\mathbf{Z}_p$ is a subring of A, A is also a $\mathbf{Z}_p$-algebra. This proves (iii).

(iv) Let M be a maximal ideal of A. Since A is commutative, A/M is a field, and so under the canonical homomorphism $A \to A/M$, $\mathbf{Z}_p$ maps onto an integral domain. Since $p\mathbf{Z}_p$ is the unique maximal ideal of $\mathbf{Z}_p$, this shows that $p\mathbf{Z}_p = \mathbf{Z}_p \cap M$. In particular, $p \in M$, and so $M \supseteq pA$. Hence we conclude that rad $A \supseteq pA$, and $\mathrm{rad}(A/pA) = (\mathrm{rad}\, A)/pA$.

On the other hand, since $p\mathbf{Z}_p \subseteq \mathrm{Ann}(A/pA)$, A/pA may be considered as a $\mathbf{Z}_p/p\mathbf{Z}_p$-module. Since $\mathbf{Z}_p/p\mathbf{Z}_p \simeq \mathbf{Z}/p\mathbf{Z}$ is a finite field, it follows from the Wedderburn structure theorem (Theorem 1.3A) that $\mathrm{rad}(A/pA) = \mathrm{rad}\, A/pA$ is a nilpotent ideal, and $A/\mathrm{rad}\, A \simeq (A/pA)/\mathrm{rad}(A/pA)$ is a direct sum of simple rings. However, since A is an integral domain, it has only one idempotent 1, and so it follows from Theorem 3.1B that $A/\mathrm{rad}\, A$ has only one idempotent; therefore $A/\mathrm{rad}\, A$ is a simple, commutative ring (and hence a field). This shows that $M := \mathrm{rad}\, A$ is a maximal ideal in A, and so (from its definition) it is the unique maximal ideal.

Finally, since $M/pA = \mathrm{rad}\, A/pA$ is nilpotent, $M^t \subseteq pA$ for some integer $t \geq 1$. Since $\bigcap_{n=1}^{\infty} p^n \mathbf{Z}_p = 0$ (Theorem 3.1A), and A is a free $\mathbf{Z}_p$-module, $\bigcap_{n=1}^{\infty} p^n A = 0$. Therefore $\bigcap_{n=1}^{\infty} M^n = 0$.

(v) We saw in the proof of (iv) that $M^t \subseteq pA$ for some integer $t \geq 1$. Choose t so that $M^t \subseteq pA$ but $M^{t-1} \not\subseteq pA$; then choose $\beta \in M^{t-1} \setminus pA$, and put $\pi = p/\beta \in K$. Since $\beta \notin pA$, $\pi^{-1} = \beta/p \notin A$; however

$$\pi^{-1} M \subseteq (p^{-1} M^{t-1}) M \subseteq p^{-1}(pA) = A.$$

We claim $\pi^{-1} M \not\subseteq M$. Otherwise take any $\alpha \neq 0$ in M. Then $\mathbf{Z}_p[\pi^{-1}]\alpha \subseteq M$, and so $\mathbf{Z}_p[\pi^{-1}] \subseteq \alpha^{-1} M \subseteq \alpha^{-1} A$. Since the last is a $\mathbf{Z}_p$-module, this shows that $\mathbf{Z}_p[\pi^{-1}]$ is a $\mathbf{Z}_p$-module, and so $\pi^{-1} \in A$, contrary to what we saw above. Thus we have shown that $\pi^{-1} M$ is an ideal of A not contained in M, and so (iv) shows that $\pi^{-1} M = A$. Hence $M = \pi A$.

The powers $M^n = \pi^n A$ $(n = 1, 2, \ldots)$ are certainly proper ideals of A. It remains to show that any nonzero proper ideal I of A equals one of these. Since $I \neq 0$, (iv) shows that for some n, $M^n \not\supseteq I$; choose n so that $M^n \supseteq I$ but $M^{n+1} \not\supseteq I$. Then $\pi^{-n}I \subseteq \pi^{-n}M^n = A$, but $\pi^{-n}I \not\subseteq \pi^{-n}M^{n+1} = M$. Thus $\pi^{-n}I$ is an ideal of A not contained in M which shows that $\pi^{-n}I = A$; and hence $I = \pi^n A$ as required.

(vi) We saw in the proof of (iv) that $\pi A = \operatorname{rad} A \supseteq pA$. Since A is a free $\mathbf{Z}_p$-algebra of rank r, say, $|A/pA| = |\mathbf{Z}_p/p\mathbf{Z}_p|^r = p^r$. Since πA is maximal, $k := A/\pi A$ is a field, and $|A/\pi A|$ divides $|A/pA| = p^r$; so k is finite of characteristic p.

EXERCISES

1. Let K_1 be a finite extension of the rational field $\mathbf{Q}$ generated by α and let A_1 be the integral closure of $\mathbf{Z}$ in K_1. Let $K = \mathbf{Q}_p(\alpha)$ and A be the integral closure of $\mathbf{Z}_p$ in K. Assume $\mathbf{Q} \subseteq \mathbf{Q}_p$. Show that $A_1 \subseteq A$. Let M_1 be a maximal ideal of A_1 such that $p \in M_1$ and $M_1 \subseteq \pi A$, where πA is the unique maximal ideal of A. Then prove that $A_1/M_1 \simeq A/\pi A = k$. (Therefore a representation over k can also be considered as a representation over A_1/M_1.) [*Hint:* One way to prove this result is to consider a p-adic completion of K_1.]
2. With the notation of Exercise 1, let $A_2 = \{\alpha/\beta \in K_1 \,|\, \alpha, \beta \in A_1, \text{ and } \beta \notin M_1\}$. Show that A_2 is an integral domain with a unique maximal ideal M_2 such that $A_1 \subseteq A_2 \subseteq A$, $M_1 \subseteq M_2 \subseteq \pi A$, and $A_1/M_1 \simeq A_2/M_2 \simeq A/\pi A = k$. Also prove that the nonzero ideals of A_2 are $\alpha^n A_2$ for some $\alpha \in A_2$ and $n = 0, 1, 2, \ldots$.

3.3 Ordinary and modular representations

We shall now apply these results on p-adic algebras (§3.2) to study the relation between representations of a group over a field K of characteristic 0 and the representations over a field of characteristic $p > 0$. At this point it is convenient to introduce notation for the various fields and rings; this notation will be used throughout the remainder of the book except in a few cases where the contrary is explicitly stated.

NOTATION Let G be a group of order g and let p be a prime. Let K be a field extension of finite degree over the p-adic field $\mathbf{Q}_p$ such that the polynomial $X^g - 1$ splits into linear factors over K. Let A be the integral closure of $\mathbf{Z}_p$ in K and let πA be the unique maximal ideal of A. Put $k := A/\pi A$. Then k is a finite field of characteristic p, and since the roots of $X^g - 1$ in K lie in A (by the definition of A), $X^g - 1$ splits into linear factors over k. Thus by §2.7, K and k are splitting fields for G, and all its subgroups and factor groups; in particular, all irreducible KG-modules and all irreducible kG-modules are absolutely irreducible (Theorems 2.7A and 2.7B).

Remark A is a principal ideal domain (Theorem 3.2), and so by the structure theory for modules over a principal ideal domain, an A-module V is free iff V is "A-torsion-free" in the sense that $\alpha v = 0$ ($\alpha \in A, v \in V$) implies $\alpha = 0$ or $v = 0$. In particular, an A-submodule U of a free A-module V is free and (again by the structure theorem) $\text{rank}_A U \leq \text{rank}_A V$. However, the homomorphic image of a free A-module need not be free. We use these observations frequently in what follows.

Our first objective is to show how to go from a KG-module V to a related kG-module in a way which preserves at least part of the structure. We shall do this in two steps; first from KG-modules to AG-modules, and then from AG-modules to kG-modules. The first step is based on the following theorem.

Theorem 3.3 (*Burnside*) Let V be a KG-module of dimension n. Then there exists an A-free AG-module W such that $V \simeq W \otimes_A K$, and $\text{rank}_A W = n$.

Proof Let $v_1, v_2, \ldots, v_n$ be a K-basis of V and put $W := v_1 AG + v_2 AG + \cdots + v_n AG \subseteq V$. The module W is finitely generated as an A-module (requiring at most ng generators), and is A-torsion-free since V is; therefore by the remark above, W is a free-A-module. Let $w_1, w_2, \ldots, w_m$ be an A-basis of W. Since no nontrivial linear combination of $w_1, w_2, \ldots, w_m$ over A is 0, and K is the field of quotients of A (Theorem 3.2), the $w_1, w_2, \ldots, w_m$ are linearly independent over K. But from the definition of W, $w_1, w_2, \ldots, w_m$ span the vector space V and so form a K-basis. Hence $\text{rank}_A W = m = n$. Finally $w_i \mapsto w_i \otimes 1$ gives a KG-isomorphism of V onto $W \otimes_A K$. This proves the theorem.

It is now clear how we can pass from a KG-module V to a kG-module. First, using Theorem 3.3, we choose an A-free AG-module W such that $V \simeq W \otimes_A K$. Then $\bar{W} := W \otimes_A k$ (where $k = A/\pi A$) is a kG-module. In terms of bases this second step can be described as follows. Let $w_1, w_2, \ldots, w_n$ be an A-basis of the free A-module W. Then $\bar{w}_i := w_i \otimes 1$ ($i = 1, 2, \ldots, n$) is a k-basis for $\bar{W}$, and the mapping from $W \to \bar{W}$ is given by

$$\sum_{i=1}^{n} \alpha_i w_i \mapsto \sum_{i=1}^{n} \bar{\alpha}_i \bar{w}_i,$$

where $\alpha_i \in A$ and $\bar{\alpha}_i := \alpha_i + \pi A \in k$. We shall call the canonical mapping $A \to A/\pi A = k$ *reduction modulo* π and denote the image of α under this mapping by $\bar{\alpha}$. More generally, if W is an A-free AG-module, then we refer to

3.3 ORDINARY AND MODULAR REPRESENTATIONS

$\bar{W} := W \otimes_A k$ as the reduction of W modulo π; note that if W is also an A-algebra, then $\bar{W}$ is a k-algebra.

Suppose ζ is the character afforded by the KG-module V. Then W affords the same character ζ (take an A-basis for W such that $w_1 \otimes 1, w_2 \otimes 1, \ldots, w_n \otimes 1$ is a K-basis of $W \otimes_A K \simeq V$). The character $\bar{\zeta}$ afforded by the kG-module $\bar{W}$ is then given by

$$\bar{\zeta}(x) = \overline{\zeta(x)} \quad \text{for each} \quad x \in G;$$

this is clear if the traces are calculated with respect to the bases $w_1, w_2, \ldots, w_n$ and $\bar{w}_1, \bar{w}_2, \ldots, \bar{w}_n$, respectively.

Note It follows that if W is an AG-module of rank n, say, and $\bar{W}$ is irreducible, then $W \otimes_A K$ is also irreducible. Indeed, let ζ be the character afforded by $W \otimes_A K$. If ζ were reducible, then ζ, and hence $\bar{\zeta}$, would be a sum of characters afforded by modules of dimensions less than n. But $\bar{\zeta}$ is afforded by the irreducible module $\bar{W}$ of dimension n; and so we have a contradiction by Corollary 1 of Theorem 2.3.

Two obvious questions arise: (a) is the kG-module $\bar{W}$ determined (up to isomorphism) by the KG-module V; and (b) is every kG-module isomorphic to a module $\bar{W}$ obtained in this way from some KG-module? In general, the answer to both these questions is no. (See Exercises 1 and 2.) However, as we shall see in the sections to follow, something can be said about the uniqueness of the $\bar{W}$ in (a), and the answer to both questions is yes whenever p does not divide $|G|$. More importantly, the process always gives considerable information about the irreducible and principal indecomposable kG-modules.

EXAMPLE 1 If W_1 and W_2 are two AG-modules such that $\bar{W}_1$ is irreducible and $W_1 \otimes_A K \simeq W_2 \otimes_A K$, then $\bar{W}_1 \simeq \bar{W}_2$. (So the answer to (a) is yes in this case.) Indeed, let ζ^1 and ζ^2 be the characters afforded by $W_1 \otimes_A K$ and $W_2 \otimes_A K$, respectively. Then by Corollary 2 of Theorem 2.3, $W_1 \otimes_A K \simeq W_2 \otimes_A K$ iff $\zeta^1 = \zeta^2$. Similarly, the irreducibility of $\bar{W}_1$ and Corollary 1 of Theorem 2.3 show that $\bar{W}_1 \simeq \bar{W}_2$ iff $\bar{\zeta}^1 = \bar{\zeta}^2$. Since $\zeta^1 = \zeta^2$ implies $\bar{\zeta}^1 = \bar{\zeta}^2$, the assertion follows. Note that in general this argument cannot be reversed to show that $\bar{W}_1 \simeq \bar{W}_2$ implies $W_1 \otimes_A K \simeq W_2 \otimes_A K$ because in general $\bar{\zeta}^1 = \bar{\zeta}^2$ does not imply $\zeta^1 = \zeta^2$ (however, see §3.5).

EXAMPLE 2 Suppose that W is an A-free AG-module. Then Ker $\bar{W}$ clearly contains Ker W. However it cannot be "too much" larger; the group Ker $\bar{W}/$Ker W is always a p-group. To prove this it is enough to show that if $x +$ Ker W has prime order q in Ker $\bar{W}/$Ker W, then $q = p$. Since $x \in$ Ker $\bar{W}$ but $x \notin$ Ker W, it follows from Theorem 3.2 that there exists an integer $n \geq 1$ such that $W(x - 1) \subseteq \pi^n W$ but $W(x - 1) \nsubseteq \pi^{n+1} W$. On the other hand, $x^q - 1 = (1 + x - 1)^q - 1 = q(x - 1) + \binom{q}{2}(x - 1)^2 + \cdots + \binom{q}{q}(x - 1)^q$ by the binomial theorem. Therefore, since $x^q \in \text{Ker } W$, $Wq(x - 1) \subseteq W(x^q - 1) + W(x - 1)^2 = W(x - 1)^2 \subseteq \pi^{2n}W \subseteq \pi^{n+1}W$ because $n \geq 1$. This implies $q \in \pi A$; thus q equals 0 in $A/\pi A = k$ and so $q = p$ by Theorem 3.2(vi).

EXERCISES

1. Let G be the alternating group of order 60 generated by $a = (123)$ and $b = (345)$. Let K be a field extension of finite degree over $\mathbf{Q}_2$ such that $X^{60} - 1$ splits into linear factors in $K[X]$, A the integral closure of $\mathbf{Z}_2$ in K, πA the unique maximal ideal in A, $k = A/\pi A$, and ω a root of $X^2 + X + 1 \in k[X]$. Prove that the kG-module V of dimension 2 over k such that for a fixed basis $\{v_1, v_2\}$ of V

$$(\alpha_1 v_1 + \alpha_2 v_2)a = \omega \alpha_1 v_1 + (1 + \omega)\alpha_2 v_2$$
$$(\alpha_1 v_1 + \alpha_2 v_2)b = \alpha_2 v_1 + (\alpha_1 + \alpha_2)v_2$$

is an irreducible kG-module. Show that there exists no AG-module U such that $\bar{U} \simeq V$.

2. Let K, $\mathbf{Q}_2$, $\mathbf{Z}_2$, A, πA, and k be as in Exercise 1, and let G be the group generated by a and b such that $a^3 = b^2 = (ab)^2 = 1$. Suppose ω is a root of the polynomial $X^2 + X + 1$ over $\mathbf{Q}_2$. Show that the matrix representations T and S of G over K defined by

$$T(a) := \begin{bmatrix} \omega & 0 \\ 0 & \omega^{-1} \end{bmatrix}, \quad T(b) := \begin{bmatrix} 0 & 1 \\ 1 & 0 \end{bmatrix}$$

and

$$S(a) := \begin{bmatrix} -1 & 1 \\ -1 & 0 \end{bmatrix}, \quad S(b) := \begin{bmatrix} 0 & 1 \\ 1 & 0 \end{bmatrix}$$

are equivalent over K, but the representations $\bar{T}$ and $\bar{S}$ over k obtained from T and S by reducing modulo π are inequivalent. However, show that $\bar{T}$ and $\bar{S}$ have the same irreducible constituents.

3. Let K, $\mathbf{Q}_2$, $\mathbf{Z}_2$, A, πA, and k be as in Exercise 1, and let G be the permutation group generated by $a = (12)$, $b = (23)$, and $c = (34)$. Show that the matrix representations T and S of G over K given by

$$T(a) := \begin{bmatrix} 1 & 0 & 0 \\ 0 & 1 & 3 \\ 0 & 0 & -1 \end{bmatrix}, \quad T(b) := \begin{bmatrix} 1 & 4 & 4 \\ 0 & -2 & -3 \\ 0 & 1 & 2 \end{bmatrix},$$

$$T(c) := \begin{bmatrix} -3 & -8 & -12 \\ 1 & 3 & 3 \\ 0 & 0 & 1 \end{bmatrix}$$

3.4 LIFTING IDEMPOTENTS

and

$$S(a) := \begin{bmatrix} 1 & 3 & 8 \\ 0 & -1 & 0 \\ 0 & 0 & -1 \end{bmatrix}, \quad S(b) := \begin{bmatrix} -2 & -3 & -8 \\ 1 & 2 & 8 \\ 0 & 0 & -1 \end{bmatrix},$$

$$S(c) := \begin{bmatrix} -1 & -1 & -4 \\ 0 & -3 & -8 \\ 0 & 1 & 3 \end{bmatrix}$$

are inequivalent over K. Show that the representations $\bar{T}$ and $\bar{S}$ of G over k obtained from T and S by reducing modulo π have the same irreducible constituents.

4. Let $\mathbf{Q}_p$ be the field of p-adic numbers and K the extension of $\mathbf{Q}_p$ generated by the roots of $X^g - 1$, where g is the order of a group G. Let A be the integral closure of $\mathbf{Q}_p$, πA the unique maximal ideal of A, and $k = A/\pi A$. Prove that two kG-modules associated with the same KG-module have the same composition factors.

3.4 Lifting idempotents

We continue with the notation of §3.3, and in this section examine the relationship between the idempotents of the group algebra AG and the group algebra kG. This will lead to a result which shows that if W is an AG-module and $\bar{W}$ is its reduction modulo π, then both W and $\bar{W}$ decompose into direct sums in essentially the same way. (See §1.6 for elementary facts about idempotents.)

Theorem 3.4A (*Lifting idempotents*) Let R be a free A-algebra.

(i) If $e_1, \ldots, e_n$ is a complete set of orthogonal idempotents in R (so $e_i e_j = 0$ for $i \neq j$ and $\sum_{i=1}^n e_i = 1$), then $\bar{e}_1, \ldots, \bar{e}_n$ (reduction modulo π) is a complete set of orthogonal idempotents of $\bar{R} = R/\pi R$.

(ii) Conversely, if $f_1, \ldots, f_n$ is a complete set of orthogonal idempotents of $\bar{R}$, then there exists a complete set of orthogonal idempotents $e_1, \ldots, e_n$ of R such that $\bar{e}_i = f_i$ for each i.

Remarks We shall principally be applying the theorem in two situations. First where $R = AG$ and $\bar{R} = kG$, and second where $R = Z(AG)$, the center of AG, and $\bar{R} = Z(kG)$, the center of kG.

Note that (ii) shows that if f is an idempotent in $\bar{R}$, then there is an idempotent e in R such that $\bar{e} = f$ [apply (ii) to the set $e, 1 - e$].

Proof (i) Since reduction modulo π is a ring homomorphism, $\bar{e}_i \bar{e}_j = 0$ for $i \neq j$, $\bar{e}_i^2 = \bar{e}_i$ for each i, and $\bar{e}_1 + \cdots + \bar{e}_n = 1$. It remains to show that

$\bar{e}_i \neq 0$ for each i. However $\bar{e}_i = 0$ would imply that $e_i \in \pi R$. But then for each integer $n \geq 1$, the idempotent $e_i = e_i^n \in \pi^n R$. Since R is a free A-module,

$$\bigcap_{n=1}^{\infty} \pi^n R = \bigcap_{n=1}^{\infty} (\pi A)^n \cdot R = 0$$

by Theorem 3.2(iv). This implies $e_i = 0$, contrary to the definition of idempotent. Hence $\bar{e}_i \neq 0$, and (i) is proved.

(ii) We prove (ii) by induction on n. For $n = 1$ the result is true (take $e_1 = 1$). For $n \geq 2$, induction shows that there exists a complete set of $n - 1$ orthogonal idempotents $e_1, \ldots, e_{n-2}, e$ of R such that under reduction modulo π, $\bar{e}_1 = f_1, \ldots, \bar{e}_{n-2} = f_{n-2}$, and $\bar{e} = f_{n-1} + f_n$.

Now define $R_0 := eRe$. Then R_0 is a ring with unity e. Moreover, since A is finitely generated as a $\mathbf{Z}_p$-module, the same is true of R, and hence of R_0 itself. Thus R_0 is a $\mathbf{Z}_p$-algebra; and R_0 is $\mathbf{Z}_p$-free because $R_0 \subseteq R$ and R is $\mathbf{Z}_p$-free. By Theorem 3.2 there exists an integer $n \geq 1$ such that the ideal $pA = \pi^n A$. Therefore $(\pi R_0)^n \subseteq \pi^n A R_0 = p R_0 \subseteq \pi R_0$, and so the hypotheses of Theorem 3.1B hold for R_0 with $J = \pi R_0$. Since $\bar{e} = f_{n-1} + f_n$ and the latter are orthogonal by hypothesis, it follows that $f_{n-1} = \bar{e} f_{n-1} \bar{e}$ and $f_n = \bar{e} f_n \bar{e}$ both lie in $\bar{R}_0 \simeq R_0/\pi R_0$. Hence by Theorem 3.1B there exists an idempotent e_{n-1} of R_0 such that $\bar{e}_{n-1} = f_{n-1}$; and then $e_n := e - e_{n-1}$ is an idempotent in R_0 such that $\bar{e}_n = f_n$ (note $e_n \neq 0$ since $f_n \neq 0$). Clearly $e_n e_{n-1} = e_{n-1} e_n = 0$. Since both e_{n-1} and e_n lie in eRe and $e = e_{n-1} + e_n$, this shows that $e_1, \ldots, e_{n-1}, e_n$ is a complete set of orthogonal idempotents of R with the required property.

As an immediate application of Theorem 3.4A we have the following result.

Theorem 3.4B (i) Let R be an A-free A-algebra. Then R is a local ring iff $\bar{R}$ is a local ring.

(ii) Let U be an A-free AG-module. Then U is indecomposable iff $\bar{U}$ is an indecomposable kG-module.

(iii) Let V be an A-free AG-module. Then the conclusion of the Krull–Schmidt theorem (the corollary of Theorem 1.6A) holds for V.

Proof (i) Recall that a ring S is local iff $S/\text{rad } S$ is a division ring (see the remark following Theorem 1.7A). From the remark following the statement of Theorem 3.1B we know that rad $R \supseteq pR$, and rad $R/pR = \text{rad}(R/pR)$ is the unique maximal nilpotent ideal in R/pR. By Theorem 3.2, $pA = \pi^n A$ for some integer $n \geq 1$, and so $(\pi R)^n \subseteq pR \subseteq \pi R$. Thus $\pi R/pR$ is a nilpotent ideal in R/pR, and therefore $\pi R \subseteq \text{rad } R$. This shows that rad $\bar{R} = \text{rad}(R/\pi R) = \text{rad } R/\pi R$; and so $R/\text{rad } R \simeq \bar{R}/\text{rad } \bar{R}$. Thus R is local iff $\bar{R}$ is local.

(ii) Put $E := \text{End}_{AG}(U) \subseteq \text{End}_A(U)$. Since U is a free A-module, $\text{End}_A(U)$ is also a free A-module; hence E is A-free. Moreover

(1) $$\bar{E} = \text{End}_{AG}(U) \otimes_A k \simeq \text{End}_{kG}(U \otimes_A k) = \text{End}_{kG}(\bar{U}).$$

By §1.6 U (respectively, $\bar{U}$) is indecomposable iff 1 is the only idempotent in E (respectively, $\bar{E}$). By Theorem 3.4A, 1 is the only idempotent of E iff 1 is the only idempotent of $\bar{E}$ (write 1 as a sum of primitive idempotents). Thus U is indecomposable iff $\bar{U}$ is indecomposable.

(iii) Any direct summand of V is a free A-module of smaller rank, and so it is clear that V is a direct sum of a finite set of indecomposable AG-modules. The uniqueness of this decomposition will follow from Theorem 1.6A when we have shown that $E := \text{End}_{AG}(U)$ is a local ring for each indecomposable A-free AG-module U. However, by (ii), $\bar{U}$ is an indecomposable kG-module, and so $\text{End}_{kG}(\bar{U})$ is local by Theorem 1.7A. Then (1) shows that $\bar{E}$ is local and (i) shows that E is also local as required.

3.5 The case where p does not divide $|G|$

We now consider the especially simple case where $p \nmid g$ (where $g := |G|$). Write the AG-module AG in the form

(1) $$AG = U_1 \oplus \cdots \oplus U_t$$

where $U_1, U_2, \ldots, U_t$ are indecomposable AG-modules. Then reduction modulo π gives

(2) $$kG = \bar{U}_1 \oplus \cdots \oplus \bar{U}_t$$

and by Theorem 3.4B we know that the $\bar{U}_i$ are indecomposable kG-modules. However, in the case where $p \nmid g$, Theorem 1.3B (Maschke's theorem) shows that every kG-module is completely reducible, and so the indecomposability of the $\bar{U}_i$ implies that each $\bar{U}_i$ is irreducible.

This leads to the following theorem.

Theorem 3.5 Suppose that $p \nmid |G|$. Then the following hold.

(i) Each kG-module is isomorphic to a module of the form $\bar{U}$ for some A-free AG-module U.

(ii) If U is an A-free AG-module, then the following three conditions are equivalent: $\bar{U}$ is an irreducible kG-module, U is an indecomposable AG-module, and $U \otimes_A K$ is an irreducible KG-module.

(iii) If $\zeta^1, \ldots, \zeta^r$ are the distinct irreducible characters of G over K, then $\bar{\zeta}^1, \ldots, \bar{\zeta}^r$ are distinct and are the irreducible characters of G over k. In particular, for two irreducible characters ζ^i, ζ^j of G over K, we have $\zeta^i = \zeta^j$ iff $\bar{\zeta}^i = \bar{\zeta}^j$.

(iv) If U and V are A-free AG-modules, then $\bar{U} \simeq \bar{V}$ iff $U \otimes_A K \simeq V \otimes_A K$.

Proof (i) By Theorem 1.3B (Maschke's theorem), every kG-module is completely reducible; therefore it is enough to show that each irreducible kG-module has the form $\bar{U}$ for some A-free AG-module U. However each irreducible kG-module is an irreducible constituent of kG (see the note following Theorem 1.3A) and so is isomorphic to one of the summands in (2). This proves (i).

(ii) If $U = U' \oplus U''$, then $U \otimes_A K = (U' \otimes_A K) \oplus (U'' \otimes_A K)$. Hence, if $U \otimes_A K$ is irreducible, then U is indecomposable. Secondly, if U is indecomposable, then $\bar{U}$ is indecomposable by Theorem 3.4B and hence irreducible because $\bar{U}$ is completely reducible (Theorem 1.3B). Finally, if $\bar{U}$ is irreducible, then the note of §3.3 shows that $U \otimes_A K$ is also irreducible. Thus the three conditions are equivalent.

(iii) From (1) we obtain

$$(3) \quad KG \simeq AG \otimes_A K = (U_1 \otimes_A K) \oplus \cdots \oplus (U_t \otimes_A K),$$

where the summands are all irreducible KG-modules by (ii). By the note following Theorem 1.3A every irreducible KG-module is isomorphic to one of the $U_i \otimes_A K$; and we may assume that the U_i are numbered so that each irreducible KG-module is isomorphic to exactly one of $U_i \otimes_A K$ $(i = 1, \ldots, r)$ and that $U_i \otimes_A K$ affords the character ζ^i. Similarly each irreducible kG-module is isomorphic to one of $\bar{U}_1, \ldots, \bar{U}_t$. Since irreducible kG-modules with the same character are isomorphic (Corollary 1 of Theorem 2.3), this implies that every irreducible kG-module is isomorphic to one of $\bar{U}_i$ $(i = 1, \ldots, r)$. Since $p \nmid |G|$, Theorem 1.5 shows that the number of irreducible kG-modules (up to isomorphism) is the same as the number of irreducible KG-modules. Therefore the modules $\bar{U}_i$ $(i = 1, \ldots, r)$ are nonisomorphic, and their characters $\bar{\zeta}^1, \ldots, \bar{\zeta}^r$ are distinct and are all the irreducible characters of G over k.

(iv) Write $U = U_1 \oplus \cdots \oplus U_m$ and $V = V_1 \oplus \cdots \oplus V_n$ as sums of indecomposable AG-modules. Then

$$\bar{U} = \bar{U}_1 \oplus \cdots \oplus \bar{U}_m, \quad \bar{V} = \bar{V}_1 \oplus \cdots \oplus \bar{V}_n,$$

$$U \otimes_A K = (U_1 \otimes_A K) \oplus \cdots \oplus (U_m \otimes_A K),$$

and

$$V \otimes_A K = (V_1 \otimes_A K) \oplus \cdots \oplus (V_n \otimes_A K).$$

By (ii), each of the summands occurring in these latter modules is irreducible. Thus the proof of (iv) reduces easily to the case where U and V are indecomposable and $\bar{U}$, $\bar{V}$, $U \otimes_A K$, and $V \otimes_A K$ are irreducible. But in the latter case the result is immediate from (iii).

3.6 MODULAR CHARACTERS

Corollary If $p \nmid |G|$ and χ is an irreducible character of G over k, then $\deg \chi$ divides $|G|$.

Proof By part (iii) of the theorem, $\deg \chi = \deg \zeta$ for some irreducible character ζ of G over K. Then the result follows from Theorem 2.4B and §3.5.

EXERCISE

Let G be the alternating group generated by $a = (123)$ and $b = (345)$, and let k be a finite field of characteristic 2 containing a root ω of $X^2 + X + 1$. Let V be a kG-module of dimension 2 such that

$$(\alpha_1 v_1 + \alpha_2 v_2)a = (1 + \omega)\alpha_1 v_1 + \omega\alpha_2 v$$
$$(\alpha_1 v_1 + \alpha_2 v_2)b = \alpha_2 v_1 + (\alpha_1 + \alpha_2)v_2$$

where $\{v_1, v_2\}$ is a fixed basis of V. Show that there is no AG-module U such that $\bar{U} \simeq V$.

3.6 Modular characters

Let W be a kG-module, let x be a p'-element of G, and put $H := \langle x \rangle$. Then the restriction W_H is a kH-module. Since $p \nmid |H|$, Theorem 3.5 shows that there exists an A-free AH-module U such that $\bar{U} \simeq W_H$ and $U \otimes_A K$ is a KH-module uniquely determined up to isomorphism by W_H. In particular, the character of $U \otimes_A K$ is uniquely determined by W_H and so we can define $\phi(x)$ as the value of this character at x, where $\phi(x)$ only depends on W.

Definition We shall denote the set of all p'-elements of G by G°. Then the function $\phi: G^\circ \to K$ defined above is called the *modular character* (or "Brauer character") afforded by W. The modular character ϕ is called *irreducible* if it is afforded by an irreducible module (see the note below).

Note If χ is the (Frobenius) character afforded by W, then it follows in the notation above that $\bar{U}$ affords the character χ_H of χ restricted to H. Therefore $\overline{\phi(x)} = \chi(x)$ for each $x \in G^\circ$. Moreover, the modular character ϕ completely determines χ. For suppose y is any element of G. Then for some integer $m \geq 1$, $y^{p^m} \in G^\circ$. In a representation of G corresponding to χ, the linear transformation corresponding to y has eigenvalues $\omega_1, \ldots, \omega_n$, say, in k, and $\chi(y) = \omega_1 + \cdots + \omega_n$. Similarly $\chi(y^{p^m}) = \omega_1^{p^m} + \cdots + \omega_n^{p^m}$. Since k has characteristic p, the Binomial theorem (Lemma 1.5A) shows that

(1) $$\chi(y)^{p^m} = \omega_1^{p^m} + \cdots + \omega_n^{p^m} = \chi(y^{p^m}).$$

Now write $y = y_p y_{p'}$, where y_p and $y_{p'}$ are the p-part and p'-part of y, respectively (see §1.5). Then $y^{p^m} = (y_{p'})^{p^m}$, and so corresponding to (1) we get $\chi(y_{p'})^{p^m} = \chi(y^{p^m})$. Hence the Binomial theorem shows that

$$\{\chi(y) - \chi(y_{p'})\}^{p^m} = \chi(y^{p^m}) - \chi(y^{p^m}) = 0,$$

and so $\chi(y) = \chi(y_{p'}) = \overline{\phi(y_{p'})}$ because $y_{p'} \in G^\circ$. Hence ϕ completely determines χ. In particular it follows from Corollary 1 of Theorem 2.3 that if ϕ is afforded by an irreducible module W, then every kG-module V that affords ϕ is isomorphic to W.

Theorem 3.6 (*Brauer and Nesbitt*) Let U and V be kG-modules that afford the Frobenius characters σ and τ and the modular characters σ_0 and τ_0, respectively. Let S and T be the representations afforded by U and V. Then the following are equivalent.

(i) $\sigma_0 = \tau_0$ (on G°).
(ii) For each $x \in G^\circ$, $S(x)$ and $T(x)$ have the same characteristic polynomial.
(iii) U and V have the same kG-composition factors occurring with the same multiplicities.

Remark The condition (ii) may be rephrased as saying that the eigenvalues of $S(x)$ and $T(x)$ are the same (with the same multiplicities).

Proof First assume that (i) holds. Let $x \in G^\circ$ and put $H := \langle x \rangle$. By Theorem 3.5 there exist A-free AH-modules U_0 and V_0 such that $\bar{U}_0 \simeq U_H$ and $\bar{V}_0 \simeq V_H$. By the definition of modular character, the modules $U_0 \otimes_A K$ and $V_0 \otimes_A K$ afford the ordinary characters $(\sigma_0)_H$ and $(\tau_0)_H$, respectively. Since these are equal by (i), Corollary 2 of Theorem 2.3 shows that $U_0 \otimes_A K \simeq V_0 \otimes_A K$. Thus by Theorem 3.5, $\bar{U}_0 \simeq \bar{V}_0$, and so $U_H \simeq V_H$. Thus the restrictions of the representations S and T to H are equivalent, and (ii) follows.

Next assume (ii) holds. Let $\chi^1, \ldots, \chi^r$ be the irreducible characters of G over k. If m_i (respectively, n_i) is the multiplicity in U (respectively, V) of the irreducible constituent that affords the character χ^i ($i = 1, \ldots, r$), then $\sigma = \sum m_i \chi^i$ and $\tau = \sum n_i \chi^i$. Since both (ii) and (iii) only deal with the composition factors of U and V, there is no loss in generality in supposing that both U and V are completely reducible. By (ii) we know that $\sigma(x) = \tau(x)$ for all $x \in G^\circ$, and so we conclude that $\sigma = \tau$ on all of G by the note above. Thus $0 = \sigma - \tau = \Sigma(m_i - n_i)\chi^i$, and so by Corollary 1 of Theorem 2.3, $m_i - n_i = 0$ in k for each i; that is, $p \mid (m_i - n_i)$ for $i = 1, \ldots, r$. It remains to show that $m_i = n_i$ for all i. We shall proceed by induction on the degree of σ, and consider two cases.

First, suppose that for some j, $p \nmid m_j$. Then $p \nmid n_j$, and so both m_j and n_j are at least 1. Since U and V are completely reducible, they have submodules U_1 and V_1, respectively, which afford the characters $\sigma - \chi_j$ and $\tau - \chi_j$. Then by the induction hypothesis it follows that $m_i = n_i$ for each i. Second, suppose that $p \mid m_i$ and $p \mid n_i$ for all i. Write $m_i = pm'_i$ and $n_i = pn'_i$ ($i = 1, \ldots, r$). Then U and V have submodules U_2 and V_2, respectively, which afford the characters $\sigma' := \Sigma m'_i \chi^i$ and $\tau' := \Sigma n'_i \chi^i$. If S_2 and T_2 are the representations afforded by the modules U_2 and V_2, then for each $x \in G°$ the characteristic polynomial for $S(x)$ is the pth power of the characteristic polynomial of $S_2(x)$, and similarly, the characteristic polynomial for $T(x)$ is the pth power of that of $T_2(x)$. Hence the condition (ii) holds for S_2 and T_2, and so by induction we conclude $m'_i = n'_i$ for all i. Hence $m_i = n_i$ for $i = 1, \ldots, r$ and thus (ii) implies (iii).

Finally, if (iii) holds, then (i) holds because the values of the characters depend only on the composition factors.

EXAMPLE Let Q be a normal p-subgroup of G. Then each irreducible modular character ϕ of G can be written in terms of some irreducible modular character ψ of G/Q as follows: $\phi(x) = \psi(Qx)$ for all $x \in G°$. For suppose W is an irreducible kG-module which affords ϕ. Then by Clifford's theorem (Theorem 2.2A), W_Q is a completely reducible kQ-module because $Q \triangleleft G$. Since Q is a p-group, the only irreducible kQ-module is the trivial module (corollary of Theorem 1.5). Therefore $Q \subseteq \operatorname{Ker} W$, and W can be considered as an (irreducible) $k(G/Q)$-module. The modular character ϕ afforded by the kG-module W is related to the modular character ψ afforded by the $k(G/Q)$-module W by the equation $\phi(x) = \psi(Qx)$ $(x \in G°)$.

3.7 Cartan invariants, decomposition numbers, and orthogonality relations

In the present section we shall examine the relationship between the principal indecomposable kG-modules and the irreducible kG-modules. Our starting point is the material of §1.8. Write the kG-module kG as a sum of indecomposable submodules (principal indecomposables)

(1) $$kG = U_1 \oplus \cdots \oplus U_t,$$

where the U_i are enumerated so that each U_i is isomorphic to exactly one of $U_1, \ldots, U_r$. Then by the Krull–Schmidt Theorem, each principal indecomposable module of kG is isomorphic to exactly one of $U_1, \ldots, U_r$. By Theorem 1.8 and its corollary, each U_i has a unique maximal submodule $U'_i = U_i \cap \operatorname{rad} kG$ and each irreducible kG-module is isomorphic to exactly one of $U_1/U'_1, \ldots, U_r/U'_r$. We shall use the notation

(2) $$U_i \sim \sum_{j=1}^{r} c_{ij}(U_j/U'_j) \qquad (i = 1, \ldots, r),$$

where c_{ij} denotes the multiplicity of U_j/U'_j as an irreducible constituent of U_i. The nonnegative integers c_{ij} are called the *Cartan invariants* of kG, and the $r \times r$ matrix $\Gamma := [c_{ij}]$ is called the *Cartan matrix* of kG.

80   III MODULAR REPRESENTATIONS AND CHARACTERS

Note 1   It follows from Theorem 1.5 that r is equal to the number of classes of p'-elements of G, and Theorem 1.9A shows that $c_{ij} = i(U_j, U_i)$ for all i, j. The Cartan matrix Γ is only determined up to a permutation of its rows and columns (depending on the numbering of $U_1, \ldots, U_r$).

Now by Theorem 3.3 there exist A-free AG-modules $V_1, \ldots, V_s$ with the property that each irreducible KG-module is isomorphic to exactly one of the KG-modules $V_i \otimes_A K$ ($i = 1, \ldots, s$). Consider the kG-modules $\bar{V}_1, \ldots, \bar{V}_s$. Let $\zeta^1, \ldots, \zeta^s$ be the (ordinary) irreducible characters of G over K such that ζ^i is afforded by $V_i \otimes_A K$ for each i. Then the restriction of ζ^i to the set G° of p'-elements of G is the modular character afforded by $\bar{V}_i$, and Theorem 3.6 shows that the kG-composition factors of $\bar{V}_i$ and their multiplicities are uniquely determined by ζ^i. We write

(3) $$\bar{V}_i \sim \sum_{j=1}^{r} d_{ij}(U_j/U'_j) \quad (i = 1, 2, \ldots, s)$$

where d_{ij} denotes the multiplicity of the irreducible kG-module U_j/U'_j as an irreducible constituent of $\bar{V}_i$. The nonnegative integers d_{ij} are called the *decomposition numbers* of kG, and the $s \times r$ matrix $\Delta := [d_{ij}]$ is the *decomposition matrix*.

Note 2   It follows from Theorem 1.5 that s is the number of conjugate classes of G, and Theorem 1.9A shows that $d_{ij} = i(U_j, \bar{V}_i)$ for all i, j. The decomposition matrix Δ is only determined up to independent permutations of its rows and columns (depending on the ordering of the characters $\zeta^1, \ldots, \zeta^s$ and the modules $U_1, \ldots, U_r$).

The first important relation between Δ and Γ is the following.

Theorem 3.7A   With the notation above $\Gamma = \Delta^T \Delta$, where Δ^T is the transpose of Δ. In particular, Γ is symmetric.

Proof   Let $f_1, \ldots, f_t$ be a complete set of orthogonal idempotents in kG such that $U_i = f_i(kG)$ in (1) (see §1.6). Then by Theorem 3.4A these can be lifted to a complete set of orthogonal idempotents $e_1, \ldots, e_t$ of AG with $\bar{e}_i = f_i$ for each i. Write

(4) $$e_i(AG) \sim \sum_{j=1}^{s} a_{ij} V_j \quad (i = 1, \ldots, t)$$

where a_{ij} denotes the multiplicity of the irreducible KG-module $V_j \otimes_A K$ as an irreducible constituent of $e_i(AG) \otimes_A K \simeq e_i(KG)$. Since KG is semisimple, each principle indecomposable KG-module is irreducible, and hence

3.7 CARTAN INVARIANTS, NUMBERS, AND RELATIONS

isomorphic to one of the irreducible KG-modules $V_j \otimes_A K$. Hence by Theorems 1.9A and 1.9C have

$$a_{lj} = i(V_j \otimes_A K, e_l(KG)) = i(e_l(KG), V_j \otimes_A K) = \dim_K(V_j e_l \otimes K).$$

Since $V_j e_l \subseteq V_j$ is an A-free AG-module, this shows that

$$a_{lj} = \operatorname{rank}_A(V_j e_l) = \dim_k \bar{V}_j \bar{e}_l = i(\bar{e}_l(kG), \bar{V}_j)$$

by Theorem 1.9C. Since $\bar{e}_i(kG) = f_i(kG) = U_i$, Note 2 above shows that $a_{ij} = d_{ji}$. If we substitute this in (4) and go over to the associated kG-modules, we finally obtain

$$U_i = \bar{e}_i(kG) \sim \sum_{l=1}^{s} d_{li} \bar{V}_l \sim \sum_{j=1}^{s} \sum_{l=1}^{r} d_{li} d_{lj}(U_j/U'_j)$$

for $i = 1, \ldots, r$ using (3). Comparing this with (2) gives $c_{ij} = \sum_{l=1}^{r} d_{li} d_{lj}$, and so $\Gamma = \Delta^T \Delta$ as asserted.

The inner product $\langle\,,\,\rangle_G$ defined on the vector space $\operatorname{Class}_K(G)$ of class functions defined from G to K was introduced in §2.4. Here we shall also be interested in the vector space $\operatorname{Class}_K(G^\circ)$ of class functions $\theta: G^\circ \to K$ (where G° is the set of p'-elements of G). In particular, each modular character of G is an element of $\operatorname{Class}_K(G^\circ)$. In the notation introduced above there are r classes of p'-elements of G, and so $\dim_K \operatorname{Class}_K(G^\circ) = r$. We define the inner product $\langle\,,\,\rangle_{G^\circ}$ on $\operatorname{Class}_K(G^\circ)$ by

$$\langle \theta, \phi \rangle_{G^\circ} := \frac{1}{g} \sum_{x \in G^\circ} \theta(x) \phi(x^{-1})$$

where $g := |G|$. Clearly $\langle\,,\,\rangle_{G^\circ}$ is a symmetric, bilinear form.

NOTATION Let $U_1, \ldots, U_r$ be a full set of nonisomorphic principal indecomposable kG-modules [see (1) above]. Then the modular characters $\eta^1, \ldots, \eta^r$ afforded by these modules are called the *principal indecomposable modular characters*. Associated with each principal indecomposable modular character η^i there is the *irreducible modular character* ϕ^i afforded by U_i/U'_i ($i = 1, \ldots, r$). By Theorem 1.7A and the note in §3.6 we know that $\phi^1, \ldots, \phi^r$ are all distinct and include every irreducible modular character of G. We shall also let $\zeta^1, \ldots, \zeta^s$ be the irreducible (ordinary) characters of G, where in the notation above ζ^i is afforded by $V_i \otimes_A K$.

Finally enumerate the conjugate classes of G, $\mathscr{C}_1, \mathscr{C}_2, \ldots, \mathscr{C}_s$ so that $\mathscr{C}_1, \mathscr{C}_2, \ldots, \mathscr{C}_r$ are the classes of p'-elements. We shall put $h_i := |\mathscr{C}_i|$ ($i = 1, 2, \ldots, s$), and if θ is a class function on G (or G°), we shall write θ_i to denote the value of θ on $\mathscr{C}_i$. We shall also use the notation $\mathscr{C}_{i^*}$ for the class $\{x^{-1} | x \in \mathscr{C}_i\}$.

Note 3 As we saw in the proof of Theorem 3.7A, each U_i can be written in the form $\overline{W}_i$, where $W_i := e_i(AG)$ is an A-free AG-module. Therefore the modular character η_i afforded by U_i is the restriction of an ordinary character (afforded by W_i) to G°.

Note 4 It follows from (2), (3), and (4) that

$$\eta^i = \sum_{j=1}^{r} c_{ij}\phi^j \quad (i = 1, \ldots, r)$$

$$\zeta^i = \sum_{j=1}^{r} d_{ij}\phi^j \quad \text{on } G^\circ \quad (i = 1, \ldots, s)$$

and

$$\eta^i = \sum_{j=1}^{s} d_{ji}\zeta^j \quad \text{on } G^\circ \quad (i = 1, \ldots, r)$$

since $a_{ij} = d_{ji}$.

Theorem 3.7B (*Orthogonality relations*) With the notation above, the following hold.

  (i) Γ is nonsingular (we shall write $\Gamma^{-1} = [c'_{ij}]$).
  (ii) For all $i, j = 1, \ldots, r$ we have

$$\langle \phi^i, \phi^j \rangle_{G^\circ} = c'_{ij}$$

$$\langle \eta^i, \eta^j \rangle_{G^\circ} = c_{ij}$$

and

$$\langle \phi^i, \eta^j \rangle_{G^\circ} = \langle \eta^j, \phi^i \rangle_{G^\circ} = \delta_{ij}.$$

Proof We have three matrices of character values on the classes in G°: the $s \times r$ matrix $Z := [\zeta^i_j]$, the $r \times r$ matrix $F := [\phi^i_j]$, and the $r \times r$ matrix $E := [\eta^i_j]$; and also the $s \times r$ decomposition matrix $\Delta = [d_{ij}]$ and the $r \times r$ Cartan matrix $\Gamma = [c_{ij}]$. Using these we can rewrite the relations in Note 4 in matrix form:

(5) $\qquad E = \Gamma F, \quad Z = \Delta F, \quad \text{and} \quad E = \Delta^T Z.$

We also have the ordinary orthogonality relations

$$\sum_{l=1}^{s} \zeta^l_i \zeta^l_{j*} = g\, \delta_{ij}/h_i \quad \text{for } i, j = 1, \ldots, r$$

by Theorem 2.4A. These can be expressed in, say, the form

(6) $\qquad Z^T Z = [g\, \delta_{ij*}/h_i] = L.$

3.7 CARTAN INVARIANTS, NUMBERS, AND RELATIONS

Note that L is nonsingular and $L^{-1} = [h_i \, \delta_{ij*}/g]$. Since

(7) $\qquad F^T \Gamma F = (\Delta F)^T (\Delta F) = Z^T Z = L$

it follows that Γ is invertible, and so (i) is proved. Moreover from (7) and (5) and the symmetry of Γ we have the relations:

(8) $\qquad F L^{-1} F^T = F(F^{-1} \Gamma^{-1} (F^T)^{-1}) F^T = \Gamma^{-1}$

(9) $\qquad E L^{-1} E^T = \Gamma (F L^{-1} F^T) \Gamma^T = \Gamma$

(10) $\qquad F L^{-1} E^T = (F L^{-1} F^T) \Gamma^T = I \qquad$ (the identity matrix).

These are precisely the matrix forms of the relations in (ii).

Corollary The irreducible modular characters $\phi^1, \ldots, \phi^r$ and the principal indecomposable modular characters $\eta^1, \ldots, \eta^r$ for G both form K-bases for $\text{Class}_K(G^\circ)$.

Proof As we saw above $\dim_k \text{Class}_k(G^\circ) = r$, so it is enough to show that the two sets of characters are linearly independent sets. In terms of the matrices defined in the proof of the theorem this is equivalent to the conditions that F and E are nonsingular; and the nonsingularity of F and E follows at once from (10).

EXAMPLE Let $W_i = e_i(AG)$ be a principal indecomposable AG-module such that $\overline{W}_i$ affords the principal indecomposable modular character η^i, and let ψ^i be the (ordinary) character afforded by W_i (see Note 3). Then $\psi^i = \eta^i$ on G° by the definition of modular character. We shall now show that $\psi^i = 0$ on $G \backslash G^\circ$. Indeed $\psi^i = \sum_{j=1}^s d_{ji} \zeta^j$ by (4) since $a_{ij} = d_{ji}$. Therefore by Theorem 2.4A, $\langle \psi^i, \psi^i \rangle_G = \sum_{j=1}^s d_{ji}^2$. On the other hand, by Note 4 above, we have on G° that

$$\eta^i = \psi^i = \sum_{j=1}^s d_{ji} \left(\sum_{l=1}^r d_{jl} \phi^l \right).$$

Therefore by Theorem 3.7B

$$\langle \psi^i, \psi^i \rangle_{G^\circ} = \sum_{j=1}^s d_{ji} \left(\sum_{l=1}^r d_{jl} \langle \eta^i, \phi^l \rangle_{G^\circ} \right) = \sum_{j=1}^s d_{ji}^2.$$

Thus

$$0 = \langle \psi^i, \psi^i \rangle_G - \langle \psi^i, \psi^i \rangle_{G^\circ} = \frac{1}{g} \sum_{x \in G \backslash G^\circ} |\psi^i(x)|^2,$$

and so $\psi^i(x) = 0$ for all x in $G \backslash G^\circ$.

EXERCISES

1. Suppose that G is a cyclic group of order n. Describe the ordinary irreducible characters, the modular irreducible characters, the principal indecomposable modular characters, the decomposition matrix, and the Cartan matrix.

2. If G is a group with a Sylow p-subgroup P of order p^m, show that p^m divides $\eta^i(1)$ for each principal indecomposable modular character η^i. [*Hint:* With the notation of the example above, calculate $\langle \psi_P^i, 1_P \rangle_P$.]

3.8 Modular characters of p-solvable groups

Recall that a group G is called *p-solvable* for a given prime p provided it has a normal series $G = G_0 \supseteq G_1 \supseteq \cdots \supseteq G_m = 1$ with each $G_i \triangleleft G$ such that each factor G_i/G_{i+1} is either a p-group or a p'-group.

EXAMPLE 1 A solvable group is p-solvable (for any prime p). Conversely, since every $2'$-group is solvable (by a theorem of Feit and Thompson [2]), every 2-solvable group is solvable.

EXAMPLE 2 Any subgroup or factor group of a p-solvable group is also p-solvable.

Note 1 In any group G there exists a unique maximal normal p-subgroup (usually denoted by $O_p(G)$) and a unique maximal normal p'-subgroup (usually denoted by $O_{p'}(G)$); and G is p-solvable iff for each proper normal subgroup N of G, at least one of $O_p(G/N)$ and $O_{p'}(G/N)$ is nontrivial.

Note 2 Suppose that G is a p-solvable group with $O_p(G) = 1$, then $N := O_{p'}(G)$ contains $C_G(N)$. Indeed otherwise, by the definition of $O_{p'}(G)$, $C_G(N)$ has a p-subgroup $Q \neq 1$ such that the normal subgroup $C_G(N)N/N$ of G/N contains the normal p-subgroup QN/N of G/N. But then $QN = Q \times N$ since $Q \subseteq C_G(N)$, and so $Q = O_p(QN)$. Thus Q is a characteristic subgroup of the normal subgroup QN and hence $Q \triangleleft G$. This contradicts the hypothesis that $O_p(G) = 1$.

The object of the present section is to prove the following theorem (Theorem 3.8) about characters of p-solvable groups. The original proof was given by Swan [1] using results of Fong [1]; our proof is based on a modification due to Serre [1].

Before stating the theorem we first shall prove a lemma which will be needed in the proof. If N is a normal subgroup of a group G, and T is a representation of N, then for each $y \in G$ we define the *conjugate representation* T^y of N by $T^y(x) := T(yxy^{-1})$ for all $x \in N$ (see §2.2).

3.8 MODULAR CHARACTERS OF p-SOLVABLE GROUPS

Lemma 3.8 Let K^* be an algebraic closure of K, let N be a normal subgroup of a group G, and let V be an irreducible K^*N-module which affords the representation T. Assume that T is equivalent to T^y for each $y \in G$. Put $D := \{\det T(x) \mid x \in N\}$ and define

$$H := \{(y, c) \in G \times \mathrm{Aut}_{K^*}(V) \mid \det c \in D$$

and $cT(x)c^{-1} = T^y(x)$ for all $x \in N\}$.

Then H is a finite subgroup of $G \times \mathrm{Aut}_{K^*}(V)$ with the following properties.

(i) There is an injective homomorphism $N \to H$ given by $x \mapsto (x, T(x))$; we shall identify N with its image under this mapping.

(ii) There is a surjective homomorphism $H \to G$, given by $(y, c) \mapsto y$, which maps N (embedded in H) onto N (embedded in G).

(iii) There is a representation of H on V given by $(y, c) \mapsto c$, and its restriction to N is equal to T.

Proof It is readily verified that H is a subgroup of $G \times \mathrm{Aut}_{K^*}(V)$, and that (i) follows immediately. The mapping in (ii) is clearly a homomorphism which maps N onto N; the only nonobvious point is that it is surjective. However, for each $y \in G$, T is equivalent to T^y, and so for some $b \in \mathrm{Aut}_{K^*}(V)$, $bT(x)b^{-1} = T^y(x)$ for all $x \in N$. Since K^* is algebraically closed, $\det b = \beta^d$ for some $\beta \in K^*$ (and $d := \dim_{K^*} V$), then $(y, \beta^{-1}b) \in H$. This proves (ii). Moreover, the kernel of the homomorphism in (ii) is $\{(1, c) \mid \det c \in D\}$. Clearly D is a finite group; let $\lambda_1 = 1, \lambda_2, \ldots, \lambda_n$ be the elements of D. If $H_0 := \{(1, c) \mid \det c = 1\}$, then H_0 is a subgroup of H and $\{(1, c) \mid \det c \in D\} = H_0 \cup H_0 c_2 \cup \cdots \cup H_0 c_n$, where c_i is an element of $\mathrm{Aut}_{K^*}(V)$ such that $\det c_i = \lambda_i$ for $i = 2, 3, \ldots, n$. To prove the finiteness of H it suffices to prove that H_0 is finite. By Schur's lemma (see §1.2),

$$H_0 = \{(1, c) \mid \det c = 1 \text{ and } c = \gamma \cdot 1 \text{ for some } \gamma \in K^*\}$$
$$= \{(1, c) \mid c = \gamma \cdot 1 \text{ and } \gamma^d = 1\},$$

and so the order of H_0 divides d. Thus H_0, and hence H, is finite.

The assertion (iii) is obvious.

Remark The proof above shows that the kernel of the homomorphism in (ii) is $\{(1, c) \mid \det c \in D\}$. This has order dividing $d|D|$, and hence dividing $d|N|$.

Theorem 3.8 (*Fong–Swan*) Let G be a p-solvable group and ψ an irreducible (Frobenius) character of G of degree d over k. Then there exists an irreducible ordinary character ζ of G of degree d such that $\bar{\zeta} = \psi$.

Remark In general ζ is not uniquely determined by ψ, nor does each irreducible ordinary character of G correspond in this way to an irreducible character over k. (See the exercises at the end of this section.) Recall that by our standing hypothesis (§3.3), all fields involved are splitting fields for G, so ψ and ζ are absolutely irreducible.

Proof We shall proceed by induction on the degree d of ψ. By §2.7 there is no loss in generality in supposing that K (and hence k) is a splitting field for $X^{g^{2d}} - 1$ where $g := |G|$, and we may also assume that G acts faithfully on a kG-module V of dimension d which affords ψ. Note that by the corollary of Clifford's theorem (Theorem 2.2A), $O_p(G) = 1$. Put $N := O_{p'}(G)$.

Case 1 ($d = 1$) In this case G is abelian. Since $O_p(G) = 1$, G is a p'-group, and so the result follows from Theorem 3.5.

Case 2 ($d > 1$ and the irreducible constituents of V_N are not all conjugate) In this case by Clifford's theorem (Theorem 2.2A) there exists a subgroup S of G and an irreducible kS-module U of dimension less than d such that $V \simeq U^G$. By the induction hypothesis there exists an irreducible ordinary character χ of S such that U affords $\bar{\chi}$. Then χ^G is an ordinary character of G such that $\overline{\chi^G} = \psi$, and χ^G is irreducible because ψ is (see the note of §3.3).

Case 3 ($d > 1$ and all irreducible constituents of V_N are isomorphic) Let V_0 be an irreducible constituent of V_N. Since N is a p'-group, it follows from Theorem 3.5 that there exists an A-free AN-module W_0 such that $\bar{W}_0 \simeq V_0$. Put $U_0 := W_0 \otimes_A K$, and $U_0^* := U_0 \otimes_K K^*$, where K^* is an algebraic closure of K. Then U_0 is an irreducible KN-module (because V_0 is an irreducible kN-module). Since K is a splitting field for KN by the standing hypothesis of §3.3, this shows that U_0^* is also an irreducible K^*N-module. Let T be the representation of N afforded by U_0^*, and χ the character afforded by U_0^*. For each $y \in G$, T^y affords the character χ^y. On the other hand, V_0 affords the character $\bar{\chi}$ and the kN-module $V_0 y$ affords the character $\overline{\chi^y}$. Since $V_0 y$ is an irreducible constituent of V_N, $V_0^y \simeq V_0$ by hypothesis, so $\bar{\chi} = \overline{\chi^y}$. Hence $\chi = \chi^y$ by Theorem 3.5. This shows that T is equivalent to T^y for all $y \in G$ (by Theorem 2.3 and its corollary). Hence with the notation of Lemma 3.8 there exists a group H in which N is embedded together with an extension of the action on U_0^* to H so that U_0^* is a K^*H-module. Since H has order dividing $g^2 d$ (see the remark following Lemma 3.8), the character afforded by the irreducible K^*H-module U_0^* is afforded by some irreducible KH-module U_1, say. Now by Theorem 3.3 there exists an A-free AH-module W such that $U_1 \simeq W \otimes_A K$. Since $W_N \otimes_A K \simeq (U_1)_N$ it follows from Theorem 3.5 that $\bar{W}_N \simeq V_0$.

3.8 MODULAR CHARACTERS OF p-SOLVABLE GROUPS

Now define on the k-space $\operatorname{Hom}_{kN}(\bar{W}, V)$ an action of H by $\bar{w}f^{(y, c)} := [(wc^{-1})f]y$ for each $w \in W, f \in \operatorname{Hom}_{kN}(\bar{W}, V)$ and $(y, c) \in H$; clearly $f^{(y, c)}$ is k-linear, and it is a kN-homomorphism because for each $x \in N$,

$$\bar{w}f^{(y, c)}x = [\overline{(wc^{-1})}f(yxy^{-1})]y = [\overline{(wc^{-1}yxy^{-1})}f]y$$
$$= [\overline{(wc^{-1}T^y(x))}f]y = [\overline{(wT(x)c^{-1})}f]y = (\bar{w}x)f^{(y, c)}$$

by the definition of H (see Lemma 3.8). Moreover, it is clear that this action of H is k-linear, and so we have defined a kH-module structure on $\operatorname{Hom}_{kN}(\bar{W}, V)$. On the other hand we can define a kH-module structure on V using the homomorphism $H \to G$ described in part (ii) of Lemma 3.8.

We now claim that $\operatorname{Hom}_{kN}(\bar{W}, V) \otimes_k \bar{W} \simeq V$ as kH-modules under the mapping θ given by $(f, \bar{w})\theta := \bar{w}f$. It is straightforward to verify that θ is a kH-homomorphism, and it is injective because $\bar{w}f = 0$ for $f \neq 0$ implies $\bar{w} = 0$ (since $\bar{W}$ is an irreducible kN-module). Finally, by Clifford's theorem (Theorem 2.2A), V_N is completely reducible, and so by hypothesis it is a direct sum, say $V_N = V_1 \oplus V_2 \oplus \cdots \oplus V_m$, where each $V_i \simeq \bar{W}_N$. Then as k-spaces $\operatorname{Hom}_{kN}(\bar{W}, V) \simeq \bigoplus_{i=1}^{m} \operatorname{Hom}_{kN}(\bar{W}, V_i)$ is isomorphic to m copies of k because k is a splitting field. Hence $\dim_k \operatorname{Hom}_{kN}(\bar{W}, V) = m = \dim V/\dim \bar{W}$. Thus the k-spaces $\operatorname{Hom}_{kN}(\bar{W}, V) \otimes_k \bar{W}$ and V have the same dimension, and so θ is a k-isomorphism. Since θ is a kH-homomorphism, this shows that θ is a kH-isomorphism as asserted. Note that $\dim_k \bar{W} > 1$. Otherwise, V_N is a sum of isomorphic one-dimensional kN-submodules and that implies N acts as a group of scalars on V. However $N = O_{p'}(G) \supseteq C_G(N)$ by Note 2 above. Since G acts faithfully on V and N acts as a group of scalars on V, this would imply $G \subseteq N$, and so G is abelian. But $d > 1$ by hypothesis, so the corollary of Theorem 1.5 gives a contradiction. Hence $\dim_k \bar{W} > 1$, and hence $\dim_k \operatorname{Hom}_{kN}(\bar{W}, V) < \dim_k V = d$.

Now the kH-module $\operatorname{Hom}_{kN}(\bar{W}, V)$ is irreducible since otherwise V would be reducible as a kH-module and hence as a kG-module, which is contrary to hypothesis. Since $\dim_k \operatorname{Hom}_{kN}(\bar{W}, V) < d$, the induction hypothesis shows that there exists an irreducible ordinary character ζ_1 of H such that $\operatorname{Hom}_{kN}(\bar{W}, V)$ affords the k-character $\bar{\zeta}_1$ of H. Again we have seen above that there exists an irreducible ordinary character ζ_2 of H (afforded by W) such that $\bar{W}$ affords the k-character $\bar{\zeta}_2$ of H. Then $\zeta := \zeta_1 \zeta_2$ is an irreducible ordinary character of H such that $V \simeq \operatorname{Hom}_{kN}(\bar{W}, V) \otimes_k \bar{W}$ affords the character $\overline{\zeta_1 \zeta_2} = \bar{\zeta}_1 \bar{\zeta}_2$. Finally, we shall show that the kernel of $\zeta_1 \zeta_2$ contains the kernel L of the homomorphism $H \to G$ defined in part (ii) of Lemma 3.8. Indeed we know from Example 2 of §3.3 that if U is any AH-module, then $\operatorname{Ker} \bar{U}/\operatorname{Ker} U$ is always a p-group. However, by the remark following Lemma 3.8, $|L|$ divides $d|N|$. Since N is a p'-subgroup of G, and $d \mid |N|$ by the corollary of Theorem 2.4B, this shows that L is a p'-subgroup of H. On the other hand Ker $V = $ Ker $\overline{\zeta_1 \zeta_2}$ (where V is considered as a kH-module) contains L by the definition of the action of H; hence $L \subseteq$ Ker $\zeta_1 \zeta_2 = $ Ker ζ. Thus we can consider ζ as an irreducible ordinary character of $G \simeq H/L$, and this completes the proof.

EXERCISES

1. Let G be an abelian group of order p^n and ζ^1, ζ^2, ..., ζ^{p^n} be the irreducible ordinary characters of G. Then show that $\overline{\zeta}^i = \overline{\zeta}^1$ for all $i = 1, 2, \ldots, p^n$.

2. Let G be the nonabelian group of order 6. Show that G has an irreducible character of degree 2 over a field of characteristic 2 and that G has no irreducible character of degree 2 over a field of characteristic 3.

3.9 Notes and comments

The connection between the ordinary and modular representations of a finite dimensional algebra was first recognized by Brauer (see Artin, et al. [1]) who developed the relation between the decomposition numbers and the Cartan invariants. This was later applied to the study of modular representations of a group by Brauer and Nesbitt in an important paper [1] which set the stage for the subsequent development of the theory. Our treatment of the relation between the ordinary and modular representations follows closely a paper of Dade [5]. For other references to this topic see Artin et al. [1], Curtis and Reiner [1], and Dornhoff [1]. Theorem 3.3 was first proved by Burnside when A is the ring of rational integers. The basic properties of modular characters were proved by Brauer and Nesbitt [1,2] and Brauer [1,8]. Note that our definition of modular characters is equivalent to the one defined by Brauer and Nesbitt.

There are a number of papers on the representations of p-solvable groups. For further results on modular representations of p-solvable groups, see Berman [4], Dade [4], Fong and Gaschutz [1], Green and Hill [1], Isaacs [2], Itô [1,2], Nagao [2], Rukolaine [1], and Winter [1].

CHAPTER **IV**

Blocks of Group Algebras

The theory of blocks of a group algebra attempts to classify the irreducible representations. This chapter presents several of the characterizations of the blocks and their properties. The main topics are blocks and their characterization in terms of ideals, irreducible modular and ordinary characters, irreducible modular and ordinary representations, and principal indecomposable modules. We then study the defect groups of a block and by making a further analysis of the Cartan matrix and the matrix of decomposition numbers we obtain bounds on the number of irreducible characters in a block of given defect. Finally, in the last section we discuss blocks of small defect.

4.1 Blocks

For the present section let k denote an arbitrary field (possibly of characteristic 0) and let G be a group. In §1.8 we considered the direct decomposition of the group algebra kG as a kG-module into principal indecomposable modules. We shall consider here the decomposition of kG as an algebra into a direct sum of indecomposable (two-sided) ideals. In many ways the situation here is simpler. We shall use it as a basic tool in the classification of kG-modules.

Definition A *block* of kG is a nonzero ideal B such that (a) $kG = B \oplus B'$ for some ideal B', and (b) B cannot be written as a direct sum of two nonzero ideals of kG.

Since kG is Artinian, we can certainly write

(1) $$kG = B_1 \oplus \cdots \oplus B_t$$

as a direct sum of blocks. Associated with (1) we have the decomposition

(2) $$1 = f_1 + \cdots + f_t,$$

where $f_1, \ldots, f_t$ is a complete set of orthogonal idempotents of kG with $f_i \in B_i$ and $B_i = f_i(kG)$ for each i (see §1.6). These idempotents are completely determined by the decomposition (1), and we call f_i the idempotent *associated* with the block B_i.

Note 1 The ideals $B_1, \ldots, B_t$ in (1) are the only blocks of kG, and so the decomposition (1) is unique. Indeed, suppose that B is a block of kG. Then since B and B_i are ideals, $BB_i \subseteq B \cap B_i$ for each i. Thus from (1) we have

$$B = BB_1 + \cdots + BB_t \subseteq (B \cap B_i) \oplus \cdots \oplus (B \cap B_t) \subseteq B.$$

Hence $B = (B \cap B_1) \oplus \cdots \oplus (B \cap B_t)$, and since B cannot be written as a sum of two nonzero ideals, there exists j such that $B \subseteq B_j$ and $B \cap B_i = 0$ for all $i \neq j$. Since B is a block, there is an ideal B' such that $kG = B \oplus B'$. Therefore $B_j = BB_j + B'B_j \subseteq (B \cap B_j) \oplus (B' \cap B_j) \subseteq B_j$, and so

$$B_j = (B \cap B_j) \oplus (B' \cap B_j).$$

Since B_j is not the sum of two nonzero ideals, this shows that $B_j = B \cap B_j = B$. Thus $B_1, \ldots, B_t$ are the only blocks of kG.

Note 2 Let $a \in kG$. Then by (2) we have

(3) $$a = af_1 + \cdots + af_t = f_1 a + \cdots + f_t a.$$

Since B_i is an ideal, af_i and $f_i a$ both lie in B_i and so by (1) and (3) we conclude $af_i = f_i a$ for each $a \in kG$. Hence the idempotents f_i in (2) all lie in the center $Z(kG)$ of kG. We shall call them the *block idempotents* (or "central primitive idempotents") of kG. Thus kG has exactly t block idempotents and these are given by (2).

Lemma 4.1 Let $f_1, \ldots, f_t$ be the block idempotents of kG. Then we have the decomposition

(4) $$Z(kG) = f_1 Z(kG) \oplus \cdots \oplus f_t Z(kG)$$

of the center $Z(kG)$ of kG into a sum of ideals. Moreover, each $f_i Z(kG)$ is

4.1 BLOCKS

indecomposable as a $Z(kG)$-module. If k is a splitting field for $Z(kG)$, then for each i we have

$$f_i Z(kG)/\text{rad}[f_i Z(kG)] \simeq k.$$

Remark Both $Z(kG)$ and $f_i Z(kG)$ are commutative k-algebras (in the latter case f_i is the unity element). Thus by Theorem 1.3A(i), their radicals consist of all their nilpotent elements. In particular, $\text{rad}[f_i Z(kG)] = f_i Z(kG) \cap \text{rad } Z(kG) = f_i \text{ rad } Z(kG)$.

Proof Since $f_i Z(kG) \subseteq f_i(kG) = B_i$, the decomposition (4) follows from (1) and (2). To show that each $f_i Z(kG)$ is indecomposable as a $Z(kG)$-module, it follows from §1.6 that it is enough to prove that the idempotent f_i of $Z(kG)$ cannot be written $f_i = e + e'$, where e and e' are idempotents of $Z(kG)$ and $ee' = 0$. Suppose on the contrary that such a decomposition existed. Then $B_i = e(kG) + e'(kG)$ is a sum of two ideals in kG (since e and e' lie in the center of kG). Moreover, $e(kG) \cap e'(kG) = 0$; indeed, if $ea = e'a'$, where $a, a' \in kG$, then $ea = e(e'a) = 0$. Thus $B_i = e(kG) \oplus e'(kG)$ is a sum of two nonzero ideals, contrary to the definition of a block. This shows that $f_i Z(kG)$ is an indecomposable $Z(kG)$-module.

Thus we are in the situation of Theorem 1.8 with $R = Z(kG)$, and the $f_i Z(kG)$ are the principal indecomposable R-modules. In particular, $\text{rad}[f_i Z(kG)]$ is the unique maximal R-submodule of $f_i Z(kG)$. Since R is commutative, this shows that $f_i Z(kG)/\text{rad}[f_i Z(kG)]$ is a simple commutative algebra over k. If k is a splitting field for $Z(kG)$, then the latter algebra has dimension 1 over k and so the final assertion of the lemma follows.

Definition A *central character* of kG is a k-algebra homomorphism $\omega: Z(kG) \to k$. [By our convention, since ω is an algebra homomorphism it maps the unity of $Z(kG)$ onto the unity of k; hence ω is surjective.]

Theorem 4.1 Let k be an arbitrary splitting field of the group algebra kG (possibly k has characteristic 0), and let $f_1, \ldots, f_t$ be the block idempotents of kG. Then kG has exactly t distinct central characters $\omega_1, \ldots, \omega_t$. These are characterized by the equations

(5) $$\omega_i(f_j) = \begin{cases} 1 & \text{if } i = j \\ 0 & \text{otherwise.} \end{cases}$$

Remark We shall call ω_i the central character *associated* with the block $f_i(kG)$.

Proof The composition ω_i of the projection $Z(kG) \to f_i Z(kG)$ and the canonical mapping $f_i Z(kG) \to f_i Z(kG)/\text{rad}[f_i Z(kG)]$ is a k-algebra homomorphism. Since k is a splitting field for kG, $\omega_i: Z(kG) \to k$ by Lemma 4.1 and so is a central character. Since the f_j are orthogonal, it is clear that (5) holds for ω_i. On the other hand, suppose that ω is any central character of kG. If $a \in \text{rad } Z(kG)$, then a is nilpotent and so $\omega(a)^m = \omega(a^m) = 0$ in k for some integer $m \geq 1$; hence $\omega(a) = 0$. Thus rad $Z(kG)$ lies in the kernel of every central character. Moreover, since k is a splitting field for kG, Lemma 4.1 shows that $f_i Z(kG) = kf_i \oplus \text{rad}[f_i Z(kG)]$ as k-spaces. Hence we have

(6) $Z(kG) = kf_1 \oplus \cdots \oplus kf_t \oplus \text{rad } Z(kG)$ as k-spaces.

Now any central character ω satisfies $\omega(1) = 1$. Since $1 = f_1 + \cdots + f_t$ in $Z(kG)$, this shows that $\omega(f_j) \neq 0$ for at least one j. Since in k we have

$$\omega(f_i)\omega(f_j) = \omega(f_i f_j) = \begin{cases} \omega(f_j) & \text{if } i = j \\ 0 & \text{otherwise,} \end{cases}$$

this shows that $\omega(f_j) = 1$ and $\omega(f_i) = 0$ for all $i \neq j$. Thus the two central characters ω_j and ω agree on $f_1, \ldots, f_t$, and each contains rad $Z(kG)$ in its kernel. Therefore it follows from (6) that $\omega = \omega_j$. Hence $\omega_1, \ldots, \omega_t$ are the only central characters of kG.

EXAMPLE 1 (*Constructing central characters from irreducible characters*) Under the hypotheses of Theorem 4.1 let ζ be an irreducible character of G over k. Let V be a kG-module which affords ζ and let T be the representation afforded by V. Let $\mathscr{C}_1, \ldots, \mathscr{C}_s$ be the conjugate classes of G, let $c_1, \ldots, c_s$ be the class sums in kG, and put $h_i := |\mathscr{C}_i|$ ($i = 1, \ldots, s$). We know from §2.3 that $c_1, \ldots, c_s$ is a k-basis of $Z(kG)$. Since c_i centralizes kG, and k is a splitting field for kG, $T(c_i) = \gamma_i 1$ for some $\gamma_i \in k$ ($i = 1, \ldots, s$). Taking traces we obtain $h_i \zeta_i = \gamma_i \zeta(1)$, where ζ_i is the value on $\mathscr{C}_i$. Now suppose that $\zeta(1) \neq 0$ in k (so either k has characteristic 0, or k has characteristic $p > 0$, where p does not divide the degree of ζ). Then $\gamma_i = h_i \zeta_i / \zeta(1)$, so we can define a k-linear mapping $\omega: Z(kG) \to k$ by $\omega(c_i) := h_i \zeta_i / \zeta(1)$ ($i = 1, \ldots, s$). Since T is a k-algebra homomorphism, it is clear that ω is a central character of kG.

EXAMPLE 2 (*The central characters and block idempotents when the characteristic is 0*) Now suppose that K is a field of characteristic 0 and that K is a splitting field for KG. In this case rad $KG = 0$ (Theorem 1.3B). Since the nilpotent elements of $Z(KG)$ all generate nilpotent ideals in KG, this means that 0 is the only nilpotent element in $Z(KG)$, and so rad $Z(KG) = 0$. Then from Lemma 4.1, since K is a splitting field for KG, each $f_i KG \simeq K$, and so $\dim_K Z(KG) = t$ (the number of block idempotents). However (see Example 1), the class sums $c_1, \ldots, c_s$ form a K-basis of $Z(KG)$, and so $t = s$ (the number of conjugate classes of G). Furthermore KG has exactly s central characters by Theorem 4.1, and exactly s irreducible characters $\zeta^1, \ldots, \zeta^s$ over K by Theorem 2.4A.

4.2 CLASSIFYING MODULES, CHARACTERS, AND IDEMPOTENTS

Using Example 1 (and the notation there), we can construct all s central characters $\omega_1, \ldots, \omega_s$. They are the K-linear mappings defined by

(7) $\qquad \omega_j(c_l) := h_l \zeta_l^j / \zeta^j(1) \qquad (l = 1, \ldots, s)$.

(Note that these are distinct because

$$0 = \sum_{l=1}^{s} h_l \zeta_l^j \zeta_{l^*}^i \neq \sum_{l=1}^{s} h_l \zeta_l^j \zeta_{l^*}^j = g$$

for $i \neq j$ by Theorem 2.4A).

It follows from Theorem 4.1 that the block idempotent f_j associated with ω_j satisfies the relations (5). Then writing $f_j = \sum_{l=1}^{s} \lambda_{jl} c_l$, say, with $\lambda_{jl} \in K$, we have $\sum_{l=1}^{s} \lambda_{jl} h_l \zeta_l^i / \zeta^i(1) = \delta_{ij}$, and so by Theorem 2.4A we find

$$\zeta^j(1) \zeta_m^j = \sum_{i=1}^{s} \delta_{ij} \zeta^i(1) \zeta_m^i = \sum_{l=1}^{s} \lambda_{jl} \sum_{i=1}^{s} h_l \zeta_l^i \zeta_m^i = \lambda_{jm^*} g.$$

Hence (in the notation of Theorem 2.4A)

(8) $\qquad f_j = \dfrac{\zeta^j(1)}{g} \sum_{l=1}^{s} \zeta_{l^*}^j c_l \qquad$ for $j = 1, \ldots, s$.

Remark The concept of block can be extended as follows. Suppose we use the notation of §3.3 and consider the group algebra AG. First note that $\overline{Z(AG)} = Z(kG)$ since in both cases the class sums form a basis. The block idempotents $f_1, \ldots, f_t$ of kG form a complete set of orthogonal idempotents of $Z(kG)$ and so may be lifted to a complete set of orthogonal idempotents $e_1, \ldots, e_t$ of $Z(AG)$ (Theorem 3.4A). It can now be proved that the $e_i Z(AG)$ ($i = 1, \ldots, t$) are uniquely determined by the f_i and are indecomposable $Z(AG)$-modules. (See the exercise below). The ideals $e_i(AG)$ ($i = 1, \ldots, t$) cannot be written as sums of two nonzero ideals of AG [since the same is true of $\overline{e_i(AG)} = f_i(kG)$] and are called the blocks of AG. However in general $e_i(KG)$ is *not* a block of KG, but splits further into a sum of nonzero ideals in KG. (This is the point of much of the next section.)

EXERCISE

With the notation of §3.3, let f be a block idempotent of kG and let e be an idempotent of AG such that $\bar{e} = f$. Prove that $eZ(AG)$ is indecomposable.

4.2 Classifying modules, characters, and idempotents into blocks

We now return to the notation of §3.3 for the rest of this chapter. In the present section we shall describe how to classify the modules and characters of G with respect to the blocks of kG.

(a) Indecomposable kG-modules lying in blocks

Let V be an indecomposable kG-module, and suppose that $f_1, \ldots, f_t$ are the block idempotents of kG (see §4.1). Since $f_1, \ldots, f_t$ is a complete set of orthogonal idempotents

$$V = Vf_1 \oplus \cdots \oplus Vf_t \quad \text{as} \quad k\text{-spaces}$$

(see §1.6), and since the f_i are all central, this is a direct sum of kG-modules. Because V is indecomposable, this means that for some j we have

(1) $\qquad Vf_j = V \quad$ and $\quad Vf_i = 0 \quad$ for all $\quad i \neq j$.

In this case we associate V with the block $B_j := f_j(kG)$ of kG, and say V *lies in the block* B_j. Clearly all kG-modules isomorphic to V lie in the same block. This gives a classification of all indecomposable (and in particular, all irreducible) kG-modules into blocks.

Note 1 It follows at once from (1) that if an indecomposable kG-module V lies in a block B_j, then so does each of its irreducible constituents. Moreover, considering any block $B_i = f_i(kG)$ of kG as a kG-module, we see that the indecomposable components and the irreducible constituents of B_i all lie in the block B_i (we have $B_i f_i = B_i$ because the block idempotent f_i is central).

Note 2 Since $1 = f_1 + \cdots + f_t$, it follows from (1) that if V lies in the block B_j, then $v = v1 = vf_j$ for all $v \in V$.

The block of kG in which the trivial kG-module lies is called the *principal block*. This block will play an especially important role; partly because it is the block which is most easily analyzed.

The relation (1) classifies indecomposable kG-modules into blocks using the block indempotents. For irreducible kG-modules it is often easier to use the following criterion, which uses the associated central character.

Theorem 4.2A Let B be a block of kG and let ω be the associated central character of kG. Then an irreducible kG-module V lies in the block B iff

(2) $\qquad vz = \omega(z)v \quad$ for all $\quad v \in V, z \in Z(kG)$.

In particular, the central character associated with the principal block of kG is given by

$$\sum_{x \in G} \alpha_x x \mapsto \sum_{x \in G} \alpha_x \quad \text{for all} \quad \sum_{x \in G} \alpha_x x \in Z(kG).$$

Proof Since k is a splitting field for kG (see §3.3), $\text{End}_{kG}(V) = k$. However each element $z \in Z(kG)$ acts as an element of $\text{End}_{kG}(V)$ on V, so there is a function $\psi: Z(kG) \to k$ such that $vz = \psi(z)v$ for all $v \in V, z \in Z(kG)$. Clearly, ψ

4.2 CLASSIFYING MODULES, CHARACTERS, AND IDEMPOTENTS

is a central character of kG; we must prove that it is the central character associated with B. Let f be the block idempotent of kG such that $B = f(kG)$. Since V lies in B, $vf = v$ for all $v \in V$ (Note 2 above). On the other hand $f \in Z(kG)$, and so $vf = \psi(f)v$ for all $v \in V$. Hence we conclude $\psi(f) = 1$. It now follows from Theorem 4.1 that $\omega(f) = 1$ and that $\omega = \psi$ as asserted.

Finally, in the case that B is the principal block, we can take V as the trivial kG-module. Then

$$v\left(\sum_{x \in G} \alpha_x x\right) = \sum_{x \in G} \alpha_x(vx) = \left(\sum_{x \in G} \alpha_x\right) v$$

for all $v \in V$ and $\sum_{x \in G} \alpha_x x \in kG$. Thus for the principal block

$$\omega\left(\sum_{x \in G} \alpha_x x\right) = \sum_{x \in G} \alpha_x \quad \text{for all} \quad \sum_{x \in G} \alpha_x x \quad \text{in } Z(kG)$$

by (2).

Corollary Suppose that the irreducible kG-module V lies in the block B and affords the character ϕ over k. Then with the notation of Example 1 of §4.1 we have from (2) that

$$h_i \phi_i = \omega(c_i)\phi(1) \quad \text{for} \quad i = 1, \ldots, r,$$

where ω is the central character of kG associated with B. In particular, if B is the principal block, then

$$h_i \phi_i = h_i \phi(1) \quad \text{for} \quad i = 1, \ldots, r.$$

(b) Characters lying in blocks

Each irreducible character χ of G over k is afforded by an irreducible kG-module, and this kG-module is uniquely determined up to isomorphism (Corollary 1 of Theorem 2.3). Since isomorphic irreducible kG-modules lie in the same block [see (a)], this means that χ determines a unique block of kG. We shall say that χ *lies in a block* B of kG if the irreducible kG-modules that afford χ lie in B.

Similarly, if ϕ is the modular character afforded by an irreducible or principal indecomposable kG-module, then ϕ determines the kG-module up to isomorphism (Theorem 3.7B). Again we shall say that ϕ *lies in the block* B of kG if the kG-modules that afford ϕ lie in this block.

To extend this classification of characters into blocks to the ordinary irreducible characters of G over K we need the following lemma.

Lemma 4.2 Let V be an irreducible KG-module, and choose an A-free AG-module W such that $V \simeq W \otimes_A K$ (see Theorem 3.3). Let $\overline{W}$ denote the reduction of W modulo π. Then the irreducible constituents of the kG-module $\bar{W}$ all lie in the same block of kG, and this block depends only on V and not on the choice of W.

Proof Let $f_1, \ldots, f_t$ be the block idempotents of kG. Since this is a complete set of orthogonal idempotents of $Z(kG)$, Theorem 3.4A shows that there exists a complete set of orthogonal idempotents $e_1, \ldots, e_t$ of $Z(AG)$ with $\bar{e}_i = f_i$ ($i = 1, \ldots, t$). [Note that $\overline{Z(AG)} = Z(kG)$ because in each case the class sums of G form a basis.] Thus we can write

$$W = We_1 \oplus \cdots \oplus We_t$$

as a direct sum of AG-modules. Since V is irreducible, W is indecomposable (Theorem 3.4B), and so for some j we have $W = We_j$ and $We_i = 0$ for all $i \neq j$. Thus $\bar{W} = \bar{W}\bar{e}_j = \bar{W}f_j$ and $0 = \bar{W}\bar{e}_i = \bar{W}f_i$ for all $i \neq j$. Hence from (1), all irreducible constituents of $\bar{W}$ lie in the block $B_j := f_j(kG)$. By Theorem 3.6 the irreducible constituents of $\bar{W}$ are uniquely determined up to isomorphism by V. Therefore the block B_j depends only on V.

It is now clear how to classify the irreducible KG-modules and the ordinary irreducible characters of G into blocks of kG. If ζ is an irreducible character of G over K, and ζ is afforded by an (irreducible) KG-module V, then choose an A-free AG-module W such that $V \simeq W \otimes_A K$. Then ζ and V will be said to *lie in the block* B containing the irreducible constituents of $\bar{W}$. The block B is well defined and uniquely determined by ζ (or V) by Lemma 4.2.

Remark This classification of characters into blocks does not really depend on the particular splitting fields K and k that we have chosen, but only on the characteristic p of k. Sometimes the terminology "p-blocks of characters" is used to describe this classification without specific reference to the fields involved.

Once again it is often easier to describe the classification of the irreducible characters of G over K into blocks in terms of the associated central characters. This is given in the following theorem.

Theorem 4.2B Let $\mathscr{C}_1, \ldots, \mathscr{C}_s$ be the conjugate classes of G with class sums $c_1, \ldots, c_s$ and put $h_i := |\mathscr{C}_i|$ ($i = 1, \ldots, s$).

(i) To each irreducible KG-module V there is associated a central character ω of KG such that

(3) $\qquad vz = \omega(z)v \qquad$ for all $\quad v \in V, z \in Z(KG)$.

If V affords the character ζ, then ω is the K-linear mapping determined by $\omega(c_i) = h_i \zeta_i / \zeta(1)$ ($i = 1, \ldots, s$), where ζ_i is the value of ζ on the class $\mathscr{C}_i$.

(ii) In (i) the values $\omega(c_i)$ of the class sums all lie in A. The k-linear

4.2 CLASSIFYING MODULES, CHARACTERS, AND IDEMPOTENTS

mapping $\bar{\omega} \colon Z(kG) \to k$ defined by $\bar{\omega}(c_i) := \overline{\omega(c_i)}$ ($i = 1, \ldots, s$) is a central character of kG. The KG-module V lies in the block of kG associated with $\bar{\omega}$.

(iii) If ζ and χ are two irreducible characters of G over K, then ζ and χ lie in the same block of kG iff

(4) $\qquad h_i \zeta_i / \zeta(1) \equiv h_i \chi_i / \chi(1) \pmod{\pi}$ ($i = 1, 2, \ldots, s$).

Remark See the corollary of Theorem 4.2D.

Proof (i) This follows from Example 1 of §4.1.

(ii) It follows from Theorem 2.4B that each $h_i \zeta_i / \zeta(1)$ is an algebraic integer in K, and so lies in A (Theorem 3.2). Thus $\omega(c_i) \in A$ for each i, and $\bar{\omega}$ is properly defined. Since the restriction of ω to $Z(AG) \subseteq Z(KG)$ is an A-algebra homomorphism, it is clear that $\bar{\omega}$ is a k-algebra homomorphism, and hence a central character of kG. Choose W as an A-free AG-module such that $V \simeq W \otimes_A K$. Then from the action of G on V we have $wc_i = \omega(c_i)w$ for all $w \in W$ and each i. Thus $\bar{w}c_i = \bar{\omega}(c_i)\bar{w}$ for all $\bar{w} \in \bar{W}$ and each i. Hence by Theorem 4.2A the irreducible constituents of $\bar{W}$ all lie in the block of kG associated with the central character $\bar{\omega}$. By definition V lies in the same block.

(iii) Let V and U be two irreducible KG-modules that afford the characters ζ and χ, respectively. Let ω and ψ be the central characters of KG associated with V and U as in (i). Then by (ii) we know that V and U lie in the same block of kG iff $\bar{\omega} = \bar{\psi}$, and the latter is equivalent to (4).

EXAMPLE Let B_0 be the principal block of kG. Then the ordinary character 1_G lies in B_0 and so it follows from (4) that an ordinary irreducible character ζ of G lies in B_0 iff $h_i \zeta_i / \zeta(1) \equiv h_i \pmod{\pi}$ for $i = 1, \ldots, s$. Similarly, it follows from Theorem 4.2A that an irreducible kG-module V lies in B_0 iff

(5) $\qquad vc_i = \bar{h}_i v \qquad$ for all $v \in V$ and $i = 1, \ldots, s$;

that is, iff in the representation afforded by V, each class sum c_i is represented by the scalar $\bar{h}_i 1$. This latter condition gives the following interesting result. Suppose that V affords the character χ over k. Let σ be a field automorphism of k, and let U be a kG-module affording the irreducible character χ^σ (see the example of §2.7). Since the values $\bar{h}_i$ ($i = 1, \ldots, s$) lie in the prime subfield of k, they are fixed under σ. Therefore it follows from (5) that if χ lies in the principal block B_0, then χ^σ also lies in B_0 for each automorphism σ of k.

Warning The corresponding result is not true for other blocks of kG.

As an example of the usefulness of classifying characters into blocks we have the following orthogonality relations due to Osima.

Theorem 4.2C (*Block orthogonality relations*) Let $\zeta^1, \zeta^2, \ldots, \zeta^m$ be the set of all ordinary irreducible characters of a group G that lie in a given block B. Suppose that x is a p'-element of G and y is *not* a p'-element of G. Then we have

$$\sum_{i=1}^{m} \zeta^i(x)\zeta^i(y) = 0.$$

Proof Let $\zeta^1, \zeta^2, \ldots, \zeta^s$ denote the set of all ordinary irreducible characters of G and let $\phi^1, \phi^2, \ldots, \phi^r$ denote the set of all irreducible modular characters of G enumerated so that $\phi^1, \phi^2, \ldots, \phi^n$ are those lying in the block B. Then from the way in which the block of characters ζ^i is defined, we have $\zeta^i = \sum_{j=1}^{r} d_{ij}\phi^j$ on G°, where $d_{ij} = 0$ if either $i \leq m$ and $j > n$ or $i > m$ and $j \leq n$. Thus the decomposition matrix $\Delta = [d_{ij}]$ has the form

$$\Delta = \begin{bmatrix} \Delta_1 & 0 \\ 0 & \Delta_2 \end{bmatrix} \begin{matrix} m \\ s-m \end{matrix}$$
$$\quad\;\; n \quad r-n$$

(see §3.7). Let $\mathscr{C}_1, \ldots, \mathscr{C}_r$ denote the classes of p'-elements of G and let $Z := [\zeta^i_j]$ and $F := [\phi^i_j]$ denote, respectively, the $s \times r$ and $r \times r$ matrices of the values of the characters on these classes. Since y is not a p'-element, the orthogonality relations for ordinary characters give

$$[\zeta^1(y), \zeta^2(y), \ldots, \zeta^s(y)]Z = 0.$$

But $Z = \Delta F$ and F is invertible (see the corollary to Theorem 3.7B). Therefore we conclude $[\zeta^1(y), \ldots, \zeta^m(y)]\Delta_1 = 0$. Multiplying this equation on the right by the column $[\phi^1(x), \phi^2(x), \ldots, \phi^n(x)]$, we obtain

$$0 = \sum_{i=1}^{m} \zeta^i(y) \sum_{j=1}^{n} d_{ij}\phi^j(x) = \sum_{i=1}^{m} \zeta^i(y)\zeta^i(x)$$

as required.

(c) Block idempotents of KG lying in blocks of kG

In Example 2 of §4.1 we constructed the block idempotents of KG and showed how each block idempotent of KG is associated with one of the irreducible characters of G over K, and can in fact be expressed in terms of that character [Eq. (8) of §4.1]. This gives a classification of the block idempotents of KG into the blocks of kG. The relation is described further in the following theorem.

4.2 CLASSIFYING MODULES, CHARACTERS, AND IDEMPOTENTS

Theorem 4.2D Let $\zeta^1, \ldots, \zeta^m$ be the irreducible ordinary characters of G over K lying in a block B, and let

(6) $$e_j := \frac{\zeta^j(1)}{g} \sum_{l=1}^{s} \zeta^j_{l*} c_l \qquad (j = 1, \ldots, m)$$

be the associated block idempotents of KG [see (8) of §4.1]. Then $e := e_1 + \cdots + e_m$ is the unique central idempotent of AG such that $B = \bar{e}(kG)$. Moreover we can write $e = \sum_{i=1}^{s} \beta_i c_i$, where

(7) $$\beta_i := \frac{1}{g} \sum_{l=1}^{m} \zeta^l(1) \zeta^l_{i*} \qquad (i = 1, \ldots, s)$$

and $\beta_i = 0$ whenever $\mathscr{C}_i$ is not a class of p'-elements.

Remark An interesting consequence of (7) is that the right-hand side lies in A because $e \in AG$.

Proof Since $\overline{Z(AG)} = Z(kG)$, it follows from Theorem 3.4A that there is a central idempotent f of AG such that $\bar{f}$ is the block idempotent associated with B. Let $e_1, \ldots, e_s$ be the block idempotents of KG (with the first m lying in B), and $\omega_1, \ldots, \omega_s$ the associated central characters of KG. Since $e_1, \ldots, e_s$ are the primitive idempotents of $Z(KG)$ and $f \in Z(AG) \subseteq Z(KG)$, we can write $f = e_{j_1} + \cdots + e_{j_n}$, say, as a sum of certain e_j. Then $\overline{\omega_j(f)} = 1$ if $j \in \{j_1, \ldots, j_n\}$ and is 0 otherwise. On the other hand $\bar{\omega}_j(\bar{f}) = \overline{\omega_j(f)} = 1$ iff the irreducible character ζ^j associated with ω_j lies in $B = \bar{f}(kG)$ (Theorem 4.2B). It follows from our choice of notation that the latter occurs only for $j = 1, \ldots, m$; therefore we conclude $n = m$ and $f = e_1 + \cdots + e_m$. This shows that $f = e$, and hence e is the unique central idempotent of AG such that $B = \bar{e}(kG)$.

Since $c_1, \ldots, c_s$ is an A-basis of $Z(AG)$, e can be written in the form $e = \sum_{i=1}^{s} \beta_i c_i$ with $\beta_i \in A$ and the values (7) of the coefficients come directly from (6).

The final assertion in the lemma follows directly from Theorem 4.2C.

Corollary In order to show that two ordinary irreducible characters of G lie in the same block of kG it is enough to verify (4) in Theorem 4.2B only in the cases where $\mathscr{C}_i$ is a class of p'-elements.

Proof With the notation of (i) of Theorem 4.2B we know that ζ lies in the block $B = \bar{e}(kG)$ iff $\bar{\omega}(\bar{e}) = 1$ (Theorem 4.1); that is, iff $\omega(e) \equiv 1 \pmod{\pi}$. The latter condition can be written in the form

$$1 \equiv \sum_{i=1}^{s} \beta_i \omega(c_i) = \sum_{i=1}^{s} \beta_i h_i \zeta_i / \zeta(1) \pmod{\pi}.$$

Similarly

$$1 \equiv \sum_{i=1}^{s} \beta_i h_i \chi_i / \chi(1) \pmod{\pi}$$

iff χ lies in B. Since $\beta_i = 0$ whenever $\mathscr{C}_i$ is not a class of p'-elements, the corollary follows.

(d) Blocks and principal indecomposable kG-modules

Each block B of kG is a kG-module and a direct summand of kG. Thus it is possible to write B as a direct sum of principal indecomposable kG-modules; and it is evident that these principal indecomposables and all their irreducible constituents "lie in" B in the sense of (a) above. It is possible to describe this relation in a different way which is sometimes illuminating. Define the relation $\sim$ on the set of all principal indecomposable kG-modules by writing $U \sim V$ iff there exist principal indecomposable kG-modules $U = U_1, U_2, \ldots, U_m = V$ (for some $m \geq 1$) such that for each $i = 1, \ldots, m-1$ the modules U_i and U_{i+1} have at least one irreducible constituent in common. Then clearly $\sim$ is an equivalence relation and the definitions in (a) show that $U \sim V$ implies that U and V lie in the same block. On the other hand, let U be a principal indecomposable kG-module and let I be the sum of all principal indecomposable kG-modules V such that $V \sim U$. Since $V \simeq xV$ as kG-modules for all $x \in G$, therefore $V \sim U$ implies $xV \sim U$; hence $xI \subseteq I$ for all $x \in G$. This shows that I is a (two-sided) ideal of kG. Since all $V \sim U$ lie in the same block, say $f(kG)$; it follows from (1) that $I \subseteq If(kG) \subseteq f(kG)$. Now let $I_1, \ldots, I_n$ be the ideals of kG that correspond in this way to the different $(\sim)$-classes of principal indecomposable kG-modules; then $kG = I_1 + \cdots + I_n$. On the other hand, it follows from the definition of $\sim$ that principal indecomposable modules in different classes have no irreducible constituents in common; therefore I_i and I_j have no common irreducible constituent when $i \neq j$. But then the Jordan–Hölder theorem shows that $(I_1 + \cdots + I_{j-1}) \cap I_j = 0$ for $j = 2, \ldots, n$, and so $kG = I_1 \oplus \cdots \oplus I_n$. However, as we saw above, each I_j lies in a block of kG. Thus by the definition of block we conclude that each I_j is a block, and $V_1 \sim V_2$ iff V_1 and V_2 lie in the same block. Hence $\sim$ is equivalent to the relation "lies in the same block of kG."

4.3 Defect groups

In this section we shall show how to associate with each conjugate class of G and each kG-block a class of conjugate p-subgroups of G called defect groups.

4.3 DEFECT GROUPS

First let $\mathscr{C}$ be a conjugate class of G. A p-subgroup D of G is a *defect group* of $\mathscr{C}$ if for some $x \in \mathscr{C}$, D is a Sylow p-subgroup of $C_G(x)$.

Note 1 If D is a Sylow p-subgroup of $C_G(x)$, then $y^{-1}Dy$ is a Sylow p-group of $y^{-1}C_G(x)y = C_G(y^{-1}xy)$. Since the Sylow p-subgroups of $C_G(x)$ are all conjugate, it follows that the set of defect groups of $\mathscr{C}$ is a single conjugate class of p-subgroups in G. In particular, the order $p^d := |D|$ is uniquely determined by $\mathscr{C}$. We call d the *defect* of $\mathscr{C}$. If Q is a p-subgroup in the center $Z(G)$ of G, then Q is contained in each defect group of every class of G.

In working with defect groups we shall frequently use the following combinatorial lemma.

Lemma 4.3A Let $\mathscr{C}_1, \ldots, \mathscr{C}_s$ be the conjugate classes of G with class sums $c_1, \ldots, c_s$ in kG. Let Q be a p-subgroup of G and suppose that the class $\mathscr{C}_i$ has a defect group D_i which is contained in Q.

(i) If $z \in G$ and $C_G(z)$ has no Sylow p-subgroup conjugate to a subgroup of Q, then for each j the set $S_z := \{(x, y) \in \mathscr{C}_i \times \mathscr{C}_j \mid xy = z\}$ has cardinality $|S_z| \equiv 0 \pmod{p}$.

(ii) For each j we have $c_i c_j = \sum_{l=1}^{s} \gamma_l c_l$ in $Z(kG)$, where $\gamma_l = 0$ if the class $\mathscr{C}_l$ has no defect group contained in Q.

Proof (i) Let P be a Sylow p-subgroup of $C_G(z)$. Since P centralizes z, it acts on S_z by conjugation $(x, y)^u := (u^{-1}xu, u^{-1}yu)$ for all $u \in P$. Moreover, for each $x \in \mathscr{C}_i$, $P \nsubseteq C_G(x)$ since the defect groups of $\mathscr{C}_i$ are conjugate to a subgroup of Q. Hence each $(x, y) \in S_z$ lies in a nontrivial orbit under P. Since P is a p-group, all nontrivial orbits have orders that are multiples of p, and so $|S_z| \equiv 0 \pmod{p}$ as asserted.

(ii) Since $c_i c_j \in Z(kG)$ and $c_1, \ldots, c_s$ is a k-basis of $Z(kG)$, we can write $c_i c_j = \sum_{l=1}^{s} \gamma_l c_l$. Since

$$c_i c_j = \left(\sum_{x \in \mathscr{C}_i} x\right)\left(\sum_{y \in \mathscr{C}_j} y\right)$$

we see that γ_l is the value of $|S_z|$ in k, where $z \in \mathscr{C}_l$. Thus (ii) follows from (i) because k has characteristic p.

In order to define what we mean by the defect group of a block we shall need the result to be proved in the next lemma. However we first introduce a little notation. Let Q be any p-subgroup of G, and let $\mathscr{C}_1, \ldots, \mathscr{C}_n$ be those conjugate classes of G that have subgroups of Q as defect groups. We denote by I_Q the k-subspace of $Z(kG)$ spanned by the corresponding class sums $c_1, \ldots, c_n$.

Note 2 It follows at once from Lemma 4.3A(ii) that I_Q is always an ideal in the ring $Z(kG)$. If Q is a Sylow p-subgroup of G, then $I_Q = Z(kG)$.

Lemma 4.3B Let f be a block idempotent of kG. Then there exists a p-subgroup D of G with the following properties.

(i) $f \in I_D$.

(ii) If Q is a p-subgroup of G then $f \in I_Q$ iff D is conjugate to a subgroup of Q.

(iii) Let $f = \sum_{i=1}^{s} \gamma_i c_i$, where $c_1, c_2, \ldots, c_s$ are the class sums of $\mathscr{C}_1, \mathscr{C}_2, \ldots, \mathscr{C}_s$, and let $\mathscr{D}^\circ$ denote the set of defect groups of all classes $\mathscr{C}_i$ for which $\gamma_i \neq 0$. Then each subgroup in $\mathscr{D}^\circ$ is conjugate to a subgroup of D and the maximal elements of $\mathscr{D}^\circ$ are precisely the subgroups of G that are conjugate to D.

Proof Let $\mathscr{C}_1, \mathscr{C}_2, \ldots, \mathscr{C}_s$ be the conjugate classes of G and $c_1, c_2, \ldots, c_s$ the corresponding class sums in kG. Since $f \in Z(kG)$ we can write $f = \sum_{i=1}^{s} \gamma_i c_i$ for some $\gamma_i \in k$. Let ω be the central character of kG associated with the block $f(kG)$. By Theorem 4.1 we have $\omega(f) = 1$, and so for some j, $\omega(c_j) \neq 0$ and $\gamma_j \neq 0$. Let D be a defect group of the class $\mathscr{C}_j$. We claim that D has the required properties.

(i) Since $\omega(fc_j) = \omega(f)\omega(c_j) \neq 0$, fc_j is not nilpotent, and so $fc_j \notin \mathrm{rad}[fZ(kG)]$. By Lemma 4.1, $fZ(kG)$ is a local ring, and therefore $fc_j Z(kG) = fZ(kG)$. But $c_j \in I_D$, so $f \in fc_j Z(kG) \subseteq I_D$.

(ii) If D is conjugate to a subgroup of Q, then $I_D \subseteq I_Q$ by definition, and so $f \in I_Q$. Conversely, suppose that $f \in I_Q$ for some p-subgroup Q of G. Since $c_1, \ldots, c_s$ are linearly independent, this implies that $c_i \in I_Q$ whenever $\gamma_i \neq 0$; in particular, $c_j \in I_Q$. Hence by the definition of I_Q, Q contains a defect group of $\mathscr{C}_j$ and hence a conjugate of D.

(iii) This is a consequence of (i) and (ii).

It is now clear how we may define the defect groups of a block B of kG. Let f be the block idempotent associated with B. Then define the *defect groups* of B to be the p-subgroups D of G that satisfy the conditions of Lemma 4.3B. It is clear from part (ii) that the defect groups of B form a single conjugacy class of p-subgroups of G. A slightly simpler description of the defect groups is given by the following theorem.

Theorem 4.3 Let $B = f(kG)$ be a block of kG and let ω be the associated central character of kG. Let $\mathscr{C}_1, \ldots, \mathscr{C}_s$ be the classes of G with class sums $c_1, \ldots, c_s$, respectively, and let $\mathscr{D}$ denote the set of all defect groups of all $\mathscr{C}_i$ for which $\omega(c_i) \neq 0$. Then each subgroup in $\mathscr{D}$ contains a defect group of B and the minimal elements of $\mathscr{D}$ are precisely the defect groups of B.

Moreover, there exists a class $\mathscr{C}_j$ of p'-elements of G such that the defect groups of $\mathscr{C}_j$ are the same as those of B.

Proof Let Q be a defect group of $\mathscr{C}_i$ and suppose that $\omega(c_i) \neq 0$. Then the

4.3 DEFECT GROUPS

proof of part (i) of Lemma 4.3B shows that fc_i is not nilpotent, and so

$$f \in fZ(kG) = fc_i Z(kG) \subseteq I_Q .$$

Hence by Lemma 4.3B, Q contains a defect group of $B = f(kG)$. This shows that each subgroup Q in $\mathscr{D}$ contains a defect group of B. On the other hand, the definition of D in the proof of Lemma 4.3B shows that D and its conjugates lie in $\mathscr{D}$, and so the defect groups of B all lie in D and are therefore the minimal elements of $\mathscr{D}$.

To prove the final statement of the theorem we use Theorem 4.2D. From that theorem we know that the block idempotent f for B has the form $f = \sum_{i=1}^{s} \bar{\beta}_i c_i$, where $\beta_i \in A$ and $\beta_i = 0$ whenever $\mathscr{C}_i$ is not a class of p'-elements of G. This shows that the class $\mathscr{C}_j$ chosen in the proof of Lemma 4.3B must be a class of p'-elements (since $\gamma_j \neq 0$), and the defect group D of B obtained there is the defect group of the class $\mathscr{C}_j$.

EXAMPLE (*Characters whose degrees are not divisible by p*) Suppose ζ is an irreducible ordinary character of G such that $p \nmid \zeta(1)$, and let B be the block of kG in which ζ lies. Let ω be the central character of KG corresponding to ζ. Then it follows from Theorem 4.2B (with the notation there) that $\omega(c_i) = h_i \zeta_i / \zeta(1)$ for $i = 1, \ldots, s$. Let P be a Sylow p-subgroup of G. If P is not a defect group of the class $\mathscr{C}_i$, then the Sylow p-subgroups of $C_G(x)$ ($x \in \mathscr{C}_i$) have orders less than $|P|$. Thus in this case $p \mid h_i$; and so $\bar{\omega}(c_i) = 0$ because $p \nmid \zeta(1)$. Thus $\bar{\omega}(c_i) = 0$ except for the classes $\mathscr{C}_i$ that have P as a defect group (or equivalently, for which $\mathscr{C}_i \cap C_G(P) \neq \varnothing$). By Theorem 4.3 this shows that B has P as a defect group.

Now suppose that ζ' is a second irreducible ordinary character with $p \nmid \zeta'(1)$. Let ζ' lie in the block B' of kG and let ω' be the central character of KG corresponding to ζ'. It then follows from Theorem 4.2B, the corollary of Theorem 4.2D, and what we have just shown that

(1) $\qquad B = B'$ iff $h_i \zeta_i / \zeta(1) \equiv h_i \zeta_i' / \zeta'(1)$ (mod π)

for all i such that $\mathscr{C}_i$ is a class of p'-elements and $\mathscr{C}_i \cap C_G(P) \neq \varnothing$. However, when $\mathscr{C}_i \cap C_G(P) \neq \varnothing$, then $p \nmid h_i$, and so we can cancel the h_i in (1). Finally, since the characters are class functions, this yields the following very simple condition [valid when $p \nmid \zeta(1)$ and $p \nmid \zeta'(1)$]:

$\qquad B = B'$ iff $\zeta(x)/\zeta(1) \equiv \zeta'(x)/\zeta'(1)$ (mod π)

for all p'-elements $x \in C_G(P)$. In particular (taking $\zeta' = 1_G$), ζ lies in the principal block of kG iff $\zeta(x) \equiv \zeta(1)$ (mod π) for all p'-elements $x \in C_G(P)$.

An analogous argument (appealing to the corollary of Theorem 4.2A in place of Theorem 4.2B) shows that if ϕ and ϕ' are two irreducible characters of G over k, and their degrees are not divisible by p, then ϕ and ϕ' lie in the same block of kG iff

$$\phi(x)/\phi(1) = \phi'(x)/\phi'(1) \quad \text{for all } p'\text{-elements } x \in C_G(P).$$

In particular, ϕ lies in the principal block of kG iff

$$\phi(x) = \phi(1) \quad \text{for all } p'\text{-elements } x \in C_G(P).$$

4.4 Further analysis of the Cartan matrix and decomposition matrix

Using the concepts of block and defect group we can now analyse more carefully the decomposition matrix Δ and the Cartan matrix $\Gamma = \Delta^T \Delta$ (see §3.7). We shall need the following construction of generalized characters.

Lemma 4.4 (i) Let $\mathscr{C}$ be a conjugate class of p'-elements of G. Then there exists $\lambda: G \to K$ which is an A-linear combination of irreducible ordinary characters of G such that λ takes integer values, $\lambda(y) \not\equiv 0 \pmod{p}$ for $y \in \mathscr{C}$, and $\lambda(y) = 0$ for each p'-element $y \notin \mathscr{C}$.

(ii) Suppose that the Sylow p-subgroups of G have order p^m. If ψ is any modular character of G then the function $\theta: G \to K$ defined by

$$\theta(y) := \begin{cases} p^m \psi(y) & \text{if } y \text{ is a } p'\text{-element} \\ 0 & \text{otherwise} \end{cases}$$

is an element of Char(G).

Proof (i) Take $x \in \mathscr{C}$ and choose Q as a Sylow p-subgroup of $C_G(x)$. Since x is a p'-element, the subgroup H generated by Q and x has the form $H = Q \times \langle x \rangle$. Let $\chi^1, \chi^2, \ldots, \chi^n$ be those irreducible characters of H over K that have Q in their kernels. Since H/Q is abelian it follows from the corollary of Theorem 1.5 that each χ^i has degree 1 and $n = |H:Q|$. We define μ as the A-linear combination of characters of H given by

$$\mu := \sum_{i=1}^{n} \chi^i(x^{-1})\chi^i$$

(see Theorem 2.4B); and then

$$\lambda := \mu^G = \sum_{i=1}^{n} \chi^i(x^{-1})(\chi^i)^G$$

is an A-linear combination of irreducible ordinary characters of G. Because $Q \subseteq \operatorname{Ker} \chi^i$ for each i, the ordinary character relations (Theorem 2.4A) show that

$$\mu(y) = \begin{cases} \sum_{i=1}^{n} \chi^i(x^{-1})\chi^i(x) = n & \text{if } y \in Qx \\ 0 & \text{otherwise.} \end{cases}$$

4.4 FURTHER ANALYSIS OF THE CARTAN DECOMPOSITION MATRIX

$(\chi^1, \ldots, \chi^n$ correspond to a complete set of irreducible characters of H/Q). This shows that $\lambda = \mu^G$ takes integer values. On the other hand, since x is the only p'-element in Qx we have $z^{-1}xz \in Qx$ only if $z \in C_G(x)$. Therefore from the definition of induced class function (§2.5)

$$\lambda(x) = |C_G(x) : H|, \quad \mu(x) = |C_G(x) : Q| \not\equiv 0 \pmod{p},$$

and

$$\lambda(y) = 0 \quad \text{if } y \text{ is not conjugate to an element of } Qx.$$

Since λ is a class function and x is the only p'-element in Qx, this proves (i).

(ii) By Brauer's theorem on the characterization of characters (Theorem 2.6B) it is enough to show that $\theta_E \in \text{Char}(E)$ for each elementary subgroup of G. In particular, it is enough to do this for each subgroup E of the form $Q \times S$, where Q is a p-subgroup and S is a p'-subgroup since each elementary q-subgroup (for any prime q) can be written in this form. However, since S is a p'-subgroup the restriction ψ_S of the modular character ψ to S is an (ordinary) character (see §3.5). Therefore, by the definition of θ

$$\theta_E = p^{m-l}(\psi_S)^E \in \text{Char}(E),$$

where $p^l := |Q| \le p^m$. This proves (ii).

EXAMPLE Let p^m be the order of the Sylow p-subgroups of G, and put $g := |G|$. Then p^m divides the degree of each principal indecomposable modular character η. Indeed, let $\zeta^1, \zeta^2, \ldots, \zeta^s$ and $\phi^1, \phi^2, \ldots, \phi^r$ be irreducible ordinary characters and irreducible modular characters, respectively. We can write $\zeta^i = \sum_{j=1}^r d_{ij} \phi^j$ on the set G° of p'-elements ($i = 1, 2, \ldots, s$), where the integers d_{ij} are the decomposition numbers. Now define λ as in Lemma 4.4(i) with $\mathscr{C} = \{1\}$. Since λ is an A-linear combination of $\zeta^1, \zeta^2, \ldots, \zeta^s$, we can write $\lambda = \sum_{j=1}^r a_j \phi^j$ on G° with the $a_j \in A$. Then by Theorem 3.7B, since λ is 0 on all p'-elements except 1,

$$\frac{1}{g}\lambda(1)\eta(1) = \langle \lambda, \eta \rangle = \sum_{j=1}^s a_j \langle \phi^j, \eta \rangle \in A.$$

Since p is not a unit in A, this shows that $p^m \mid \eta(1)\lambda(1)$. But $p \nmid \lambda(1)$ by the construction of λ, so $p^m \mid \eta(1)$. (See also Exercise 2 of §3.7.)

NOTATION Let $\zeta^1, \zeta^2, \ldots, \zeta^s$ be the irreducible (ordinary) characters of the group G over K, $\phi^1, \phi^2, \ldots, \phi^r$ the irreducible modular characters of G, and $\Delta = [d_{ij}]$ the $s \times r$ decomposition matrix (so the d_{ij} are nonnegative integers and $\zeta^i = \sum_{j=1}^r d_{ij} \phi^j$ on the set G° of p'-elements of G). Let $\Gamma = \Delta^T \Delta$ be the Cartan matrix (Δ^T denotes the transpose of Δ). Moreover, suppose B is a block of kG and that the characters have been enumerated so that ζ^1,

$\zeta^2, \ldots, \zeta^{s_1}$ and $\phi^1, \phi^2, \ldots, \phi^{r_1}$ are those characters that lie in B. This means that Δ can be written

$$\begin{bmatrix} \Delta_1 & 0 \\ 0 & \Delta_2 \end{bmatrix},$$

where Δ_1 is an $s_1 \times r_1$ submatrix. We shall use $\bar{\Delta}$ and $\bar{\Delta}_1$ to denote the matrices obtained from Δ and Δ_1 by reducing the entries modulo π. Note that since $\Gamma = \Delta^T \Delta$, Γ is an $r \times r$ matrix of the form

$$\begin{bmatrix} \Gamma_1 & 0 \\ 0 & \Gamma_2 \end{bmatrix},$$

where Γ_1 is an $r_1 \times r_1$ submatrix equal to $\Delta_1^T \Delta_1$.

Note Let ϕ^i and ϕ^j be irreducible modular characters of G, and let η^i and η^j be the corresponding principal indecomposable modular characters. Suppose that ϕ^i and ϕ^j lie in different blocks. Then from the above decomposition of Γ we see that $c_{ij} = c'_{ij} = 0$ [where $\Gamma = [c_{ij}]$ and $\Gamma^{-1} = [c'_{ij}]$]. Therefore by Theorem 3.7B, $\langle \phi^i, \phi^j \rangle_{G^\circ} = \langle \eta^i, \eta^j \rangle_{G^\circ} = 0$. Similarly, if ζ^i and ζ^j are irreducible ordinary characters of G lying in different blocks, then $\langle \zeta^i, \zeta^j \rangle_{G^\circ} = 0$. (Express ζ^i and ζ^j in terms of the ϕ^l and the decomposition numbers.)

Theorem 4.4 With the notation above,

(i) The $s \times r$ matrix $\bar{\Delta}$ has rank r (the largest possible), and $\bar{\Delta}_1$ has rank r_1;
(ii) $|\det \Gamma|$ is a power of p;
(iii) There exist rational integers b_{ji} such that for $j = 1, 2, \ldots, r_1$ we have $\phi^j = \sum_{i=1}^{s_1} b_{ji} \zeta^i$ on G°.

Proof (i) Let $\mathscr{C}_1, \mathscr{C}_2, \ldots, \mathscr{C}_r$ be the conjugate classes of p'-elements of G. We first show that the restrictions of $\bar{\zeta}^1, \bar{\zeta}^2, \ldots, \bar{\zeta}^s$ to G° span the full r-dimensional space of k-valued class functions on G°. Indeed, for each $i = 1, 2, \ldots, r$ we can construct by Lemma 4.4(i), $\lambda_i \in \text{Char}(G)$ such that

$$\overline{\lambda_i(y)} \neq 0 \quad \text{for} \quad y \in \mathscr{C}_i$$
$$\overline{\lambda_i(y)} = 0 \quad \text{for all} \quad y \in G^\circ \backslash \mathscr{C}_i.$$

Clearly the restrictions of $\bar{\lambda}_1, \bar{\lambda}_2, \ldots, \bar{\lambda}_r$ form a basis of the space of k-valued class functions on G°. Since each λ_i is an A-linear combination of $\zeta^1, \zeta^2, \ldots, \zeta^s$, this means that $\bar{\zeta}^1, \bar{\zeta}^2, \ldots, \bar{\zeta}^s$ also span the space of k-valued class functions on G°.

Now let $[\zeta_j^i]$ and $[\phi_j^i]$ denote respectively the $s \times r$ and $r \times r$ matrices of character values on $\mathscr{C}_1, \mathscr{C}_2, \ldots, \mathscr{C}_r$. By definition we have $[\zeta_j^i] = \Delta[\phi_j^i]$;

hence $[\bar{\zeta}_j^i] = \bar{\Delta}[\bar{\phi}_j^i]$. But from what we have just shown, the rows of $[\bar{\zeta}_j^i]$ span an r-dimensional space. Hence $[\bar{\zeta}_j^i]$, and consequently $\bar{\Delta}$ as well, has rank r.

Since $\bar{\Delta}$ has rank r, its columns are linearly independent, hence so are the columns of $\bar{\Delta}_1$; so $\bar{\Delta}_1$ has rank r_1.

(ii) For each modular character ϕ^i, we define θ^i as in Lemma 4.4(ii) taking $\psi = \phi^i$. Since $\theta^i \in \text{Char}(G)$, $\langle \theta^i, \theta^j \rangle = p^{2m} \langle \phi^i, \phi^j \rangle_{G^\circ}$ is an integer. By Theorem 3.7B we know that $\Gamma^{-1} = [c'_{ij}]$, where $\langle \phi^i, \phi^j \rangle_{G^\circ} = c'_{ij}$. Hence if L denotes the $r \times r$ matrix $[\langle \theta^i, \theta^j \rangle_{G^\circ}]$, then $L\Gamma = p^{2m} \cdot 1$, and so $\det L \det \Gamma = p^{2mr}$. Since $\det L$ is an integer, $|\det \Gamma|$ is a power of p.

(iii) By (i), Δ_1 has an $r_1 \times r_1$ submatrix Δ_0 with $\det \bar{\Delta}_0 \neq 0$; we claim that $\det \Delta_0 = \pm 1$. Indeed otherwise there exists a prime $q \neq p$ such that $q \mid \det \Delta_0$. But then the rank of $\Delta(\bmod q)$ is less than or equal to $r - 1$, and so the rank of the $r \times r$ matrix $\Gamma = \Delta^T \Delta \pmod{q}$ is less than or equal to $r - 1$. But by (ii) $\det \Gamma \not\equiv 0 \pmod{q}$, so we have a contradiction. Hence $\det \Delta_0 = \pm 1$ and $\Delta_0^{-1} = [b_{ij}]$ has integer entries (by Cramer's rule). Without loss in generality we may suppose that Δ_0 consists of the first r_1 rows of Δ_1. Then we have

$$\zeta^i = \sum_{j=1}^{r_1} d_{ij} \phi^j$$

on G° for $i = 1, 2, \ldots, r_1$, which yields

$$\phi^j = \sum_{i=1}^{r_1} b_{ji} \zeta^i$$

on G° for $j = 1, 2, \ldots, r_1$ as required.

4.5 The characters in a block of given defect

Recall that a block of kG has defect d if p^d is the order of its defect groups. In the present section we shall show how to characterize the defect d in terms of the degrees of the irreducible characters lying in the block, and how to find an upper bound for the number of ordinary (and modular) irreducible characters in a block of given defect.

Theorem 4.5A *Suppose that the Sylow p-groups of G have order p^m, and let B be a block of defect d of kG. Then*

(i) $p^{m-d} \mid \zeta(1)$ *for every irreducible ordinary character ζ lying in B, and for at least one of those characters $p^{m-d+1} \nmid \zeta(1)$;*

(ii) $p^{m-d} \mid \phi(1)$ *for every irreducible modular character ϕ lying in B, and for at least one of those characters $p^{m-d+1} \nmid \phi(1)$.*

Proof Let $\zeta^1, \zeta^2, \ldots, \zeta^{s_1}$ and $\phi^1, \phi^2, \ldots, \phi^{r_1}$ be the irreducible ordinary characters and the irreducible modular characters, respectively, lying in B. Let η^i be the principal indecomposable modular character associated with ϕ^i ($i = 1, 2, \ldots, r_1$). Then from §3.7B and Theorem 4.4

(1) $$\zeta^i = \sum_{j=1}^{r_1} d_{ij} \phi^j \quad \text{on } G° \quad (i = 1, 2, \ldots, s_1)$$

(2) $$\eta^j = \sum_{i=1}^{s_1} d_{ij} \zeta^i \quad \text{on } G° \quad (j = 1, 2, \ldots, r_1)$$

(3) $$\phi^j = \sum_{i=1}^{s_1} b_{ji} \zeta^i \quad \text{on } G° \quad (j = 1, 2, \ldots, r_1),$$

where the d_{ij} and b_{ij} are integers. It follows from (1) and (3) that the largest power of p dividing all $\zeta^i(1)$ ($i = 1, 2, \ldots, s_1$) is equal to the largest power of p dividing all $\phi^i(1)$ ($i = 1, 2, \ldots, r_1$). Hence (i) implies (ii) (and conversely). Thus it will be sufficient to prove (i).

Using the notation of Theorem 4.2D, let e be the central idempotent of AG such that $\bar{e}$ is the block idempotent associated with B. Then Theorem 4.2D shows that $e = \sum_{i=1}^{s} \beta_i c_i$, where

(4) $$\beta_i := \frac{1}{g} \sum_{l=1}^{s_1} \zeta^l(1) \zeta^l_{i*} \quad (i = 1, 2, \ldots, s)$$

and $\beta_i = 0$ whenever $\mathscr{C}_i$ is not a class of p'-elements. Suppose that $\mathscr{C}_i$ is a class of p'-elements. Then so is $\mathscr{C}_{i*}$, and hence by (4), (1), and (2) we obtain

$$\beta_i = \frac{1}{g} \sum_{l=1}^{s_1} \zeta^l(1) \sum_{j=1}^{r_1} d_{lj} \phi^j_{i*} = \frac{1}{g} \sum_{j=1}^{r_1} \eta^j(1) \phi^j_{i*}.$$

Since $p^m | \eta^j(1)$ by the example of §4.4, this shows that $\eta^j(1)/g \in A$ for each j, and so $\beta_i = \sum_{j=1}^{r_1} a_j \phi^j_{i*}$ for suitable $a_j \in A$. In particular, since $\bar{e} \neq 0$, there exists a class $\mathscr{C}_i$ of p'-elements with $\bar{\beta}_i \neq 0$, and some j such that $\phi^j_{i*} \notin \pi A$; it then follows from (3) that there exists some l such that $\zeta^l_{i*} \notin \pi A$ ($1 \leq l \leq s_1$). Choose this class $\mathscr{C}_i$ so that the defect of $\mathscr{C}_i$ is as large as possible. Then Lemma 4.3B(iii) shows that $\mathscr{C}_i$ has the same defect d as B has, and so the largest power of p dividing $h_i := |\mathscr{C}_i| = |\mathscr{C}_{i*}|$ is p^{m-d} from the definition of defect group of $\mathscr{C}_i$. Since $h_i \zeta^l_{i*}/\zeta^l(1) \in A$ (Theorem 2.4B) and $\zeta^l_{i*} \notin pA$, we conclude that $p^{m-d+1} \nmid \zeta^l(1)$.

Finally we show that $p^{m-d} | \zeta(1)$ for each irreducible ordinary character lying in B. By Theorem 4.3 we may choose j so that $\omega(c_j) \neq 0$ and the defect groups of the class $\mathscr{C}_j$ are the defect groups of B; this implies $p^{m-d} | h_j$ but $p^{m-d+1} \nmid h_j$. However, since ζ lies in B, Theorem 4.2B shows that $\omega(c_j) = \overline{h_j \zeta_j/\zeta(1)}$, and so $h_j \zeta_j/\zeta(1) \notin \pi A$. Since $\zeta_j \in A$ by Theorem 2.4B, and

4.5 THE CHARACTERS IN A BLOCK OF GIVEN DEFECT

$p^{m-d} \mid h_j$, we have $p^{m-d} \mid \zeta(1)$. This completes the proof of (i) and so the theorem is proved.

EXAMPLE Suppose that ζ is an irreducible ordinary character of G such that $p \nmid \zeta(1)$. Then it follows from Theorem 4.5A that ζ lies in a block B of defect m where p^m is the order of the Sylow p-subgroup of G. The defect groups of B are the Sylow p-subgroups of G. Such a block is sometimes said to have *full defect*. In particular, taking $\zeta = 1_G$ we see that the principal block of kG has full defect.

In order to estimate the number of irreducible characters in a block of given defect we shall use the following result.

Lemma 4.5 (i) Let $\eta^1, \eta^2, \ldots, \eta^r$ be the principal indecomposable modular characters of G. Suppose $\theta \in \text{Char}(G)$ has the property that $\theta(x) = 0$ for all $x \in G \backslash G°$. Then there exist integers $a_1, a_2, \ldots, a_r$ such that $\theta = \sum_1^r a_i \eta^i$ on $G°$.

(ii) Suppose that the Sylow p-subgroups of G have order p^m and let B be a block of defect d of kG. Let ζ be an irreducible ordinary character lying in B. Define the class function $\theta: G \to K$ by

$$\theta(x) := \begin{cases} p^d \zeta(x) & \text{if } x \in G° \\ 0 & \text{otherwise.} \end{cases}$$

Then $\theta \in \text{Char}(G)$, and if $p^{m-d+1} \nmid \zeta(1)$, then $(1/p)\theta \notin \text{Char}(G)$. Moreover, θ is a **Z**-linear combination of irreducible characters lying in B.

Proof (i) By the corollary of Theorem 3.7B, $\eta^1, \eta^2, \ldots, \eta^r$ is a K-basis for the vector space $\text{Class}_K(G°)$. Thus there exist $a_i \in K$ such that $\theta = \sum_{i=1}^r a_i \eta^i$. It remains to show that the a_i are integers. Let $\zeta^1, \zeta^2, \ldots, \zeta^s$ and $\phi^1, \phi^2, \ldots, \phi^r$ be the irreducible ordinary characters and the irreducible modular characters, respectively, of G. Then by Theorem 3.7B and Theorem 4.4(iii) there exist integers b_{jl} such that for each j

$$a_j = \sum_{i=1}^r a_i \langle \eta^i, \phi^j \rangle_{G°} = \langle \theta, \phi^j \rangle_{G°} = \sum_{l=1}^s b_{jl} \langle \theta, \zeta^l \rangle_{G°}.$$

However $\langle \theta, \zeta^l \rangle_{G°} = \langle \theta, \zeta^l \rangle_G$, which is an integer since $\theta \in \text{Char}(G)$. Therefore a_j is an integer as asserted.

(ii) Let $\mathscr{C}_1, \mathscr{C}_2, \ldots, \mathscr{C}_r$ be the classes of p'-elements of G and put $h_i := |\mathscr{C}_i|$ and $g := |G|$. Then for each irreducible ordinary character ζ^l of G we have

(1) $$\langle \theta, \zeta^l \rangle = \frac{p^d}{g} \sum_{j=1}^r h_j \zeta_j \zeta^l_{j*} = \frac{p^d \zeta(1)}{g} \sum_{j=1}^r \zeta^l_{j*} \omega(c_j),$$

where ω is the central character of KG associated with ζ (see Example 2 of

§4.1). Now $p^{m-d}\theta \in \mathrm{Char}(G)$ by Lemma 4.4 because the restriction of ζ to G° is a modular character; therefore $n_l := \langle p^{m-d}\theta, \zeta^l \rangle$ is an integer. Also $p^d\zeta(1)/g \in A$ because $p^{m-d} \mid \zeta(1)$, and $\omega(c_j) = h_j \zeta_j/\zeta(1)$ and ζ^l_{j*} lie in A; so by (1), $n_l/p^{m-d} \in A$. This shows that $p^{m-d} \mid n_l$, and so $\theta = \sum_{l=1}^{s} \langle \theta, \zeta^l \rangle \zeta^l = \sum_{l=1}^{s} (n_l/p^{m-d})\zeta^l \in \mathrm{Char}(G)$.

On the other hand, if $p^{m-d+1} \nmid \zeta(1)$ then $(1/p)\theta \notin \mathrm{Char}(G)$. Indeed, otherwise it follows from (i) that $(1/p)\theta = \sum_{i=1}^{r} a_i \eta^i$ on G° for some integers a_i. But by the example of §4.4, $p^m \mid \eta^i(1)$ for each i, and so $p^m \mid \theta(1)/p$. Since $\theta(1) = p^d\zeta(1)$, this implies $p^{m-d+1} \mid \zeta(1)$ contrary to the choice of ζ.

Finally, suppose that the irreducible character ζ^l does not lie in the block B. Then by the note preceding Theorem 4.4 we know that $\langle \zeta, \zeta^l \rangle_{G^\circ} = 0$. Hence $\langle \theta, \zeta^l \rangle = p^{m-d}\langle \zeta, \zeta^l \rangle_{G^\circ} = 0$, and so ζ^l is not one of the constituents of θ. This completes the proof of (ii).

Theorem 4.5B Suppose that the Sylow p-subgroups of G have order p^m. Let B be a block of kG of defect d. Then

(i) at most $\frac{1}{4}p^{2d} + 1$ irreducible ordinary characters lie in B;

(ii) if ζ is an irreducible ordinary character lying in B, then for $d \leq 2$ we have $p^{m-d} \mid \zeta(1)$ but $p^{m-d+1} \nmid \zeta(1)$, and for $d > 2$ we have $p^{m-1} \nmid \zeta(1)$;

(iii) an irreducible ordinary character ζ of G lies in a block of defect 0 if $p^m \mid \zeta(1)$ and in a block of defect 1 if $p^{m-1} \mid \zeta(1)$ and $p^m \nmid \zeta(1)$.

Proof (i) Let $\zeta^1, \zeta^2, \ldots, \zeta^{s_1}$ be the irreducible ordinary characters that lie in B; and for $i = 1, 2, \ldots, s_1$, define $\theta^i \in \mathrm{Char}(G)$ as in Lemma 4.5, taking $\zeta = \zeta^i$. Then for $i, j = 1, 2, \ldots, s_1$ we have

$$p^d \langle \theta^i, \zeta^j \rangle_G = \langle \theta^i, \theta^j \rangle_G = p^{2d}\langle \zeta^i, \zeta^j \rangle_{G^\circ},$$

and using Theorem 2.4B we see

(1) $\qquad \langle \zeta^i, \zeta^j \rangle_{G^\circ} = \dfrac{\zeta^j(1)}{g} \sum_{t=1}^{r} \{h_t \zeta^j_t/\zeta^j(1)\} \zeta^i_{t*} \in \dfrac{\zeta^j(1)}{g} A.$

Since ζ^i and ζ^j lie in the same block

$$h_t \zeta^i_t/\zeta^i(1) \equiv h_t \zeta^j_t/\zeta^j(1) \pmod{\pi} \qquad \text{for all } t$$

by Theorem 4.2B. Therefore

(2) $\qquad \dfrac{p^{m-d}\langle \theta^l, \zeta^i \rangle}{\zeta^i(1)} \equiv \dfrac{p^{m-d}\langle \theta^l, \zeta^j \rangle}{\zeta^j(1)} \pmod{\pi}$

for all $i, j, l = 1, 2, \ldots, s_1$. [Note that both sides of (2) lie in A by (1).]

Using Theorem 4.5A we may assume that the ζ^i have been enumerated so that $p^{m-d+1} \nmid \zeta^1(1)$. We claim that $\langle \theta^1, \zeta^1 \rangle \notin \pi A$. Indeed otherwise, $p^{m-d}\langle \theta^1, \zeta^1 \rangle/\zeta^1(1) \in \pi A$ because $p^{m-d+1} \nmid \zeta^1(1)$. Then by (2) we have

4.5 THE CHARACTERS IN A BLOCK OF GIVEN DEFECT

$p^{m-d}\langle\theta^1, \zeta^j\rangle/\zeta^j(1) \in \pi A$; and so p divides the integer $\langle\theta^1, \zeta^j\rangle$ for $j = 1, 2, \ldots, s_1$ because $p^{m-d} \mid \zeta^j(1)$ by Theorem 4.5A. But by Lemma 4.5 we have

$$\theta^1 = \sum_{j=1}^{s_1} \langle\theta^1, \zeta^j\rangle \zeta^j$$

and so we conclude $(1/p)\theta^1 \in \text{Char}(G)$. This contradicts Lemma 4.5, and we have $\langle\theta^1, \zeta^1\rangle \notin \pi A$.

Since $\langle\theta^1, \zeta^1\rangle \notin \pi A$, (2) shows that for each $j = 1, 2, \ldots, s_1$, $p^{m-d}\langle\theta^1, \zeta^j\rangle/\zeta^j(1) \notin \pi A$; in particular, $a_j := \langle\theta^1, \zeta^j\rangle \neq 0$. Then each of the ζ^j lying in B is a constituent of θ^1. Hence we conclude that

$$p^d a_1 = p^d\langle\theta^1, \zeta^1\rangle = \langle\theta^1, \theta^1\rangle = \sum_{i=1}^{s_1}\langle\theta^1, \zeta^j\rangle^2 = \sum_{i=1}^{s_1} a_j^2.$$

Since the right-hand side is at least $(s_1 - 1) + a_1^2$, therefore $0 \geq (s_1 - 1) + a_1^2 - p^d a_1 = (s_1 - 1) - \frac{1}{4}p^{2d} + (a_1 - \frac{1}{2}p^d)^2$. Hence $s_1 - 1 \leq \frac{1}{4}p^{2d}$ and this proves (i).

(ii) Now we turn to (ii). If $s_1 = 1$ this follows at once from Theorem 4.5A; so suppose that $s_1 > 1$. Suppose $p^{m-n} \mid \zeta^j(1)$, where $n < d$. Then since $p^{m-n}\langle\theta^1, \zeta^j\rangle/\zeta^j(1) \in A$, we conclude as above that $\langle\theta^1, \zeta^j\rangle_G \equiv 0 \pmod{p}$. Since $\langle\theta^1, \zeta^j\rangle_G = \langle\theta^j, \zeta^1\rangle_G$ by definition, (2) now yields (with $l = j$, $i = 1$)

$$0 \equiv p^{m-d}\langle\theta^j, \zeta^j\rangle/\zeta^j(1) \pmod{\pi}$$

and so $p^{d-n+1} \mid \langle\theta^i, \zeta^j\rangle$. As we saw in the proof (i), $\langle\theta^j, \zeta^1\rangle = \langle\theta^1, \zeta^j\rangle \neq 0$, and so $\theta^j = \Sigma_i b_i \zeta^i$, say, where $b_1 \geq 1$. Since $\sum_{i=1}^{s_1} b_i^2 = \langle\theta^j, \theta^j\rangle_G = p^{2d}\langle\zeta^j, \zeta^j\rangle_{G^\circ} \leq p^{2d}\langle\zeta^j, \zeta^j\rangle = p^{2d}$, this shows that $b_j < p^d$. But $b_j = \langle\theta^j, \zeta^j\rangle$, and so $p^{d-n+1} < p^d$; hence $n \geq 2$ if $s_1 > 1$. Thus $n < d$ implies $n \geq 2$, and hence $d \geq 3$. Since $m - n < m - d$, this proves (ii).

(iii) This follows immediately from (ii).

Corollary 1 With the notation of the theorem, at most $\frac{1}{4}p^{2d} + 1$ irreducible modular characters lie in B.

Proof This is a consequence of Theorems 4.4(i) and 4.5B(i).

Corollary 2 If ζ is an irreducible ordinary character of G such that $\zeta(x) = 0$ for each p-element x of G with $x \neq 1$, then ζ lies in a block of defect 0.

Proof Let P be a Sylow p-subgroup of G with $|P| = p^m$. Then

$$\langle\zeta_P, 1_P\rangle = \frac{1}{p^m}\sum_{x\in P}\zeta(x) = \zeta(1)/p^m$$

is an integer and the conclusion follows from the theorem.

EXERCISES

1. Let ζ be an irreducible ordinary character of the group G. Show that ζ lies in the principal block of kG iff $\langle \zeta, 1_G \rangle_{G^\circ} \neq 0$.

2. In contrast to Theorem 4.5B(ii), if $d > 2$ then a block of defect d may have an ordinary irreducible character lying in it such that $p^{m-d+1} \mid \zeta(1)$; give an example for each $d \geq 3$. [*Hint*: If G is a nonabelian p-group of order p^d and $d \geq 3$, then G has an ordinary irreducible character of degree p.]

4.6 Blocks of small defect

The situation where a block has defect 0 is especially simple to describe.

Theorem 4.6A Let B be a block of defect 0 of kG. Then there is exactly one irreducible ordinary character ζ, one irreducible modular character ϕ, and one principal indecomposable modular character η of G lying in B. Moreover, $\phi = \eta$ and

$$\zeta(x) = \begin{cases} \phi(x) & \text{for } x \in G^\circ \\ 0 & \text{otherwise.} \end{cases}$$

Proof Theorem 4.5B(i) shows that B contains a single irreducible ordinary character ζ and Corollary 1 of Theorem 4.5B shows that B contains a single irreducible modular character ϕ, and so $\zeta = b\phi$ on G° for some integer $b > 0$. The decomposition matrix and the Cartan matrix of B are the one-by-one matrices $\Delta_1 = [b]$ and $\Gamma_1 = [b^2]$, respectively. By Theorem 4.4(ii), $|\Gamma_1|$ is a power of p, and so b is a power of p. But Theorem 4.5A(ii) then shows that $b = 1$ and hence $\zeta = \phi$ on G°, $\Gamma_1 = [1]$, and so $\phi = \eta$. Since ζ is irreducible, we have using Theorem 3.7B that $\langle \zeta, \zeta \rangle_{G^\circ} = \langle \phi, \eta \rangle_{G^\circ} = 1$. Hence

$$0 = \langle \zeta, \zeta \rangle_G - \langle \zeta, \zeta \rangle_{G^\circ} = \frac{1}{g} \sum_{x \in G \setminus G^\circ} |\zeta(x)|^2,$$

and so $\zeta(x) = 0$ for all $x \in G \setminus G^\circ$. This proves the theorem.

The situation where a block has defect 1 is considerably more complicated. However this case has been analyzed (unlike the cases of defects greater than 1) by Brauer. In order to state the result of Brauer on blocks of defect 1, we define the concept of p-conjugate characters. Let $g := |G| = p^m g'$, where $(p, g') = 1$, and let E and F be the extensions of the rational field generated by a primitive gth root of unity and a primitive g'th root of unity, respectively. Then $K \supseteq E \supseteq F$ and by §2.7, E is a splitting field for G. The Galois group $\text{Gal}(E/F)$ of E over F has order $p^m(p-1)$, and it can be considered as a group of automorphisms of the F-algebra E.

If ζ is a character of G over E, then for any σ in $\text{Gal}(E/F)$, ζ^σ is also a

4.6 BLOCKS OF SMALL DEFECT

character of G over E (see §2.7). Two characters ζ and χ are called *p-conjugate* if $\zeta = \chi^\sigma$ for some σ in $\mathrm{Gal}(E/F)$. Clearly $\zeta^\sigma(x) = \zeta(x)$ for all p'-elements x of G and hence two p-conjugate irreducible ordinary characters lie in the same block of G and they have the same irreducible modular characters as constituents.

Let B be a p-block of G and let Δ_1 be the decomposition matrix corresponding to B. Let $\zeta^1, \zeta^2, \ldots, \zeta^u$ be a full set of representatives of the p-conjugate classes of characters lying in B. The u-rowed submatrix Δ_1° of Δ_1 corresponding to the characters ζ^i is called the *reduced decomposition matrix* of B. Brauer [2] has proved the following result on blocks of defect 1.

Theorem 4.6B Let B be a block of defect 1, $\zeta^1, \zeta^2, \ldots, \zeta^{s_1}$ the irreducible ordinary characters, and Δ_1 the decomposition matrix with Δ_1° being the reduced decomposition matrix. Suppose t_i is the number of p-conjugates of ζ^i and d_i the degree of ζ^i. If p^m is the order of the Sylow p-subgroups of G, then

(i) the irreducible ordinary characters lying in B can be partitioned into two disjoint classes S and T such that $t_i d_i \equiv t_1 d_1 \pmod{p^m}$ for all ζ^i in S and $t_i d_i \equiv -t_1 d_1 \pmod{p^m}$ for all ζ^i in T;

(ii) if ζ^i and ζ^j are not p-conjugate and belong to the same partition class in (i), then they have no irreducible modular character in common;

(iii) each coefficient of Δ_1 is 1 or 0;

(iv) each column of Δ_1° contains exactly two nonzero coefficients, one in a row corresponding to a member of S, and the other corresponding to a member of T.

The proof (which we shall not give) may be found in Brauer [2]. However see Theorem 8.5 of Chapter 8 for a special case.

We conclude this section by giving examples of how the theorems of this chapter can be applied to deduce the existence of irreducible characters of certain degrees, and an elementary way in which they help classification of simple groups.

EXAMPLE 1 Suppose G is a group of order $g = p^m g_0$, where $m = 1$ or 2 and $p \nmid g_0$. Suppose further that g_0 has exactly two different prime divisors q_1 and q_2. Then G has nontrivial irreducible ordinary characters (in the principal block of kG) whose degrees divide g_0 and are powers of q_1 and q_2, respectively. For, suppose that $\zeta^1 = 1_G, \zeta^2, \ldots, \zeta^{s_1}$ are the irreducible ordinary characters of G lying in the principal block B_0 of kG. By the example of §4.5 the block B_0 has defect m. Since $m \leq 2$, Theorem 4.5B shows that $p \nmid \zeta^i(1)$ for $i = 1, 2, \ldots, s_1$. However, by Theorem 4.2C we have

(1) $$1 + \sum_{i=2}^{s_1} \zeta^i(1)\zeta^i(y) = 0$$

for each $y \in G \backslash G^\circ$ (since $\zeta^1(1) = \zeta^1(y) = 1$); in particular, $s_1 > 1$. If q_1 and q_2 are the prime factors of g_0, then it follows from (1) that for some j ($2 \le j \le s_1$) we have $q_2 \nmid \zeta^j(1)$. However $\zeta^j(1) \mid g$ by Theorem 2.4B; so we conclude from the hypothesis on g that $\zeta^j(1)$ is a power of q_1 dividing g_0. Similarly there is a character ζ^i lying in B_0 whose degree is a power of q_2.

EXAMPLE 2 It follows from a classical theorem of Burnside that a simple group whose order is divisible by at most two primes is cyclic of prime order (see Curtis and Reiner [1]). On the other hand, there are eight known simple groups whose orders are divisible by exactly three distinct primes. Under further restriction it is possible to make a partial classification of these latter groups. Let G be a simple group whose order g is divisible by exactly three primes and which has one of the forms $3p^m q^n$, $4p^m q^n$, $5p^m q^n$, or $7p^m q^n$ (p, q primes). Suppose further that $m = 1$ or 2. In these cases it follows from Example 1 that G has a nontrivial irreducible ordinary character whose degree divides 3, 4, 5, or 7, respectively. Since G is simple, the character is faithful. It is not difficult to show that no simple group has an irreducible ordinary character of degree 2, and so in the respective cases G has a character of degree equal to 3, 4, 5, or 7; equivalently, G is isomorphic to a subgroup of $GL(d, \mathbf{C})$, where $d = 3, 4, 5$, or 7. The (finite) subgroups of these general linear groups have all been classified and consequently have lead to a classification of the simple groups of these types.

EXERCISES

1. Suppose that the group G has a faithful irreducible ordinary character of degree 2. Prove that either $G \ne G'$ (the commutator subgroup) or $Z(G) \ne 1$; in particular, G is not simple. [Hint: $2 \mid |G|$ by Theorem 2.4B. Show that if z is an element of order 2 in G, then either $z \notin G'$ or $z \in Z(G)$.]
2. Let G be a group of order pg_0, where $(p, g_0) = 1$, and let k_0 be the prime subfield of the field k. Show that each irreducible (Frobenius) character of G over k lying in the principal block of kG is afforded by some $k_0 G$-module.
3. Show that the number of irreducible ordinary characters lying in a block B is equal to the number of irreducible modular characters lying in B iff B is a block of defect 0.

4.7 Notes and comments

There are purely ring theoretic proofs available for some of the results of this chapter. For example see Michler [1,2,3]. The central characters of kG are the irreducible representations of $Z(kG)$ and easily generalize for any finite dimensional algebra over a field. In this connection see Michler [1].

4.7 NOTES AND COMMENTS

Theorem 4.2B was proved by Brauer and Nesbitt [1] and the Theorems 4.2C and 4.2D are essentially due to Osima [1]. The bound $\frac{1}{4}p^{2d} + 1$ for the number of irreducible ordinary characters in a p-block of defect d was obtained by Brauer and Feit [1]. It is conjectured that there are at most p^d irreducible ordinary characters in a p-block of defect, and this has been shown to be true in case the defect groups are cyclic (Dade [2]).

The blocks of defect 1 were analyzed by Brauer [2] and the information was applied to determine the simple groups of orders $4p^a q^b$ ($a \leq 2$) and $3p^a q^b$ ($a \leq 2$). The blocks of defect 1 bear a close resemblance to the blocks with cyclic defect groups (which will be studied in Chapter 8). For further results in this direction, see Brauer [5,8], Brauer and Fowler [1], and Carleson [1].

CHAPTER V

The Theory of Indecomposable Modules

Further applications of modular representation theory depend on the work of Green [1,2], which makes an incisive analysis of the indecomposable modules of a group algebra. The purpose of this chapter is to develop this theory of the structure of indecomposable modules. The topics covered include the following: relatively projective modules, vertices and sources, absolutely indecomposable induced modules, the degrees of indecomposable modules, the relation between vertices and defect groups, and the restriction of indecomposable modules to subgroups. Finally, we use some of these results to prove a theorem of Brauer and Feit on the existence of normal abelian subgroups in linear groups.

5.1 Relatively projective modules

The Frobenius reciprocity theorem (Theorem 2.5A) is a basic tool in the theory of ordinary characters. Unfortunately, in the modular case the full reciprocity relation no longer holds. However as Green first showed in [1,2], something can be salvaged, and this is the theory of vertices and sources which we shall look at in the next few sections.

Let G be a group and A a Noetherian integral domain. The cases in which we shall be interested later will be where A is a p-adic algebra of the type we considered in §3.3 or a field. Consider a short exact sequence

(1) $$0 \to U \xrightarrow{f} V \xrightarrow{g} W \to 0$$

5.1 RELATIVELY PROJECTIVE MODULES

of AG-modules (so $\text{Ker } f = 0$, $\text{Im } f = \text{Ker } g$, and $\text{Im } g = W$). Recall that (1) *splits* if there exists an AG-homomorphism $h: W \to V$ such that hg is the identity 1_W on W. An equivalent condition is that (1) splits if $V = U_1 \oplus W_1$, where $U_1 := \text{Im } f$ (which is isomorphic to U) and W_1 is a complementary AG-submodule (which is isomorphic to W).

NOTATION If U and V are AG-modules and V can be written as a direct sum $V = V_1 \oplus V_2$ of AG-submodules with $U \simeq V_1$, then we write $U \mid V$ (as AG-modules).

Definition Let W be an AG-module and H a subgroup of G. Then W is called *H-projective* if, whenever there is an exact sequence of the form (1) with W as the right-hand AG-module such that under restriction to H the exact sequence

(2) $$0 \to U_H \xrightarrow{f} V_H \xrightarrow{g} W_H \to 0$$

splits, then (1) also splits. In other words, the existence of an AH-homomorphism $h: W \to V$ with $hg = 1_W$ (the identity on W) implies that there is an AG-homomorphism with the same property.

The following characterization of H-projectivity is due to Higman [1,2].

Theorem 5.1 Let H be a subgroup of the group G and let $x_1 = 1, x_2, \ldots, x_n$, be a right transversal of H in G. Then each of the following properties of an AG-module W implies the others:

(a) W is H-projective;
(b) $W \mid (W_H)^G$;
(c) $W \mid U^G$ for some AH-module U; and
(d) there exists $f \in \text{End}_{AH}(W)$ such that $\sum_{i=1}^{n} x_i^{-1} f x_i = 1_W$.

Remark In (d), $\sum_{i=1}^{n} x_i^{-1} f x_i$ should be interpreted as the endomorphism of W given by $w \mapsto \sum_{i=1}^{n} w x_i^{-1} f x_i$.

Proof Assume that (a) holds; we shall prove (b). By definition $(W_H)^G = W_H \otimes_{AH} AG$ and each element of $(W_H)^G$ can be written uniquely in the form $\sum_{i=1}^{n} w_i \otimes x_i$ for suitable $w_i \in W$. Define $g: (W_H)^G \to W$ by

$$\left(\sum_{i=1}^{n} w_i \otimes x_i\right) g := \sum_{i=1}^{n} w_i x_i.$$

Clearly g is a surjective AG-homomorphism. Moreover the exact sequence

(3) $$0 \to \text{Ker } g \to (W_H)^G \to W \to 0$$

splits as a sequence of AH-modules; the required AH-homomorphism $h: W \to (W_H)^G$ is given by $wh := w \otimes 1$. Since W is H-projective by (a), we conclude that (3) splits as a sequence of AG-modules. Thus $(W_H)^G =$ Ker $g \oplus W_1$, where $W_1 \simeq W$. Thus $W \mid (W_H)^G$ and (b) is proved.

Trivially (b) implies (c).

Assume (c) holds; we shall prove (d). Without loss in generality we may assume that $U^G = W \oplus V$ for some AG-modules U and V. Let h denote the projection of U^G onto W with kernel V, and define $g: U^G \to U^G$ by

$$\left(\sum_{i=1}^n w_i \otimes x_i\right) g := w_1 \otimes x_1 = w_1 \otimes 1$$

(with each $w_i \in W$). It is readily verified that g is an AH-homomorphism. Moreover for each $v \in U^G$ we can write $v = \sum_{j=1}^n v_j \otimes x_j$ for some $v_j \in W$ and then

$$v \sum_{i=1}^n x_i^{-1} g x_i = \sum_{j=1}^n \sum_{i=1}^n \{(v_j \otimes x_j x_i^{-1})g\} x_i$$
$$= \sum_{i=1}^n (v_i \otimes 1) x_i = \sum_{i=1}^n v_i \otimes x_i = v$$

since $x_j x_i^{-1} \notin H$ if $i \neq j$. Thus $\sum_{i=1}^n x_i^{-1} g x_i$ acts as the identity on U^G. Finally since h is an AG-homomorphism, for all $w \in W$ we have

$$w \sum_{i=1}^n x_i^{-1}(gh) x_i = w\left\{\sum_{i=1}^n x_i^{-1} g x_i\right\} h = wh = w.$$

Thus if we take f as the restriction of gh to W, then $f \in \text{End}_{AH}(W)$ and satisfies (d).

To complete the proof, assume that (d) holds; we shall prove that (a) holds. Let

$$0 \to U \to V \xrightarrow{g_1} W \to 0$$

be an exact sequence of AG-modules which splits as a sequence of AH-modules; and let $h: W \to V$ be an AH-homomorphism for which $hg_1 = 1_W$. We claim that $h_1: W \to V$ defined by $h_1 := \sum_{i=1}^n x_i^{-1} f h x_i$, where $f \in \text{End}_{AH}(W)$ satisfies $\sum_{i=1}^n x_i^{-1} f x_i = 1_W$, is an AG-homomorphism such that $h_1 g_1 = 1_W$; this will prove (a). Indeed, since g_1 is an AG-homomorphism,

$$h_1 g_1 = \left(\sum_{i=1}^n x_i^{-1} f h x_i\right) g_1 = \sum_{i=1}^n x_i^{-1} (f h g_1) x_i = \sum_{i=1}^n x_i^{-1} f x_i = 1_W,$$

since Im $f \subseteq W$ and $hg_1 = 1_W$. On the other hand, for each $y \in G$ and each x_i

5.1 RELATIVELY PROJECTIVE MODULES

there exists $z_i \in H$ and $x_{i'}$ such that $x_i y = z_i x_{i'}$. Then for all $w \in W$

$$w(h_1 y) = w \sum_{i=1}^{n} x_i^{-1} f h z_i x_{i'} = w \sum_{i=1}^{n} x_i^{-1} z_i f h x_{i'}$$

$$= wy \sum_{i=1}^{n} x_{i'}^{-1} f h x_{i'} = (wy) h_1,$$

since $i \mapsto i'$ is a permutation of $\{1, 2, \ldots, n\}$. Thus h_1 is an AG-homomorphism and (a) is proved.

Corollary Let H and L be subgroups of the group G.

(i) If $x^{-1} H x \subseteq L$ for some $x \in G$, then each H-projective AG-module is L-projective.

(ii) If $H \subseteq L$, V is an H-projective AL-module, and W is an AG-module such that $W \mid V^G$, then W is an H-projective AG-module.

Proof (i) Let W be an H-projective AG-module. By part (c) of the theorem there exists an AH-module U such that $W \mid U^G$. Then $U \otimes x$ is an $A(x^{-1} H x)$-submodule of U^G and it is easily seen that $(U \otimes x)^G \simeq U^G$ as AG-modules. Let V be the induced AL-module $(U \otimes x)^L$. Since inducing is a transitive operation, $V^G = (U \otimes x)^G \simeq U^G$ and so $W \mid V^G$. Thus the theorem shows that W is L-projective.

(ii) Since V is an H-projective AL-module, part (c) of the theorem shows that $V \mid U^L$ for some AH-module U. Then $V^G \mid (U^L)^G = U^G$, so $W \mid U^G$ and hence W is H-projective.

EXAMPLE 1 Let $A = k$ be a field. Then every exact sequence (1) of kG-modules splits as a sequence of k-modules (that is, vector spaces). Thus over a field 1-projectivity of W simply means that each exact sequence (1) of AG-modules with W as the right-hand term splits; in other words, W is a projective kG-module in the usual sense of projectivity.

EXAMPLE 2 Suppose A has the property that each integer not divisible by a prime p has an inverse in A. (This is the case if A is a p-adic algebra described in §3.3, or A is a field of characteristic p.) Suppose H contains a Sylow p-group of G. Then $p \nmid |G : H|$ so there exists $\alpha \in A$ such that $\alpha |G : H| = 1$. Taking f as the scalar $\alpha \cdot 1$ in part (d) of the theorem we see that $\sum_{x \in T} x^{-1} f x = \alpha |G : H| = 1$, where T is a right transversal of H in G, and so in this case every AG-module W is H-projective.

EXAMPLE 3 Suppose that A is either a field or the ring of p-adic integers described in §3.3. Then an indecomposable A-free AG-module W is 1-projective iff W is isomorphic to a principal indecomposable AG-module. Indeed, first suppose that W is an indecomposable 1-projective AG-module, then by Theorem 5.1, $W \mid W_1^G$ where W_1 is the restriction of W to 1. Since the Krull–Schmidt theorem holds because of our hypothesis on A (see Theorem 1.6A and Theorem 3.4B) we can choose an indecomposable component U of W_1 such that $W \mid U^G$. Because W_1 is A-free, U is therefore the unique A-free indecomposable (trivial) $A\{1\}$-module. Thus $U^G = U \otimes_A AG \cong AG$ as AG-modules. Hence $W \mid AG$, and so is isomorphic to a principal indecomposable AG-module. Conversely, if $W \mid AG$, then $W \mid U^G$, where U is the trivial A-free $A\{1\}$-module. Therefore Theorem 5.1 shows that W is 1-projective.

EXAMPLE 4   Suppose U and V are A-free AG-modules. If V is H-projective for some subgroup H of G, then $V \mid W^G$ for the A-free AH-module $W := V_H$. Then by Theorem 2.1B, $(U_H \otimes W)^G \simeq U \otimes W^G$. Hence $U \otimes V \mid (U_H \otimes W)^G$, which shows that $U \otimes V$ is also H-projective.

EXERCISE

Let k be a field and G a group. If the characteristic of k is either 0 or a prime not dividing $|G|$, show that for each subgroup H of G, each kG-module is H-projective.

5.2   Vertices and sources

We shall now restrict ourselves to the case where A is a Noetherian integral domain such that each A-submodule of an A-free A-module is A-free, and such that the conclusion of the Krull–Schmidt theorem holds for A-free AH-modules for all subgroups H of G. The cases in which we are really interested are (a) A is a field, and (b) A is a the p-adic algebra described in §3.3. It follows from Theorem 1.6 and 3.4B and Example 4 of §1.1 that the hypotheses hold in these two cases.

Note 1   Under these additional hypotheses the AH-module U in Theorem 5.1(c) may be taken as an indecomposable A-free AH-module whenever W is A-free (for example, take U as a suitable indecomposable summand of W_H).

Let V be an indecomposable AG-module. Then a subgroup Q of G is a *vertex* of V if V is Q-projective but V is not H-projective for any proper subgroup H of Q. It follows from Theorem 5.1(c) that corresponding to any vertex Q of V there is at least one indecomposable AQ-module U such that $V \mid U^G$; such an AQ-module is called a *source* of V.

Note 2   Part (i) of the corollary to Theorem 5.1 shows that if Q is a vertex of V, then so is any conjugate $x^{-1}Qx$ $(x \in G)$.

These definitions of vertex and source and their basic properties (which

5.2 VERTICES AND SOURCES

we shall give in this and the following sections) were given in a paper of Green [1]. In order to prove the latter properties we shall need a simple lemma.

Lemma 5.2 Suppose that A satisfies the hypotheses stated above. Let H and L be subgroups of the group G and let V be an A-free L-projective AG-module; so $V \mid U^G$ for some A-free AL-module U. Write $V_H = W_1 \oplus W_2 \oplus \cdots \oplus W_n$ as a direct sum of indecomposable AH-submodules. Then there exist elements $x_1, x_2, \ldots, x_n$ of G such that for each i, $W_i \mid (U \otimes x_i)^H$, where $U \otimes x_i$ is considered as an $A(x_i^{-1}Hx_i)$-module. In particular, W_i is $(x_i^{-1}Lx_i \cap H)$-projective.

Proof By Mackey's theorem (Theorem 2.1A)

$$(U^G)_H \simeq (U \otimes y_1)^H \oplus (U \otimes y_2)^H \oplus \cdots \oplus (U \otimes y_m)^H,$$

where $y_1 = 1, y_2, \ldots, y_m$ is a set of representatives for the (L, H)-double cosets and $U \otimes y_j$ is considered as an $A(y_j^{-1}Ly_j \cap H)$-module. Since $V \mid U^G$, it follows that each $W_i \mid (U^G)_H$. Because of the Krull–Schmidt theorem and the indecomposability of W_i, we get $W_i \mid (U \otimes y_j)^H$ for some j. The lemma now follows.

Our first main result is to prove that both vertices and sources are uniquely determined up to conjugacy.

Theorem 5.2A Suppose that A satisfies the hypotheses stated above. Let V be an indecomposable A-free AG-module, and let Q be a vertex of V.

(i) If V is H-projective for some subgroup H of G, then $H \supseteq x^{-1}Qx$ for some $x \in G$. In particular, every vertex of V is conjugate to Q in G.

(ii) Let U_1 and U_2 be A-free AQ-modules that are both sources of V. Then $U_2 \simeq U_1 \otimes z$ for some $z \in N_G(Q)$.

Remark By Note 2 above, every conjugate of a vertex of V is also a vertex for V. Thus the vertices for V form a single class of conjugate subgroups of G.

Proof (i) By Lemma 5.2, $V_H = W_1 \oplus W_2 \oplus \cdots \oplus W_n$, where each W_i is an indecomposable AH-module which is $(x_i^{-1}Qx_i \cap H)$-projective for some $x_i \in G$. Since V is H-projective we also have $V \mid (V_H)^G = W_1^G \oplus W_2^G \oplus \cdots \oplus W_n^G$; and hence by the Krull–Schmidt theorem, $V \mid W_i^G$ for some i because V is indecomposable. Hence by part (ii) of the corollary of Theorem 5.1 we conclude that V is $(x_i^{-1}Qx_i \cap H)$-projective, and so $x_i^{-1}Qx_i \cap H = x_i^{-1}Qx_i$ because $x_i^{-1}Qx_i$ is a vertex of V. Thus $H \supseteq x_i^{-1}Qx_i$ as asserted.

(ii) Write $V_Q = W_1 \oplus W_2 \oplus \cdots \oplus W_n$, where the W_i are indecomposable AQ-modules. Since V is Q-projective, $V \mid (V_Q)^G = W_1^G \oplus W_2^G \oplus \cdots \oplus W_n^G$, and so $V \mid W_i^G$ for some i because V is indecomposable. Since $V \mid U_1^G$,

Lemma 5.2 shows that there exists $x \in G$ such that $W_i \mid (U_1 \otimes x)^G$, and that W_i is $(x^{-1}Qx \cap Q)$-projective. But part (ii) of the corollary to Theorem 5.1 now implies that V is $(x^{-1}Qx \cap Q)$-projective, so $x^{-1}Qx \cap Q = Q$ because Q is a vertex of V. This shows that $x \in N_G(Q)$, and that $W_i \mid (U_1 \otimes x)$. Since U_1 is an indecomposable AQ-module, the same is true of $U_1 \otimes x$; hence $W_i \simeq U_1 \otimes x$. A similar argument now shows that $W_i \simeq U_2 \otimes y$ for some $y \in N_G(Q)$. Thus $U_2 \simeq U_1 \otimes z$, where $z = xy^{-1} \in N_G(Q)$, and the theorem is proved.

EXAMPLE 1   Let $A = k$ be a field of characteristic p and let P be a Sylow p-subgroup of the group G. Then by Example 2 of §5.1 every kG-module is P-projective. Thus in this case the vertex of each indecomposable kG-module V is a p-group, and the set of vertices of V is a class of conjugate p-subgroups of G.

Theorem 5.2B   Suppose that A satisfies the hypotheses above. Let H be a subgroup of the group G and let W be an indecomposable A-free AH-module with a vertex Q and a corresponding A-free source U. Then there exists an indecomposable A-free AG-module $V \mid W^G$ such that V also has vertex Q and source U.

Proof   Since $W \otimes 1 \simeq W$, we have $W \mid (W^G)_H$ (as AH-modules). Therefore, since W is indecomposable, there exists an indecomposable AG-module V such that $W \mid V_H$ and $V \mid W^G$; in particular, V is H-projective. By hypothesis $W \mid U^H$, and so $V \mid (U^H)^G = U^G$. Since U is an A-free AQ-module, Theorem 5.2A now shows that Q contains a vertex Q_0 of V; let U_0 be a source for V corresponding to Q_0. Since W is an indecomposable direct summand of V_H and $V \mid U_0^G$, Lemma 5.2 shows that $W \mid (U_0 \otimes x)^H$ for some $x \in G$, where $U_0 \otimes x$ is considered as an $A(x^{-1}Q_0 x \cap H)$-module. Hence W is $x^{-1}Q_0 x \cap H$-projective. Since Q is a vertex of W, $|Q_0| \geq |x^{-1}Q_0 x \cap H| \geq |Q|$. But $Q_0 \subseteq Q$, so $Q = Q_0$ and Q is a vertex of V. On the other hand, U is an indecomposable AQ-module since it is a source for W and we saw above that $V \mid U^G$. Thus U is a source for V corresponding to Q, and the theorem is proved.

EXAMPLE 2   Let H be a subgroup of the group G. Let V be a principal indecomposable kG-module, where k is a field of characteristic p. Then $V \mid U_0^G$, where U_0 is the trivial module for the group 1. (See Example 3 of §5.1.) Hence by Lemma 5.2, $V_H = W_1 \oplus W_2 \oplus \cdots \oplus W_n$, where the W_i are indecomposable kH-modules such that $W_i \mid (U_0 \otimes x_i)^H$ for some $x_i \in G$ and $U_0 \otimes x_i \simeq U_0$ is considered as a $k \cdot 1$-module. Since $U_0^H = U_0 \otimes_k kH \simeq kH$, each W_i is isomorphic to a principal indecomposable kH-module. In particular, in the case where $H = P$ is a Sylow p-group there is only one principal indecomposable kP-module, namely the regular module kP (see the

5.3 GREEN'S THEOREM

Example of §1.9). This implies that $|P| = \dim_k kP$ divides $\dim_k V$ for each principal indecomposable kG-module. (See also the example of §4.4).

EXAMPLE 3   The result noted in Example 2 leads to an interesting observation about the submatrix Γ_1 of the Cartan matrix Γ for kG corresponding to a block B_1 of kG. (See the notation of §4.4.) Suppose that B_1 has defect d. Let $\eta^1, \eta^2, \ldots, \eta^{r_1}$ and $\phi^1, \phi^2, \ldots, \phi^{r_1}$ denote, respectively, the principal indecomposable modular characters and the irreducible modular characters lying in B_1. Then Γ_1 is the $r_1 \times r_1$ matrix $[c_{ij}]$ with integer entries where

$$\eta^i = \sum_{j=1}^{r_1} c_{ij} \phi^j \qquad (i = 1, 2, \ldots, r_1).$$

In particular,

$$\eta^i(1) = \sum_{j=1}^{r_1} c_{ij} \phi^j(1) \qquad (i = 1, 2, \ldots, r_1),$$

and we can solve these equations using Cramer's rule and obtain

$$\phi^j(1)(\det \Gamma_1) = \sum_{i=1}^{r_1} c'_{ij} \eta^i(1)$$

for suitable integers c'_{ij}. By Example 2 above, the right-hand side of this equation is divisible by p^m, the order of a Sylow p-subgroup of G. By Theorem 4.5A there exists some j such that $p^{m-d+1} \nmid \phi^j(1)$. Hence $p^d \mid \det \Gamma_1$.

EXERCISES

[*For these exercises suppose that the notation of §3.3 holds.*]

1. Let W be an indecomposable kG-module. Suppose that for some subgroup H of G, W_H is isomorphic to a sum of principal indecomposable kH-modules. Show that W is isomorphic to a principal indecomposable kG-module. Hence there is an AG-module W_1 such that $\overline{W}_1 \simeq W$.

2. Show that every p-subgroup of a group G is the vertex of some indecomposable kG-module.

5.3 Green's theorem

In general, inducing from an indecomposable module does not yield another indecomposable module. However, in an especially important case, *absolute* indecomposability is preserved under induction. This is the content of Green's theorem.

V THE THEORY OF INDECOMPOSABLE MODULES

In the present section k will denote an arbitrary field of characteristic p. Recall that if V is a kG-module, and $E := \text{End}_{kG}(V)$, then V is absolutely indecomposable iff $E/\text{rad } E = k$ (Theorem 1.7).

We begin with a lemma which serves as the crucial step in the proof of Green's theorem.

Lemma 5.3 Let H be a normal subgroup of index p in the group G, and suppose that k is a perfect field of characteristic p. Let W be an absolutely indecomposable kH-module, and suppose that the indecomposable components of $(W^G)_H$ are all isomorphic to W. Then W^G is an absolutely indecomposable kG-module.

Remark Recall that a field k of characteristic p is perfect iff each element in k is a pth power of an element in k. In particular, it is well known that k is perfect when it is finite or algebraically closed.

Proof Let $y \in G \backslash H$. Then $1, y, \ldots, y^{p-1}$ is a right transversal of H in G and

(1) $$W^G = (W \otimes 1) \oplus \cdots \oplus (W \otimes y^{p-1}).$$

By hypothesis, $W \otimes y \simeq W$ as kH-modules, and so there exists an invertible $\phi \in \text{End}_k(W)$ such that $w \otimes y \mapsto w\phi$ is a kH-homomorphism from $W \otimes y$ to W. By definition, $(w \otimes y)x = wyxy^{-1} \otimes y$ for all $x \in H$, so the condition that the mapping from $W \otimes y$ to W commutes with the action of H is given by $yxy^{-1}\phi = \phi x$ for all $x \in H$. More generally, since $H \triangleleft G$, induction on i shows that

(2) $$y^i x y^{-i} \phi^i = \phi^i x \quad \text{for all} \quad x \in H, i = 0, 1, 2, \ldots.$$

In particular, we have

(3) $$y^p \phi = \phi y^p$$

since $y^p \in H$.

Let R be the k-algebra $\text{End}_k W$. Then we can construct a k-linear mapping ψ from $\text{End}_k(W^G)$ into the matrix ring $\text{Mat}(p, R)$ by defining $\psi(a) := [\alpha_{ij}]$ ($i, j = 0, 1, \ldots, p-1$), where the $\alpha_{ij} \in \text{End}_k(W)$ are uniquely determined by the conditions

(4) $$(w \otimes y^i)a = \sum_{j=0}^{p-1} (w\phi^i \alpha_{ij} \phi^{-j} \otimes y^j) \quad (i = 0, 1, 2, \ldots, p-1; w \in W).$$

It is readily verified that $\psi : \text{End}_k(W^G) \simeq \text{Mat}(p, R)$ is a k-algebra isomorphism.

Now $y^p \in H$, and for all $w \in W$ we have

$$(w \otimes y^i)y = \begin{cases} w \otimes y^{i+1} & \text{if } i = 0, 1, \ldots, p-2 \\ wy^p \otimes 1 & \text{if } i = p-1. \end{cases}$$

5.3 GREEN'S THEOREM

On the other hand, for $x \in H$, (2) shows that for all $w \in W$ we have

$$(w \otimes y^i)x = wy^i xy^{-i} \otimes y^i = w\phi^i x \phi^{-i} \otimes y^i.$$

Therefore we conclude from (4) that

$$(5) \quad \psi(y) = \begin{bmatrix} 0 & \phi & 0 & \cdots & 0 \\ 0 & 0 & \phi & \cdots & 0 \\ \hline 0 & & 0 & 0 & \cdots & \phi \\ \phi^{-p+1} y^p & 0 & 0 & \cdots & 0 \end{bmatrix} \quad \text{and} \quad \psi(x) = \begin{bmatrix} x & 0 & \cdots & 0 \\ 0 & x & \cdots & 0 \\ \hline 0 & 0 & \cdots & x \end{bmatrix}$$

for all $x \in H$. (Here the matrix entries shall be interpreted as elements of $\text{End}_k(W) = R$.)

Let $E := \text{End}_{kH}(W) \subseteq R$. Then the image of $\text{End}_{kH}(W^G)$ under ψ consists of all elements of $\text{Mat}(p, R)$ that commute with $\psi(x)$ ($x \in H$); thus $\psi(\text{End}_{kH}(W^G)) = \text{Mat}(p, E)$. Put $S := \psi(\text{End}_{kG}(W^G))$. Then $S \simeq \text{End}_{kG}(W^G)$ as k-algebras because ψ is a k-algebra isomorphism. On the other hand, S consists of all matrices in $\text{Mat}(p, E)$ that commute with $\psi(y)$, so we must consider this latter condition.

First note that we can write $\psi(y) = ab$, where

$$a = \begin{bmatrix} \phi & 0 & \cdots & 0 \\ 0 & \phi & \cdots & 0 \\ \hline 0 & 0 & \cdots & \phi \end{bmatrix} \quad \text{and} \quad b = \begin{bmatrix} 0 & 1 & 0 & \cdots & 0 \\ 0 & 0 & 1 & \cdots & 0 \\ \hline 0 & 0 & 0 & \cdots & 1 \\ \beta & 0 & 0 & \cdots & 0 \end{bmatrix},$$

with $\beta := \phi^{-p} y^p \in E$ by (2). Note that (3) also shows that a and b commute. Since W is absolutely indecomposable, $E/\text{rad } E = k$, and so there is a canonical homomorphism $\text{Mat}(p, E) \to \text{Mat}(p, k)$ (taking the entries modulo rad E); we shall denote the latter homomorphism by $c \mapsto \bar{c}$. Note that the kernel of this homomorphism is $\text{Mat}(p, \text{rad } E)$, which is a nilpotent ideal of $\text{Mat}(p, E)$ (in fact, the radical).

We now claim that for each $c \in \text{Mat}(p, E)$, $\overline{a^{-1}ca} = \bar{c}$. This is equivalent to showing that for each $\gamma \in E$, $\phi^{-1}\gamma\phi - \gamma \in \text{rad } E$. However, since $E/\text{rad } E = k$, for each $\gamma \in E$ there exists $\lambda \in k$ such that $\gamma - \lambda \cdot 1 \in \text{rad } E$. Then $\phi^{-1}\gamma\phi - \gamma \in \phi^{-1}(\lambda \cdot 1)\phi - \lambda \cdot 1 + \text{rad } E = \text{rad } E$ because ϕ commutes with λ. This proves our claim.

Now consider S and $\bar{S} := \{\bar{c} \mid c \in S\}$. To show that W^G is absolutely indecomposable we must show that $S/\text{rad } S = k$; and since $S \cap \text{Mat}(p, \text{rad } E) \subseteq \text{rad } S$, this is equivalent to showing that $\bar{S}/\text{rad } \bar{S} = k$. Since $\overline{a^{-1}ca} = \bar{c}$ for all $c \in \text{Mat}(p, E)$, and S is the centralizer of $\psi(y) = ab$, it follows that $\bar{S}$ is the centralizer in $\text{Mat}(p, k)$ of $\bar{b}$ (recall that $b \in \text{Mat}(p, E)$

because $\beta \in E$). Let $\bar{c} = [\gamma_{ij}] \in \text{Mat}(p, k)$. Then the condition $\bar{b}\bar{c} = \bar{c}\bar{b}$ is equivalent to the conditions

$$\gamma_{ij} = \gamma_{i+1, j+1}, \qquad \gamma_{i, p-1} = \gamma_{i+1, 0}\beta_0,$$

$$\gamma_{p-1, j}\beta_0 = \gamma_{0, j+1}, \qquad \gamma_{p-1, p-1}\beta_0 = \gamma_{00}\beta_0$$

for $i, j = 0, 1, \ldots, p - 2$, where $\beta_0 := \beta + \text{rad } E$. Hence $\bar{c} \in \bar{S}$ iff it has the form

$$\bar{c} = \begin{bmatrix} \gamma_0 & \gamma_1 & \cdots & \gamma_{p-1} \\ \beta_0\gamma_{p-1} & \gamma_0 & \cdots & \gamma_{p-2} \\ \hdashline \beta_0\gamma_1 & \cdots & \beta_0\gamma_{p-1} & \gamma_0 \end{bmatrix} = \gamma_0 \cdot 1 + \gamma_1\bar{b} + \cdots + \gamma_{p-1}\bar{b}^{p-1},$$

where $\gamma_0, \gamma_1, \ldots, \gamma_{p-1} \in k$. Since k is perfect there exists $\lambda \in k$ such that $\beta_0 = \lambda^p$. Then by the binomial theorem we have $(\bar{b} - \lambda \cdot 1)^p = \bar{b}^p - \lambda^p \cdot 1 = \beta_0 1 - \beta_0 \cdot 1 = 0$; and so $\bar{b} - \lambda \cdot 1$ is nilpotent. Since $\bar{S}$ is commutative, and $\bar{b} - \lambda \cdot 1 \in \bar{S}$, this shows that $\bar{S}(\bar{b} - \lambda \cdot 1) \subseteq \text{rad } \bar{S}$. But $\bar{S}(\bar{b} - \lambda \cdot 1)$ is the kernel of the algebra homomorphism $\bar{S} \to k$ defined by

$$\gamma_0 \cdot 1 + \gamma_1 \cdot \bar{b} + \cdots + \gamma_{p-1}\bar{b}^{p-1} \mapsto \gamma_0 + \gamma_1\lambda + \cdots + \gamma_{p-1}\lambda^{p-1}.$$

Therefore $\bar{S}/\text{rad } \bar{S} = \bar{S}/\bar{S}(\bar{b} - \lambda \cdot 1) = k$, as required. This proves the lemma.

Theorem 5.3 Let H be a normal subgroup of index p in a group G. Let k be a field of characteristic p, and let W be an absolutely indecomposable kH-module. Then W^G is an absolutely indecomposable kG-module.

Proof First note that it is enough to prove the theorem for the case where k is algebraically closed (and then we can drop the word "absolutely" by Theorem 1.7A). Indeed, let k^* be an algebraic closure of k. Then $W^G \otimes_k k^* = (W \otimes_k k^*)^G$, and W (respectively, W^G) is absolutely indecomposable if and only if $W \otimes_k k^*$ (respectively, $W^G \otimes_k k^*$) is indecomposable.

Consider the set $I(W) = \{x \in G \mid W \otimes x \simeq W \text{ as } kH\text{-modules}\}$. Clearly $I(W)$ is a subgroup of G and $H \subseteq I(W)$. [$I(W)$ is called the "inertia group" of W.] Since H has index p in G, we have two cases.

Case 1 ($I(W) = H$) Let $y \in G\backslash H$. Then $(W^G)_H = W \otimes 1 \oplus W \otimes y \oplus \cdots \oplus (W \otimes y^{p-1})$ as a sum of kH-modules. Using the Krull–Schmidt theorem we know that W^G has an indecomposable component V such that $W \otimes 1 \mid V_H$. But then $W \otimes y^i \mid V_H y^i$, and the latter equals V_H since V is a kG-module. Since W is indecomposable, the kH-modules $W \otimes y^i$ ($i = 1, 2, \ldots, p-1$) are indecomposable, and they are mutually nonisomorphic because $I(W) = H$. Therefore the Krull–Schmidt theorem shows that the sum of these modules divides V_H; that is, $(W^G)_H \mid V_H$. But this implies $W^G = V$, so W^G is indecomposable.

Case 2 $(I(W) = G)$ In this case we are in the situation of Lemma 5.3 and the result was proved there.
This proves the theorem.

Remark In Case 1 the proof does not depend on *absolute* indecomposability, but in Case 2 this is important. The normality of H is also important. In Green [1,2] there are examples that show that Theorem 5.3 fails to remain true if H is not assumed to be normal in G or if the words "absolutely indecomposable" are replaced by "indecomposable."

Corollary Let G be a p-group and H any subgroup. Let k be a field of characteristic p and let W be an absolutely indecomposable kH-module. Then W^G is an absolutely indecomposable kG-module.

Proof Since G is a p-group, there is a normal series of subgroups of the form $G = G_0 \supset G_1 \supset \cdots \supset G_r = H$ with $|G_i/G_{i+1}| = p$ for each i. Then the corollary follows from the theorem using induction on r.

EXAMPLE Let G be a p-group and k a field of characteristic p. Let H be a subgroup of G and W_0 the trivial kH-module. Then W_0^G is an absolutely indecomposable kG-module. Note that this module affords the transitive permutation representation of G with H as a stabilizer.

EXERCISES

1. Let H be a normal subgroup of a group G such that G/H is a cyclic group whose order m is prime to p. Let k be an algebraically closed field of characteristic p, and let W be an indecomposable kH-module such that the indecomposable components of $(W^G)_H$ are all isomorphic to W. Show that W^G is a direct sum of m nonisomorphic indecomposable kG-modules, say $W_1, \ldots, W_m$, such that $(W_i)_H \cong W$ for each i.

2. Let G be a p-group of linear transformations on a vector space V over a field k of characteristic p. If there exists a basis for V on which G acts as a transitive permutation group, show that V is an indecomposable kG-module.

5.4 The degrees of indecomposable modules

Green's theorem (Theorem 5.3) leads at once to useful facts about the degrees of indecomposable modules and the characters that they afford.

Theorem 5.4A Let k be a field of characteristic p and let P be a Sylow p-subgroup of the group G. Let V be an indecomposable kG-module with vertex $Q \subseteq P$. Then $|P : Q|$ divides $\dim_k V$.

Remark We know from Example 1 of §5.2 that V has at least one vertex contained in P. As a special case, it follows that if $p \nmid \dim_k V$, then P itself is a vertex of V.

Proof First consider the case where k is algebraically closed (and so V is *absolutely* indecomposable). Let U be a source of V corresponding to the vertex Q. By Lemma 5.2 we have $V_P = W_1 \oplus W_2 \oplus \cdots \oplus W_n$, where the W_i are indecomposable kP-modules and for some $x_1, x_2, \ldots, x_n$ in G we have $W_i | (U \otimes x_i)^P$, where $U \otimes x_i$ is viewed as a $k(x_i^{-1} Q x_i \cap P)$-module. By the corollary of Theorem 5.3, the indecomposable components of $(U \otimes x_i)^P$ are modules induced from the indecomposable components of $U \otimes x_i$ restricted to $x_i^{-1} Q x_i \cap P$; in particular, $W_i = U_i^P$ for some $k(x_i^{-1} Q x_i \cap P)$-module U_i. Thus $\dim_k W_i = |P : x_i^{-1} Q x_i \cap P| \dim_k U_i$ and so is divisible by $|P : Q|$. Since $\dim_k V = \dim_k W_1 + \cdots + \dim_k W_n$, we conclude that $|P : Q|$ divides $\dim_k V$ as required. This proves the theorem in the case k is algebraically closed.

In general, let k^* denote the algebraic closure of k. Then $V \otimes_k k^* = V_1 \oplus V_2 \oplus \cdots \oplus V_m$, where the V_i are indecomposable k^*G-modules. Since for all i we have $V_i | V \otimes_k k^*$ and $V \otimes_k k^* | U^G \otimes_k k^* = (U \otimes_k k^*)^G$, therefore $V_i | (U \otimes_k k^*)^G$ for the k^*Q-module $U \otimes_k k^*$. This shows that each V_i is Q-projective and so has a vertex $Q_i \subseteq Q$. By the algebraically closed field case considered above, $|P : Q_i|$ divides $\dim_k V_i$ for each i. Since $|P : Q|$ divides each $|P : Q_i|$ and $\dim_k V = \dim_k V_1 + \cdots + \dim_k V_m$, we conclude that $|P : Q|$ divides $\dim_k V$. This proves the theorem in general.

Lemma 5.4 Let k be a field of characteristic p and let W be an absolutely indecomposable kG-module. If x is an element of $E = \mathrm{End}_{kG}(W)$ of finite order relatively prime to p, then $x = \lambda \cdot 1$ for some $\lambda \in k$.

Proof Since W is absolutely indecomposable, $E/\mathrm{rad}\, E \simeq k$ (Theorem 1.7A), and so x may be written in the form $x = \lambda \cdot 1 + r$ for some r in rad E and λ in k. Now rad E is nilpotent and there is an integer n such that $r^n = 0$. Choose an integer m so that $p^m \geq n$. Then $x^{p^m} = (\lambda \cdot 1 + r)^{p^m} = \lambda^{p^m} \cdot 1$, since k has characteristic p. As x has order prime to p, it is a power of x^{p^m}, and the result follows.

Theorem 5.4B Let G be a group, x a p'-element of G, and P a Sylow p-subgroup of $C_G(x)$. Let k be a field of characteristic p and let V be an indecomposable kG-module affording the character ϕ over k. Suppose P properly contains a vertex Q of V. Then $\phi(x) = 0$.

Proof First consider the case where k is algebraically closed. Let $H := \langle P, x \rangle$ and let $V_H = W_1 \oplus W_2 \oplus \cdots \oplus W_n$, where each W_i is an indecomposable kH-module. Lemma 5.2 shows that there is an element x_i of

5.5 VERTICES AND DEFECT GROUPS

G such that W_i is $x_i^{-1}Qx_i \cap H = x_i^{-1}Qx_i \cap P$-projective. As $Q \neq P$, it follows that p divides $|P: x_i^{-1}Qx_i \cap P|$, and we deduce from Theorem 5.4A that p divides $\dim_k W_i$.

Now since x is central in H, we may consider x as an element of $\operatorname{End}_{kH}(W_i)$. Thus we see from Lemma 5.4 that x acts as a scalar multiple of the identity on W_i. In particular, $\phi(x) = 0$, since the multiplicity of each eigenvalue is divisible by p.

In general, let k^* be the algebraic closure of k. Then $V^* = V \otimes_k k^*$ is a k^*G-module. Let $V^* = V_1 \oplus V_2 \oplus \cdots \oplus V_m$ be a direct sum decomposition of V^* into indecomposable modules V_i. We know from the proof of Theorem 5.4A that each V_i has a vertex contained in Q. Thus our previous reasoning applied to each V_i gives the required conclusion in the general case.

EXERCISE

Let k be a field of characteristic p and let U be a principal indecomposable kG-module. Suppose η is the modular character of G associated with U. If x is a p'-element of G with $|C_G(x)| = p^m h$, where $(p, h) = 1$, then show that $\eta(x) = p^m \lambda(x)$, where λ is a character of $\langle x \rangle$.

5.5 Vertices and defect groups

Suppose k is a field of characteristic p satisfying the hypothesis of §3.3. If V is an indecomposable kG-module, then the vertices of V and the defect groups of the block in which V lies both form conjugate classes of subgroups of G. There is an interesting relation between these two classes.

Theorem 5.5 Let V be an indecomposable kG-module lying in a block B of kG. If D is a defect group of B, then V is D-projective, and hence D contains a vertex of V.

Proof Let ω be the central character associated with B. Then by Theorem 4.3 there exists a class $\mathscr{C}$ of G with class sum c in kG such that D is a defect group of $\mathscr{C}$ and $\bar{\omega}(c) \neq 0$. Choose $x \in \mathscr{C}$ so that D is a Sylow p-subgroup of $C_G(x)$, and let $x_1, x_2, \ldots, x_n$ be a right transversal of D in G. Then

$$\sum_{i=1}^{n} x_i^{-1} x x_i = mc \quad \text{where} \quad m := |C_G(x):D| \not\equiv 0 \pmod{p}.$$

Now mc acts on V as the nonzero scalar $m\bar{\omega}(c) \cdot 1$ (see Theorem 4.2B). Define $f \in \operatorname{End}_k(V)$ by $vf := vx\{m\bar{\omega}(c)\}^{-1}$. Then the condition (d) of Theorem 5.1 is satisfied for V and so V is D-projective.

EXAMPLE If B is a block of defect 0 (so $D = 1$), then each indecomposable kG-module V lying in B has vertex 1 and so is isomorphic to a principal indecomposable kG-module (Example 3 of §5.1). This gives further information about blocks of defect 0 (see also §4.6).

5.6 Restriction of indecomposable modules

In the next section we shall be proving a theorem of Brauer and Feit on the existence of normal abelian subgroups. This will require a description of the way in which certain indecomposable kG-modules decompose under restriction to subgroups of G.

The example of §4.3 shows how in the case that a character has degree not divisible by p, a knowledge of a rather few values of the character is sufficient to determine the block in which it lies. This motivates our first result.

Theorem 5.6A Suppose that the hypotheses of §3.3 hold. Let P be a Sylow p-subgroup of the group G, and put $N := N_G(P)$. Let V be an irreducible kG-module, and suppose that $p \nmid \dim_k V$. Then

(i) $V_N = W_1 \oplus W_2 \oplus \cdots \oplus W_m$ is a direct sum of indecomposable kN-modules such that $V \mid W_1^G$ and P is a vertex for W_1.

(ii) $(W_1^G)_N$ has exactly one indecomposable component with P as a vertex. Moreover, for the components of V_N we have $p \nmid \dim W_1$ but $p \mid \dim W_i$ for $i = 2, 3, \ldots, m$.

(iii) Let ϕ and θ be the characters over k afforded by V and W_1, respectively; then $\phi(x) = \theta(x)$ for all elements $x \in C_G(P)^\circ$.

(iv) If W_1 lies in the principal block of kN, then V lies in the principal block of kG; moreover, W_1 is trivial only if V is trivial.

Proof (i) Since N contains a Sylow p-subgroup, it contains a vertex of V; hence V is N-projective. Therefore by Theorem 5.1, $V \mid (V_N)^G$. Writing $V_N = W_1 \oplus W_2 \oplus \cdots \oplus W_m$ as a sum of indecomposable kN-modules, it follows from the Krull–Schmidt theorem that $V \mid (W_j)^G$ for some j, and without loss in generality we can take $j = 1$. Let Q be a vertex of W_1 with $Q \subseteq P$, and let U be a source for W_1 at Q. Then $W_1 \mid U^N$, and so $V \mid (U^N)^G = U^G$. This implies that Q contains a vertex of V. However, since $p \nmid \dim_k V$, P is a vertex for V by Theorem 5.4A. Thus $Q = P$, and P is also a vertex for W_1.

(ii) Since P is the only Sylow p-subgroup of N, $x^{-1}Px \nsubseteq N$ for each $x \in G \setminus N$; thus $N \cap x^{-1}Nx$ does not contain a Sylow p-subgroup of G if $x \notin N$. By the Mackey theorem (Theorem 2.1A)

$$(W_1^G)_N = \bigoplus_{i=1}^{n} (W_1 \otimes x_i)^N,$$

where $W_1 \otimes x$ is to be viewed as a $k(x^{-1}Nx \cap N)$-module and $x_1 = 1, x_2, \ldots, x_n$ is a set of representatives for the (N, N) double cosets of G. In

5.6 RESTRICTION OF INDECOMPOSABLE MODULES

particular, $(W_1 \otimes x_1)^N = (W_1 \otimes 1)^N \simeq W_1$ and has P as a vertex by (i). On the other hand, for $i \geq 2$ we have $x_i \notin N$, and the indecomposable components of $(W_1 \otimes x_i)^N$ are all $(x_i^{-1}Nx_i \cap N)$-projective. Since $x_i \notin N$, therefore $x_i^{-1}Nx_i \cap N$ does not contain a Sylow p-subgroup of G. This means that, except for the single component $W_1 \otimes 1$, the indecomposable components of $(W_1^G)_N$ all have vertices that are proper subgroups of the Sylow p-subgroups of G. Since $V \mid (W_1)^G$ by (i), and $V_N = W_1 \oplus W_2 \oplus \cdots \oplus W_m$, we can also conclude that each W_i for $i \geq 2$ has as its vertices proper subgroups of the Sylow p-subgroups of G. By Theorem 5.4A this implies that $p \mid \dim W_i$ for $i = 2, 3, \ldots, m$, and $p \nmid \dim W_1$ because $p \nmid \dim_k V$.

(iii) Let θ_i be the character of N over k afforded by W_i ($i = 1, 2, \ldots, m$). Our hypothesis on x shows that we can apply Theorem 5.4B to W_i with N in place of G, and it follows from (ii) that $\theta_i = 0$ on the set $C_G(P)^\circ$ of p'-elements of $C_G(P)$ for $i = 2, \ldots, m$. Since $\theta = \theta_1$, this shows that $\phi = \theta_1 + \theta_2 + \cdots + \theta_m = \theta$ on $C_G(P)^\circ$ as required.

(iv) Since the degree of ϕ is prime to p, it lies in the principal block iff $\phi(x) = \phi(1)$ for all $x \in C_G(P)^\circ$ (by the example of §4.3). On the other hand, if θ lies in the principal block, then so do all of its constituents. Let ψ be one of the irreducible constituents. For $x \in C_G(P)^\circ$ let $\mathscr{C}$ be the conjugate class of G containing x, c the corresponding class sum, and $h := |\mathscr{C}|$. Since $P \subseteq C_G(x)$, and P is a Sylow p-subgroup, $p \nmid h$. Let Y be a kN-module which affords ψ. Then by Theorem 4.2A we have $vc = hv$ for all $v \in Y$, and so taking traces we obtain $h\psi(x) = h\psi(1)$. Since $p \nmid h$, this shows that $\psi(x) = \psi(1)$ for each $x \in C_G(P)^\circ$. Adding these relations for the various irreducible constituents of θ we obtain $\theta(x) = \theta(1)$ for all $x \in C_G(P)^\circ$ whenever θ lies in the principal block. Hence it follows from (iii) that ϕ lies in the principal block of kG whenever θ lies in the principal block of kN. Finally, suppose that W_1 is trivial. Then $W_1 = (U_0)_N$ when U_0 is the trivial kG-module, and so $U_0 \mid W_1^G$ by Theorem 5.1. If $V \not\simeq U_0$, then $V \oplus U_0 \mid W_1^G$ by Krull–Schmidt theorem, and so $V_N \oplus W_1 \mid (W_1^G)_N$. This implies that $(W_1^G)_N$ has at least two components (isomorphic to W_1) with vertex P, and that is contrary to (ii). Hence $V \simeq U_0$, and the proof of (iv) is complete.

Definition Under the hypothesis of Theorem 5.6A we shall call the uniquely determined indecomposable kN-module W_1 the *derivative* of the irreducible kG-module V.

Theorem 5.6B Let N be a group containing a normal Sylow p-subgroup P and a normal p'-group H. Let k be a splitting field of characteristic p for N and its subgroups, and let U be an indecomposable kN-module. Then we have a decomposition of the form

(1) $$U_{PH} \simeq (X_1 \otimes Y_1) \oplus (X_2 \otimes Y_2) \oplus \cdots \oplus (X_m \otimes Y_m),$$

where the $X_i \otimes Y_i$ are indecomposable $k(PH)$-modules and

(i) H acts trivially on each X_i, so $(X_i)_P$ is an indecomposable kP-module;

(ii) P acts trivially on each Y_i, so $(Y_i)_H$ is an indecomposable kH-module (and therefore irreducible by Maschke's theorem);

(iii) the modules $X_i \otimes Y_i$ ($i = 1, 2, \ldots, m$) are conjugate under the action of N (see §2.2).

Proof Since PH contains a Sylow p-subgroup of N, PH contains a vertex of U, and so U is PH-projective. Therefore by Lemma 5.2 there exists an indecomposable $k(PH)$-module W such that $U \mid W^N$ and $W_{PH} = W_1 \oplus W_2 \oplus \cdots \oplus W_m$, where the W_i are indecomposable and $W_i \mid W \otimes x_i$ for some $x_i \in N$ ($i = 1, 2, \ldots, m$). (Note that in this case; $x_i^{-1} PH x_i \cap PH = PH$ for all x_i.) This implies that each $W_i \simeq W \otimes x_i$, and so the W_i are conjugate under N. It remains to prove that each W_i has the form $X_i \otimes Y_i$ described in (i) and (ii).

Dropping indices, let W be any indecomposable $k(PH)$-module. Since $p \nmid |H|$, W_H is completely reducible and hence a direct sum of irreducible KH-modules. Write $W_H = V_1 \oplus V_2 \oplus \cdots \oplus V_n$ as a direct sum of its homogeneous components (see §2.2). Since P centralizes H, $V_i x = V_i$ for all i and all $x \in P$. Thus the V_i are $k(PH)$-components of W, and so $n = 1$ by the indecomposability of W. This shows that all irreducible constituents of W_H are isomorphic to Y, say. We extend Y to a $k(PH)$-module by defining the action of P on Y as the trivial action. Now put $X := \operatorname{Hom}_{kH}(Y, W_H)$. Since each irreducible constituent of W_H is isomorphic to Y, Note 2 of 1.9 shows that $\dim_k X = \dim_k W / \dim_k Y$ because k is a splitting field for kH. Hence $\dim_k (X \otimes Y) = \dim_k W$. Define an action of PH on X pointwise; namely, $w f^{xy} := (wf) x$ for all $f \in X$, $x \in P$, $y \in H$, and $w \in X$. This defines X as a $k(PH)$-module on which H acts trivially. Now the mapping $f \otimes w \mapsto wf$ of $X \otimes Y$ into W is clearly a $k(PH)$-homomorphism; and since W_H is a direct sum of copies of Y, the mapping is surjective. But we showed that $\dim_k W = \dim_k (X \otimes Y)$, and so this surjective k-linear mapping must in fact be bijective. Hence $X \otimes Y \simeq W$ as a $k(PH)$-module. Clearly X is indecomposable as a kP-module (for otherwise W will not be indecomposable). This completes the proof.

Corollary If, under the hypothesis of the theorem, U_P has at least one component that is the trivial kP-module, then P acts trivially on U. In this case U is irreducible and $p \nmid \dim_k U$.

Proof Let $n_i := \dim_k Y_i$. Then from (1) we have $U_P \simeq n_1 X_1 \oplus n_2 X_2 \oplus \cdots \oplus n_m X_m$. By hypothesis, one of the indecomposable X_i is the trivial kP-module. Hence by (iii) of the theorem, each X_i is the trivial kP-module, so $U_{PH} \simeq Y_1 \oplus Y_2 \oplus \cdots \oplus Y_m$ and P acts trivially on U. Thus U can be

5.7 JORDAN'S THEOREM IN CHARACTERISTIC p

considered as a $k(N/P)$-module and since $p \nmid |N/P|$, the indecomposable module U is in fact irreducible (see Theorem 1.3B). Moreover, $\dim_k U$ divides $|N/P|$ by the corollary of Theorem 3.5, and so $p \nmid \dim_k U$.

5.7 Jordan's theorem in characteristic p

A classical theorem due to C. Jordan (and proved in full generality by I. Schur) states the following:

There exists a constant v_n (depending only on n) such that if G is a group with a faithful ordinary representation of degree n, then G possesses a normal abelian subgroup H such that $|G : H| \le v_n$.

For a proof of this result see Curtis and Reiner [1], §36, or Dixon [1], §5.7.

Note 1 It follows from §3.5 that the same conclusion holds if G is a p'-group and G has a faithful representation of degree n over a field k of characteristic p. The result, however, is no longer true if we drop the hypothesis that G is a p'-group (see the exercise at the end of this section).

The object of the present section is to prove an analog of Jordan's theorem for fields of characteristic p when G is not necessarily a p'-group. This result (Theorem 5.7) is due to Brauer and Feit [2].

Before we state the main theorem of this section we prove three lemmas that are needed in the proof.

We shall write $\mathrm{Triv}(H)$ to denote the trivial irreducible kH-module.

Lemma 5.7A *Suppose that the hypotheses of §3.3 hold, and let U_1 and U_2 be irreducible kG-modules that lie in the same block B of kG.*

(i) *If $p \nmid \dim_k U_i$ ($i = 1, 2$), then some irreducible constituent of $U_1^* \otimes U_2$ lies in the principal block of kG (see §1.4 for the definition of U_1^*).*

(ii) *In the case where $U_1 = U_2$, the trivial kG-module appears as an irreducible constituent of $U_1^* \otimes U_1$; and if $p \mid \dim_k U_1$, then it has multiplicity at least 2.*

Proof (i) Let ϕ^i and η^i ($i = 1, 2, \ldots, r$) be the irreducible modular characters and the principal indecomposable modular characters of G, respectively, where they are enumerated in such a way that ϕ^i and η^i ($i = 1, 2, \ldots, r_1$) are the characters in the block B. Let $\Gamma_1 = [c_{ij}]$ be the $r_1 \times r_1$ Cartan submatrix corresponding to the block B (see §4.4), and put $\Gamma_1^{-1} = [c'_{ij}]$. Then by Theorem 3.7B (using the notation above) we have

(1) $$gc'_{ij} = g\langle \phi^i, \phi^j \rangle_{G^\circ} = \sum_{l=1}^{r} h_l \phi_l^i \phi_{l^*}^j$$

for $i, j = 1, 2, \ldots, r_1$, where $g := |G|$. In particular, since the left-hand side of (1) is rational and the right-hand side of (1) is an algebraic integer, gc'_{ij} is an integer for all i, j. Let ω be the central character of kG associated with B, and let ω_l denote the value of ω on the lth class sum. Then by the definition of the modular characters, reduction modulo π of any of the irreducible modular characters ϕ^i gives a character which is equal (on G°) to an irreducible character $\bar\phi^i$ over k in the same block. Since $\bar\phi^i$ ($i = 1, 2, \ldots, r_1$) lie in B, the corollary of Theorem 4.2A therefore shows that $h_l \bar\phi^i_l = \bar\phi^i(1)\omega_l$ for $l = 1, 2, \ldots, r$ and all ϕ^i lying in B. Hence from (1) we have

$$(2) \qquad \overline{gc'_{ij}} = \phi^i(1) \sum_{l=1}^{r} \omega_l \bar\phi^j_{l*} = \phi^j(1) \sum_{l=1}^{r} \omega_l \bar\phi^i_{l*}$$

for $i, j = 1, 2, \ldots, r_1$. Without loss of generality, suppose that ϕ^1 is the irreducible modular character afforded by U_1. Then $p \nmid \phi^1(1)$ by hypothesis, and so there exists $\alpha \in k$ such that $\alpha\phi^1(1) = \sum_{l=1}^{r} \omega_l \bar\phi^1_{l*}$. Now (2) shows that $\alpha\phi^i(1) = \sum_{l=1}^{r} \omega_l \bar\phi^i_{l*}$ for $i = 1, 2, \ldots, r_1$. Hence it follows from (2) that $\overline{gc'_{ij}} = \phi^i(1)\phi^j(1)\alpha$ for all $i, j = 1, 2, \ldots, r_1$.

We now show that $\alpha \ne 0$. Indeed, from the definition of Γ_1 we have

$$\eta^i(1) = \sum_{j=1}^{r_1} c_{ij}\phi^j(1) \qquad \text{for} \quad i = 1, 2, \ldots, r_1$$

and hence

$$\phi^i(1) = \sum_{j=1}^{r_1} c'_{ij}\eta^j(1) \qquad \text{for} \quad i = 1, 2, \ldots, r_1.$$

However, if p^m is the order of the Sylow p-subgroups of G, then $p^m \mid \eta^j(1)$ for all j (see Example 2 of §5.2), and so $\eta^j(1)/g \in A$. But then we have

$$\overline{\phi^1(1)} = \sum_{j=1}^{r_1} \overline{gc'_{1j}}(\overline{\eta^j(1)/g}),$$

where the left-hand side is nonzero because $p \nmid \phi^1(1)$. Hence $\phi^1(1)\phi^j(1)\alpha = \overline{gc'_{1j}} \ne 0$ for some j, and so we conclude $\alpha \ne 0$. In particular, if ϕ^i is the character afforded by U_2, then $p \nmid \phi^i(1)$ by hypothesis and so $\overline{gc'_{1i}} = \phi^1(1)\phi^i(1)\alpha \ne 0$; this shows that $c'_{1i} \ne 0$.

Now consider the module $U_1^* \otimes U_2$. The modular character afforded by U_1^* is $(\phi^1)^*$ where $(\phi^1)^*(x) := \phi^1(x^{-1})$ for all $x \in G^\circ$. Therefore the modular character afforded by $U_1^* \otimes U_2$ is $(\phi^1)^*\phi^i$, which we can write $(\phi^1)^*\phi^i = \sum_{l=1}^{r} m_l \phi^l$ for some integers m_l. Suppose that $\phi^r = 1_G$ is the trivial modular character. Then by Theorem 3.7B

$$c'_{1i} = \langle \phi^i, \phi^1 \rangle_{G^\circ} = \langle (\phi^1)^*\phi^i, \phi^r \rangle_{G^\circ} = \sum_{l=1}^{r} m_l c'_{lr}.$$

5.7 JORDAN'S THEOREM IN CHARACTERISTIC p

Since $c'_{1i} \neq 0$, there exists l such that $m_l c'_{lr} \neq 0$. Let V be an irreducible kG-module which affords the modular character ϕ^l. Since $m_l \neq 0$, V occurs as an irreducible constituent of $U_1^* \otimes U_2$; and since $c'_{lr} \neq 0$, V lies in the same block as ϕ^r does, namely the principal block (see §4.4).

(ii) Suppose dim $U_1 = n$ and that T is the representation of G afforded by U_1. Then by Example 5 of §1.4, $U_1^* \otimes U_1$ is isomorphic to the kG-module $V := \text{End}_k(U_1)$. The set of scalars in V forms a submodule V_1 of dimension 1 on which G acts trivially so $\text{Triv}(G)$ has multiplicity greater than or equal to 1. Moreover, the elements of V with trace 0 in V form a submodule V_0 of dimension $n^2 - 1 = \dim_k V - 1$, and G has trivial action on V/V_0. If $p \mid n$, then $V_1 \subseteq V_0$, so in this case $\text{Triv}(G)$ has multiplicity greater than or equal to 2.

NOTATION Let P be a Sylow p-subgroup of G. We shall put $N := N_G(P)$ and $C := C_G(P)$. By the Schur–Zassenhaus theorem, $Z(P)$ has a p-complement H in C. Clearly $H \lhd C$, and so $C = H \times Z(P)$ and $PC = H \times P$.

Let V be a kG-module. Then we define $\text{Inv } V := \{v \in V \mid vx = v$ for all $x \in G\}$ as the set of invariant elements of V. Clearly $\text{Inv } V$ is a kG-module which is the sum of all trivial kG-submodules of V.

Note 2 If U and V are kG-modules, then $\text{Inv}(U^* \otimes V) \simeq \text{Hom}_{kG}(U, V)$ as k-spaces. Indeed, if S and T are the representations of G afforded by U and V, then $U^* \otimes V \simeq \text{Hom}_k(U, V)$ with the action $wx := S(x^{-1})wT(x)$ $(x \in G)$. Then $w \in \text{Inv } \text{Hom}_k(U, V)$ iff for all $x \in G$, $S(x)w = wT(x)$. This is equivalent to $w \in \text{Hom}_{kG}(U, V)$.

Lemma 5.7B Suppose that the hypotheses of §3.3 hold. Let U be an irreducible kG-module of dimension $n > 1$ and let B_0 be the principal block of kG. Suppose that $p \nmid |G : G'|$, where G' is the commutator subgroup of G. Then either

(i) some nontrivial irreducible constituent of $U^* \otimes U \otimes U^* \otimes U$ lies in B_0; or

(ii) there exists a nontrivial irreducible constituent U_0 of $U^* \otimes U$ with $p \nmid \dim U_0$ such that the derived kN-module W_0 contains in its kernel a subgroup $L \lhd H$ with $|H : L| \leq v_n$ (where v_n is the constant defined at the beginning of this section).

Proof Suppose (i) does not hold; we must show that (ii) holds. Let U_1, $U_2, \ldots, U_s$ be the distinct irreducible constituents of $U^* \otimes U$ with $U_1 = \text{Triv}(G)$ (see Lemma 5.7A). Let Ω be the class of kG-modules whose irreducible constituents all occur as constituents of $U_i^* \otimes U_j$ $(i, j = 1, 2, \ldots, s)$. Since (i) does not hold, U_1 is the only irreducible module from Ω that lies in B_0. More generally, if V is an indecomposable module from Ω that lies in B_0, then V has a composition series $V = V_0 \supset V_1 \supset \cdots \supset V_m = 0$ such that G

acts trivially on each factor V_i/V_{i+1}, and so the example of §1.5 shows that $G/\mathrm{Ker}\ V$ is a p-group. But $p \nmid |G:G'|$ by hypothesis, and so $\mathrm{Ker}\ V = G$. Thus G acts trivially on V. Hence the multiplicity of $\mathrm{Triv}(G)$ in $W \in \Omega$ is equal to $\dim(\mathrm{Inv}\ W)$. We shall use this observation repeatedly.

Step 1 $U^* \otimes U \simeq \bigoplus_{i=1}^{s} m_i U_i$ *is completely reducible with* $m_1 = 1$; *each of the* U_i *lies in a different block of* kG; *and* $p \nmid \dim U$ *and* $p \nmid \dim U_i$ *for* $i = 1, 2, \ldots, s$.

Let $V := U_i^* \otimes U_j$. Then $\mathrm{Inv}\ V \simeq \mathrm{Hom}_{kG}(U_i, U_j)$. Then by Schur's lemma, $\dim(\mathrm{Inv}\ V) = 1$ if $i = j$ and 0 otherwise. Hence $\mathrm{Triv}(G)$ has multiplicity 1 in $U_i^* \otimes U_i$ and 0 in $U_i^* \otimes U_j$ when $i \neq j$. Lemma 5.7A(ii) then shows that $p \nmid \dim U_i$. Since (i) does not hold, Lemma 5.7A(i) shows that the U_i lie in different blocks.

Now we show that $U^* \otimes U$ is completely reducible. Let W be an indecomposable submodule of $U^* \otimes U$ and suppose W has a composition series of length h. Since W is indecomposable, its irreducible constituents all lie in the same block and belong to Ω, so by what we have just shown they are all isomorphic to U_i for some i. Hence the multiplicity of $\mathrm{Triv}(G)$ as an irreducible constituent of $U_i^* \otimes W$ is equal to the number h of composition factors of W by the first part of this step. Thus, $h = \dim_k \mathrm{Inv}(U_i^* \otimes W) = \dim_k \mathrm{Hom}_{kG}(U_i, W)$. Now let W_0 denote the maximal completely reducible submodule of W; so W_0 is a sum of all irreducible submodules of W. If W_0 has a composition series of length h_0, then we find as before that $h_0 = \dim_k \mathrm{Hom}_{kG}(U_i, W_0)$. However, each kG-homomorphism of U_i into W maps U_i onto 0 or else onto an irreducible submodule of W (because U_i is irreducible). Therefore $\mathrm{Hom}_{kG}(U_i, W) \simeq \mathrm{Hom}_{kG}(U_i, W_0)$, which implies that $h = h_0$ and $W = W_0$. Thus each indecomposable component of $U^* \otimes U$ is completely reducible; and so $U^* \otimes U$ is completely reducible.

Finally $m_1 = 1$ and $p \nmid \dim U$. Indeed, $U^* \times U \simeq U_1 \times (U^* \times U)$ belongs to Ω and so the multiplicity m_1 of $U_1 = \mathrm{Triv}(G)$ in $U^* \otimes U$ equals $\dim_k \mathrm{Inv}(U^* \otimes U) = \dim_k \mathrm{Hom}_{kG}(U, U)$. Therefore since k is a splitting field, $m_1 = 1$. By Lemma 5.7A, $p \nmid \dim_k U$.

We now let W and W_i $(i = 1, 2, \ldots, s)$ denote the derivations of U and U_i $(i = 1, 2, \ldots, s)$, respectively, in the sense of Theorem 5.6A. (Step 1 shows that the hypotheses of Theorem 5.6A are satisfied.) Theorem 5.6A shows that W is not the trivial kN-module and that none of the W_i $(i > 1)$ lie in the principal block of kN. Moreover, $p \nmid \dim W_i$ for each i, and since $\dim_k U = n > 1$, we have $s \geq 2$.

Step 2 *There exist* kN-*modules* W' *and* W'' *which are sums of indecomposables with vertices properly contained in* P *such that* $(W^* \otimes W) \oplus W' \simeq (\bigoplus_{i=1}^{s} m_i W_i) \oplus W''$.

5.7 JORDAN'S THEOREM IN CHARACTERISTIC p

By Theorem 5.6A, $U_N = W \oplus V$, where V is a sum of indecomposable modules that are Q-projective for certain proper subgroups Q of P. Then $(U^* \otimes U)_N = (W^* \otimes W) \oplus W'$, where $W' := W^* \otimes V \oplus W \otimes V^* \oplus V \otimes V^*$. It follows from Example 4 of §5.1 that each indecomposable component of W' is Q-projective for some proper subgroup Q of P; so these components have vertices that are properly contained in P. On the other hand, Theorem 5.6A shows that $(U_i)_N = W_i \oplus V_i$, where the indecomposable components of V_i all have vertices properly contained in P. Thus Step 2 follows from Step 1 with $W'' := \bigoplus_i V_i$.

Step 3 $W_{PH} \simeq X \otimes Y$, where X is a $k(PH)$-module on which H acts trivially and Y is an irreducible $k(PH)$-module on which P acts trivially. Moreover, $\mathrm{Triv}(PH) \mid (W^* \otimes W)_{PH}$ and dim $Y > 1$.

It follows from Theorem 5.6B (using the notation there) that W_{PH} has the form $W_{PH} \simeq \bigoplus_{i=1}^m X_i \otimes Y_i$ for some $m \geq 1$. Collecting like terms we may suppose that the Y_i are mutually nonisomorphic (but the X_i may be decomposable). Clearly the $X_i \otimes Y_i$ are still conjugate under N. We now have

$$(W^* \otimes W)_{PH} = \bigoplus_{i,j=1}^m (X_i^* \otimes X_j) \otimes (Y_i^* \otimes Y_j)$$

and $Y_i^* \otimes Y_j$ has an irreducible constituent $\mathrm{Triv}(PH)$ iff $i = j$. This shows that there is a $k(PH)$-module W'''' such that

$$(W^* \otimes W)_{PH} = \bigoplus_{i=1}^m (X_i^* \otimes X_i) \oplus W'''',$$

where $\mathrm{Triv}(PH)$ is not an indecomposable component of W''''. Now $\mathrm{Triv}(N) = W_1$ has P as a vertex, so Step 2 shows that $\mathrm{Triv}(N) \mid W^* \otimes W$. Hence $\mathrm{Triv}(PH) \mid (W^* \otimes W)_{PH}$, which shows that for some i, $\mathrm{Triv}(PH) \mid X_i^* \otimes X_i$. Since the $X_i \otimes Y_i$ are conjugate under N, it follows that $\mathrm{Triv}(PH) \mid X_i^* \otimes X_i$ for each i, and so $m\,\mathrm{Triv}(PH) \mid (W^* \otimes W)_{PH}$.

Now suppose $m > 1$. If V is an indecomposable kN-module such that $\mathrm{Triv}(PH) \mid V_{PH}$, then it follows from the corollary of Theorem 5.6B that $p \nmid \dim V$. In particular, P is a vertex of V by Theorem 5.4A. Thus if $m > 1$, then by Step 2, $\mathrm{Triv}(PH) \mid (W_j)_{PH}$ for some $j \geq 2$ (recall that $m_1 = 1$). Then the corollary of Theorem 5.6B shows again that W_j is irreducible. Since $(W_j)_H$ has at least one constituent that is trivial, it follows from the corollary of Clifford's theorem (Theorem 2.2A) that H is in the kernel of W_j. Since $p \nmid \dim W_j$ by the definition of the derivative, the example of §4.3 shows that W_j lies in the principal block of kN. But then U_j lies in B_0 by Theorem 5.6A, which is impossible by Step 1 for $j \geq 2$. This shows that $m = 1$. We showed above that $\mathrm{Triv}(PH) \mid (W^* \otimes W)_{PH}$.

Finally, if dim $Y_1 = 1$, then $Y_1^* \otimes Y_1 \simeq \mathrm{Triv}(H)$, and so H is contained in the kernel of $W^* \otimes W$. But by Step 2 each $W_i \mid W^* \otimes W$, and so H lies in the kernel of each W_i. As we saw above this is impossible for $i \geq 2$ because U_i does not lie in B_0. Thus we conclude dim $Y_1 \geq 2$.

By the Schur–Zassenhaus theorem the normal Sylow p-subgroup P has a complement L_0 in N. Let K_0 be the kernel of the action of L_0 on W. Then L_0/K_0 is a p'-group with a faithful representation of degree less than or equal to n over k, and so by the classical theorem of Jordan stated in the beginning of this section, L_0/K_0 has a normal abelian subgroup L_1/K_0 of index v_n. Note that $L_1 P \triangleleft N$. Put $L := L_1 P \cap H$. Then $|H:L| \leq v_n$ and $L \triangleleft N$. Moreover if K is the kernel of the action of N on W, then $L/L \cap K$ is an abelian p'-group.

Step 4 *For some $j \geq 2$, L is contained in the kernel of W_j.*

Let $n_1 = \dim Y$ (where we write Y for Y_1); then $n_1 \geq 2$ by Step 3. Since $L/L \cap K$ is an abelian p'-group, Y_L is a direct sum of n_1 components each of dimension 1, and so $Y_L^* \otimes Y_L$ has the constituent Triv(L) with multiplicity greater than or equal to $n_1 > 1$. Since Y_H is irreducible, Triv(H) has multiplicity 1 in $(Y^* \otimes Y)_H$; therefore there is an irreducible constituent V of $Y^* \otimes Y$ such that Triv(H) is not a constituent of V_H, but Triv(L) is a constituent of V_L. Since P lies in the kernel of Y, P also lies in the kernel of V, and by Clifford's theorem (Theorem 2.2A), L is contained in the kernel of V. Note that $V \mid Y^* \otimes Y$ because $Y^* \otimes Y$ is completely reducible (P acts trivially on Y).

Now Triv$(P) \mid (W^* \otimes W)_P \simeq (\dim Y)^2 (X^* \otimes X)$, so Triv$(P) \mid (X^* \otimes X)$ by Step 3, and $V \simeq \text{Triv}(P) \otimes V \mid (X^* \otimes X) \otimes (Y^* \otimes Y) \simeq (W^* \otimes W)_{PH}$. Since P lies in the kernel of V, V has P as a vertex, and so by Step 2, $V \mid (W_j)_{PH}$ for some j; by hypothesis, $j \geq 2$ because $V \neq \text{Triv}(PH)$. However Triv$(L) \mid V_L$, and so Triv(L) is a constituent of $(W_j)_L$. Since $L \triangleleft N$, Clifford's theorem shows that L is in the kernel of W_j.

The proof of the lemma is now complete if we take U_j for U_0 and W_j for W_0.

Lemma 5.7C *Suppose the hypotheses of Lemma 5.7B hold, and that the Sylow p-subgroup P of G has order p^d. Put $\lambda_{n, p^d} := |GL(p^{2d}v_n n^4, p)|$. Then there exists a nontrivial irreducible constituent V of $U^* \otimes U \otimes U^* \otimes U$ such that $|G:\text{Ker } V| \leq \lambda_{n, p^d}$.*

Proof Suppose that V is any irreducible kG-module with $m := \dim_k V$. Let ϕ be the character afforded by V over k. Suppose that ϕ has l algebraic conjugate characters (see the example of §2.7). Then the values of the character ϕ must all lie in a field k_1 which has degree l over the prime subfield of k; in particular, $|k_1| = p^l$. Hence by Theorem 2.7B, ϕ is afforded by a $k_1 G$-module of dimension m, and so

$$|G/\text{Ker } V| \leq |GL(m, p^l)| \leq |GL(ml, p)|.$$

5.7 JORDAN'S THEOREM IN CHARACTERSTIC p

Then to prove the lemma, it is sufficient to show that there exists a nontrivial irreducible constituent V of $U^* \otimes U \otimes U^* \otimes U$ which affords a character ϕ with l algebraically conjugate characters, where $l \leq p^{2d}v_n$ (since $\dim_k V \leq (\dim_k U)^4 = n^4$). Since the hypotheses of Lemma 5.7B hold, at least one of the conclusions (i) and (ii) is true; we consider these two possibilities.

Suppose conclusion (i) of Lemma 5.7B holds. Then choose V as a nontrivial irreducible constituent of $U^* \otimes U \otimes U^* \otimes U$ from the principal block B_0. By the example of §4.2 all algebraic conjugate characters of ϕ also lie in B_0. By Corollary 1 of Theorem 4.5B this means that ϕ has at most $\frac{1}{4}p^{2d} + 1 \leq p^{2d}$ algebraic conjugates, and so the lemma is proved in this case.

Next suppose that conclusion (ii) of Lemma 5.7B holds, and choose $V = U_0$. Since $\mathrm{Triv}(G)$ and U_0 are both constituents of $U^* \otimes U$, this shows that V is a constituent of $U^* \otimes U \otimes U^* \otimes U$. Let ϕ be the character afforded by V and θ be the character of $N_G(P)$ afforded by the derivative W_0 of V. Then $\phi = \theta$ on $C_G(P)^\circ$ by Theorem 5.6A. In the notation of Lemma 5.7B, $C_G(P) = H \times Z(P)$, and so $C_G(P)^\circ = H$. Put $t := |H : H \cap \mathrm{Ker}\ W_0|$. Then the values of θ (and hence of ϕ) on H will be sums of tth roots of unity in k and hence all lie in a subfield k_2 of k of degree less than or equal to t over the prime subfield; in particular, $|k_2| \leq p^t$. Now let $\phi = \phi^1, \ldots, \phi^l$ be the algebraic conjugate characters of ϕ. If $\phi_H^i = \phi_H^j$, then ϕ^i and ϕ^j lie in the same block (see the example of §4.3), and so not more than p^{2d} of the ϕ^i can have the same restriction to H by the corollary of Theorem 4.5B. On the other hand, since the values of ϕ_H all lie in k_2, ϕ_H has at most t algebraic conjugates; hence the number of different restrictions ϕ_H^i ($i = 1, 2, \ldots, l$) is at most t. This shows that $l \leq p^{2d}t$. Since $t \leq v_n$ by (ii) of Lemma 5.7B, the lemma is proved in this case as well.

Theorem 5.7 There exists a constant $v_{n,\ p^d}$ (depending only on n and p^d) such that if G is a group with a Sylow p-subgroup of order p^d and G has a faithful representation of degree $d \leq n$ over a field k of characteristic p, then G possesses a normal abelian subgroup of index $\leq v_{n,\ p^d}$.

Proof Without loss in generality we can suppose that k satisfies the hypotheses of §3.3. Let U be a faithful kG-module of dimension n. For each subgroup H of G we define the "type" $t(H)$ to be the pair (d', m'), where $p^{d'}$ is the order of the Sylow p-subgroup of H and m' is the multiplicity of $\mathrm{Triv}(H)$ in $U_H^* \otimes U_H \otimes U_H^* \otimes U_H$. We order the set of types by writing $(d', m') < (d'', m'')$ to mean either $d' < d''$ or $d' = d''$ and $m' > m''$. Since $m' \leq n^4$, there are at most dn^4 types.

We shall first prove that if H is a subgroup of G and $t(H) = (d', m')$ with $d' > 0$, then there is a subgroup L of H such that $t(L) < t(H)$ and $|H : L| \leq \lambda_{n,\ p^d}$ (where $\lambda_{n,\ p^d}$ is defined in Lemma 5.7C). If $p \mid |H : H'|$, then H has a normal subgroup L of index p, and L clearly satisfies the requirements. Thus suppose $p \nmid |H:H'|$. If the irreducible constituents of U_H all have dimension 1, then we have a kH-composition series

$$U_H = W_0 \supset W_1 \supset \cdots \supset W_n = 0$$

with $\dim(W_{i-1}/W_i) = 1$ for $i = 1, 2, \ldots, n$. Then $x^{-1}y^{-1}xy$ acts trivially on each W_{i-1}/W_i for all $x, y \in H$, and so the same is true for each $z \in H'$. Now the example of §1.5 shows that $H'/\mathrm{Ker}\, U_{H'}$ is a p-group. Since H acts faithfully on U, therefore $\mathrm{Ker}\, U_{H'} = 1$, and so H' is a p-group. Thus we conclude that the Sylow p-subgroup of H is normal in H. Hence by the Schur–Zassenhaus theorem, H has a p-complement L. Because L is a p'-group, $t(L) < t(H)$, and because L is a p-complement, $|H:L| = p^{d'} < \lambda_{n,p^d}$. This settles the case where all constituents of U_H are one-dimensional. If U_H has an irreducible constituent of dimension greater than 1, we can apply Lemma 5.7C to this constituent to obtain a nontrivial irreducible kH-constituent V of $U_H^* \otimes U_H \otimes U_H^* \otimes U_H$ with $|H : \mathrm{Ker}\, V| \le \lambda_{n,p^{d'}} \le \lambda_{n,p^d}$. Taking $L = \mathrm{Ker}\, V$, it follows that $t(L) < t(H)$, and the assertion is proved in this case as well.

Using what we have just proved, we know that there exists a chain of subgroups $G = H_0 \supset H_1 \supset \cdots \supset H_l$ such that $t(H_{i-1}) > t(H_i)$ and $|H_{i-1} : H_i| \le \lambda_{n,p^d}$ for each i. Since there are at most dn^4 types of subgroups, there exists $l \le dn^4$ such that $t(H_l) = (0, m)$ for some $m \le n^4$. Then H_l is a p'-group, and so by the classical theorem of Jordan (see the beginning of this section), H_l has an abelian subgroup S_0 of index $\le v_n$. Then $|G : S_0| \le (\lambda_{n,p^d})^4 v_n = \mu_{n,p^d}$, say. Finally, the subgroup S_0 has at most μ_{n,p^d} conjugates in G and $S := \bigcap_{x \in G} x^{-1} S_0 x$ has index less than or equal to $(\mu_{n,p^d})!$ in G. Putting $v_{n,p^d} := (\mu_{n,p^d})!$, the proof is completed.

EXERCISE

Let k be an algebraically closed field of characteristic $p > 0$ and set $G_m := SL(2, p^m)$, the special linear group of degree 2 over the finite field of p^m elements. Let $GL(2, k)$ be the general linear group of degree 2 over k. Then $G_m \subseteq GL(2, k)$. Show that a normal abelian subgroup of G_m has order at most 2 while the order $|G_m|$ of G_m is arbitrarily large.

5.8 Notes and comments

The theory of indecomposable kG-modules as presented here is essentially due to Green [1,2]. A general setting for a part of the representation theory of finite groups, namely the part that includes the theory of vertices of modular representations and defect groups of blocks, is given in Green [3]. This theory has proved to be very useful in the theory of classification of indecomposable modules as well as in the structure theory of groups; for example, see Dade [2] and Thompson [1]. One of its uses will become apparent in Chapter 8, where the structure of a block with a cyclic defect group is studied in detail. A detailed treatment of the theory of indecomposable modules is given in Feit [2]. The material of the Section 5.7 is due to Brauer and Feit [2].

CHAPTER VI

The Main Theorems of Brauer

Throughout this chapter we shall keep the notation introduced in §3.3. In particular, we have an integral domain A with a unique maximal ideal πA, the field K of quotients of A of characteristic 0, and the residue class field $k := A/\pi A$ of characteristic p. Moreover, A is a principal ideal domain and both K and k are splitting fields for the group G and all of its subgroups.

The theorems of this chapter will investigate various connections between the theory of blocks of kG and the theory of blocks of kH for certain subgroups and quotient groups H of G.

6.1 The Brauer homomorphism

The Brauer homomorphism which we describe now is an important tool in our investigation. Recall that for any subgroup Q of G, $C_G(Q) \triangleleft N_G(Q)$. This observation will be used repeatedly in the following sections.

Lemma 6.1A Let Q be a p-subgroup of the group G, and put $N := N_G(Q)$. Then the mapping σ defined by

(1) $$\sigma\left(\sum_{x \in G} \alpha_x x\right) := \sum_{x \in C_G(Q)} \alpha_x x$$

is a k-algebra homomorphism from $Z(kG)$ into $Z(kN)$.

Remark We shall call σ the *Brauer homomorphism* with respect to Q. It is clear from the proof below that the analogous function from $Z(AG)$ into

6.1 THE BRAUER HOMOMORPHISM

$Z(AN)$ can also be defined and is an A-linear mapping; but the latter is not usually an algebra homomorphism. Note that when we interpret the elements $\sum_{x \in G} \alpha_x x$ of $Z(kG)$ as class functions $x \mapsto \alpha_x$ from G into k, then σ is simply restriction to $C_G(Q)$.

Proof We first show that σ maps $Z(kG)$ into $Z(kN)$. Let $\mathscr{C}_i$ be any conjugate class of G with class sum c_i. Then $\mathscr{C}_i^* := \mathscr{C}_i \cap C_G(Q)$ is a (possibly empty) normal subset of N since $C_G(Q) \triangleleft N$. Thus $\mathscr{C}_i^*$ is a union of conjugate classes of N, and so $\sigma(c_i) = \sum_{x \in \mathscr{C}_i^*} x \in Z(kN)$. Since $c_1, c_2, \ldots, c_s$ is a k-basis of $Z(kG)$, this shows that σ is a function of $Z(kG)$ into $Z(kN)$.

Clearly, σ is a k-linear mapping and $\sigma(1) = 1$, so it remains to show that for any class sums c_i, c_j we have $\sigma(c_i c_j) = \sigma(c_i)\sigma(c_j)$. However, $\sigma(c_i c_j) = \sum_{z \in G} v_z z$ and $\sigma(c_i)\sigma(c_j) = \sum_{z \in G} v_z^* z$, where v_z is the number of elements of the set $S_z := \{(x, y) \in \mathscr{C}_i \times \mathscr{C}_j \mid xy = z\}$ and v_z^* is the number of elements of the set $S_z^* := \{(x, y) \in \mathscr{C}_i^* \times \mathscr{C}_j^* \mid xy = z\}$. Since $S_z^* \subseteq S_z$ and

$$S_z \setminus S_z^* = \{(x, y) \in (\mathscr{C}_i \setminus \mathscr{C}_i^*) \times \mathscr{C}_j \mid xy = z\} \cup \{(x, y) \in \mathscr{C}_i^* \times (\mathscr{C}_j \setminus \mathscr{C}_j^*) \mid xy = z\}$$

as a disjoint union, it follows from Lemma 4.3A that $v_z = |S_z| \equiv |S_z^*| = v_z^*$ (mod p) for all $z \in G$. Since k has characteristic p, this shows that $\sigma(c_i c_j) = \sigma(c_i)\sigma(c_j)$ as required.

Lemma 6.1B *Let Q be a p-subgroup of the group G and $\mathscr{C}$ a conjugate class in G. Put $\mathscr{C}^* := \mathscr{C} \cap C_G(Q)$, $N := N_G(Q)$ and let σ denote the Brauer homomorphism with respect to Q. Then*

(i) $\mathscr{C}^* \neq \phi$ iff Q is contained in some defect group of $\mathscr{C}$.
(ii) $\mathscr{C}^*$ is a single conjugate class in N if Q is a defect group of $\mathscr{C}$.
(iii) Suppose that $\mathscr{C}'$ is a conjugate class in N with $\mathscr{C}' \subseteq \mathscr{C}$. If Q is a defect group of $\mathscr{C}'$, then Q is a defect group of $\mathscr{C}$ and $\mathscr{C}' = \mathscr{C} \cap C_G(Q)$.
(iv) Let f be a block idempotent of kG with a defect group containing Q, then $\sigma(f) \neq 0$ and is a central idempotent of kN. If f' is a second block idempotent of kG with a defect group containing Q, and $f \neq f'$, then $\sigma(f)$ and $\sigma(f')$ are orthogonal idempotents in $Z(kN)$.

Proof (i) $\mathscr{C}^* \neq \phi$ iff $x \in C_G(Q)$ for some $x \in \mathscr{C}$. The latter condition is equivalent to $Q \subseteq C_G(x)$ for some $x \in \mathscr{C}$, and this holds iff Q is contained in some defect group of $\mathscr{C}$.

(ii) If Q is a defect group for $\mathscr{C}$, then Q is a Sylow p-subgroup of $C_G(x)$ for some $x \in \mathscr{C}$; in particular, $x \in C_G(Q) \cap \mathscr{C} = \mathscr{C}^*$. Suppose that y also lies in $\mathscr{C}^*$. Then $y = z^{-1}xz$ for some $z \in G$, and $Q \subseteq C_G(y)$, so $z^{-1}Qz \subseteq C_G(zyz^{-1}) = C_G(x)$. Since both Q and $z^{-1}Qz$ are Sylow p-subgroups of $C_G(x)$, there exists $w \in C_G(x)$ such that $wzQz^{-1}w^{-1} = Q$. Then $wz \in N = N_G(Q)$, and $(wz)^{-1}x(wz) = y$, so y is conjugate to x in N. Thus $\mathscr{C}^*$ is a single conjugate class of N.

(iii) Suppose that Q is not a defect group of $\mathscr{C}$. Then for some $x \in \mathscr{C}'$ we have Q as a Sylow p-subgroup of $C_N(x)$, but $C_G(x)$ has a Sylow p-subgroup $D \supset Q$. Now in any p-group each proper subgroup is properly contained in its normalizer; therefore there exists $y \in D\backslash Q$ with $y \in N$. Then $\langle Q, y \rangle \subseteq C_N(x)$, and so Q is not a Sylow p-subgroup of $C_N(x)$, contrary to our hypothesis. This shows that Q must be a defect group of $\mathscr{C}$. Since $\mathscr{C}^* \supseteq \mathscr{C}'$, it now follows from (ii) that $\mathscr{C}' = \mathscr{C}^*$.

(iv) Let $\mathscr{C}_1, \mathscr{C}_2, \ldots, \mathscr{C}_s$ be the conjugate classes of G and $c_1, c_2, \ldots, c_s$ the corresponding class sums. Since $f \in Z(kG)$ we can write $f = \sum_{i=1}^{s} \gamma_i c_i$ for some $\gamma_i \in k$, and by Lemma 4.3B we know that for some j, $\gamma_j \neq 0$ and the class $\mathscr{C}_j$ has a defect group containing Q. Then $\sigma(c_j) \neq 0$ by (i) and it follows (since the c_j are linearly independent) that $\sigma(f) \neq 0$. Because σ is an algebra homomorphism, this shows that $\sigma(f)$ is an idempotent in $Z(kN)$. Similarly, $\sigma(f')$ is an idempotent in $Z(kN)$. Since f and f' are distinct block idempotents, then $ff' = 0$ and therefore $\sigma(f)\sigma(f') = 0$ so $\sigma(f)$ and $\sigma(f')$ are orthogonal.

6.2 Blocks with normal p-subgroups

In the present section we shall examine the situation where the group G contains a normal p-subgroup Q and relate the set of blocks of kG with the set of blocks of $k\tilde{G}$ where $\tilde{G} := G/Q$. In the following section we shall show how to reduce the general case to this situation.

Remark The natural mapping $G \to G/Q = \tilde{G}$ induces a ring homomorphism $\beta: kG \to k\tilde{G}$. We shall use this frequently and it is important to note that Ker $\beta \subseteq \text{rad}(kG)$ and so is a nilpotent ideal. Indeed, by the definition of $\text{rad}(kG)$ it is sufficient to show that $V \text{Ker } \beta = 0$ for each irreducible kG-module V (see §1.3). But when V is an irreducible kG-module, Q acts trivially on the kQ-module V_Q, since Q is a normal p-subgroup of G and k has characteristic p (see the corollary of Theorem 1.5 and the corollary of Theorem 2.2A). Thus $V(x - 1) = 0$ for $x \in Q$. Since Ker β is generated as an ideal by the set $\{(x - 1) \mid x \in Q\}$, $V \text{Ker } \beta = 0$ as required.

Note In the case that Q is a normal p-subgroup of G, $C_G(Q) \triangleleft G$, and so for each conjugate class $\mathscr{C}$ of G, $\mathscr{C} \cap C_G(Q) = \mathscr{C}$ or $\varnothing$. In this situation the results of Lemma 6.1B take an especially simple form.

Lemma 6.2A Let Q be a normal p-subgroup of the group G and let I_Q be the ideal in kG generated by all class sums c of classes with defect groups conjugate to subgroups of Q (see §4.3). Let J_Q be the k-subspace of I_Q spanned by those c corresponding to classes that have Q (and its conjugates) as defect groups. Then

6.2 BLOCKS WITH NORMAL p-SUBGROUPS

(i) $J_Q J_Q \subseteq J_Q$;

(ii) if $\mathscr{C}$ is any class in G which has a defect group D not containing Q, the corresponding class sum c lies in rad $Z(kG)$;

(iii) $I_Q \subseteq J_Q + \text{rad } Z(kG)$;

(iv) Q is contained in each defect group of each block of kG;

(v) if f is a block idempotent of kG, and the block $f(kG)$ has Q as its (unique) defect group, then $f \in J_Q$.

Proof Let σ denote the Brauer homomorphism with respect to Q. Then σ is an endomorphism of the algebra $Z(kG)$. By Lemma 6.1B [parts (i) and (ii)], for any class sum $c \in I_Q$, $\sigma(c) = c$ if the corresponding class $\mathscr{C}$ has Q as a defect group and $\sigma(c) = 0$ otherwise (see the note above).

(i) $J_Q = \sigma(I_Q)$ and so $J_Q J_Q = \sigma(I_Q^2) \subseteq \sigma(I_Q) = J_Q$.

(ii) By hypothesis there exists $x \in \mathscr{C}$ such that Q is not contained in $C_G(x)$. (Actually, if this happens for one $x \in \mathscr{C}$ it happens for all $x \in \mathscr{C}$ because Q is normal.) Since $Q \triangleleft G$, $C_G(x)Q$ is a subgroup of G and $|C_G(x)Q : C_G(x)| = |Q : C_G(x) \cap Q| = q$, say, is a nontrivial power of p. Choose a right transversal of $C_G(x)$ in G of the form $y_i z_j$ ($i = 1, 2, \ldots, q$; $j = 1, 2, \ldots, n$), where $y_1, y_2, \ldots, y_q$ is a right transversal for $C_G(x)$ in $C_G(x)Q$, and $z_1, z_2, \ldots, z_n$ is a right transversal of $C_G(x)Q$ in G. Then

$$c = \sum_{j=1}^{n} \sum_{i=1}^{q} z_j^{-1} y_i^{-1} x y_i z_j.$$

If we apply the homomorphism β described in the remark above we obtain

$$\beta(c) = \sum_{j=1}^{n} q\beta(z_j^{-1})\beta(x)\beta(z_j) = 0,$$

since $p \mid q$. Thus $c \in \text{Ker } \beta \subseteq \text{rad } kG$. Since $c \in Z(kG)$, this shows that $c \in \text{rad } Z(kG)$.

(iii) This follows immediately from (ii) and the definition of I_Q and J_Q.

(iv) Let ω be the central character kG associated with a block B of kG. By Theorem 4.3 there exists a class $\mathscr{C}$ in G with class sum c such that $\omega(c) \neq 0$ and the defect groups of $\mathscr{C}$ are the defect groups of B. Since $\omega(c)$ is a nonzero element of k, this implies that c is not nilpotent and so $c \notin \text{rad } Z(kG)$. Hence by (ii) the defect groups of $\mathscr{C}$ (and hence of B) all contain Q.

(v) Let $\mathscr{C}_1, \mathscr{C}_2, \ldots, \mathscr{C}_s$ be the conjugate classes of G and $c_1, c_2, \ldots, c_s$ their class sums. By Lemma 4.3A we can write $f = \sum_{i=1}^{s} \gamma_i c_i$, where $\gamma_i = 0$, unless some defect group of $\mathscr{C}_i$ contains the defect group Q of the block $f(kG)$. Hence by definition of I_Q, $f \in I_Q$. By (iii) we can write $f = a + b$, where $a \in J_Q$, and $b \in \text{rad } Z(kG)$. Since f is an idempotent and $Z(kG)$ is commutative, the binomial theorem shows that $f = f^{p^t} = a^{p^t} + b^{p^t}$ for each integer $t \geq 0$. Since $b \in \text{rad } Z(kG)$, it is nilpotent, and so for sufficiently large t, $f = a^{p^t} \in J_Q$ by (i).

Lemma 6.2B Let Q be a normal p-subgroup of a group G and let $\tilde{G} := G/Q$. For each conjugate class $\mathscr{C}$ in G we define $\tilde{\mathscr{C}} := \{Qx \mid x \in \mathscr{C}\}$ in $\tilde{G}$. Put $H := QC_G(Q)$ and $\tilde{H} := H/Q$. Then

(i) Let $\mathscr{C}$ be a conjugate class of p'-elements in G and suppose a defect group D of $\mathscr{C}$ contains Q. Then D/Q is a defect group of $\tilde{\mathscr{C}}$.

(ii) The mapping $\mathscr{C} \mapsto \tilde{\mathscr{C}}$ gives a bijection from the set of all classes of p'-elements in G with Q as defect group onto the set of all classes of p'-elements in G that lie in H and have 1 as defect group.

Proof (i) Choose $x \in \mathscr{C}$ such that D is a Sylow p-subgroup in $C_G(x)$. Then D/Q is a Sylow p-subgroup in $C_G(x)/Q$ and clearly $C_G(x)/Q \subseteq C_{\tilde{G}}(Qx)$. We claim that $C_{\tilde{G}}(x)/Q$ contains all p-elements of $C_{\tilde{G}}(Qx)$; it will then follow that D/Q is a Sylow p-subgroup of $C_{\tilde{G}}(Qx)$ and hence a defect group of $\tilde{\mathscr{C}}$. Let Qy be any p-element in $C_{\tilde{G}}(Qx)$. Then Qy commutes with Qx, and so $x^{-1}y^{-1}xy \in Q$. This means $y^{-1}xy = xz$ for some $z \in Q$, and $zx = xz$ since $Q \subseteq D \subseteq C_G(x)$. But x and $y^{-1}xy$ are p'-elements, and z is a p-element, so we conclude that $z = 1$. Hence $y \in C_G(x)$. This shows that $Qy \in C_G(x)/Q$ as claimed, and (i) is proved.

(ii) Part (i) shows that the given mapping maps a class $\mathscr{C}$ of p'-elements of G with a defect group Q onto a class $\tilde{\mathscr{C}}$ of p'-elements of $\tilde{G}$ with defect group 1. Moreover, since $Q \subseteq C_G(x)$ for all $x \in \mathscr{C}$, therefore $\mathscr{C} \subseteq C_G(Q)$ and $\tilde{\mathscr{C}} \subseteq \tilde{H}$. It remains to prove that the mapping gives a bijection between the two sets.

Let $\mathscr{C}_1$, $\mathscr{C}_2$ be two classes of p'-elements of G with Q as defect group. Since $Q \triangleleft G$, Q is the unique defect group, and so for each $x \in \mathscr{C}_i$, $Q \subseteq C_G(x)$; hence x is the unique p'-element in the coset Qx. Thus if $\mathscr{C}_1 \neq \mathscr{C}_2$, then $\tilde{\mathscr{C}}_1 \neq \tilde{\mathscr{C}}_2$; hence the map is injective. On the other hand, let $\mathscr{C}'$ be a class of p'-elements in $\tilde{G}$ with 1 as a defect group with $\mathscr{C}' \subseteq \tilde{H}$. Each element of $\mathscr{C}'$ can be written in the form Qx, where x is a p'-element of G and $x \in H$ because $\mathscr{C}' \subseteq \tilde{H}$. If $p^t := |Q|$, then $x^{p^t} \in C_G(Q)$, since $C_G(Q)$ is a normal subgroup of index dividing p^t in H. Again x is a p'-element and $x \in \langle x^{p^t} \rangle$, so $x \in C_G(Q)$. Thus the class $\mathscr{C}$ of G containing x has a defect group $D \supseteq Q$ because $Q \subseteq C_G(x)$. Now clearly $\tilde{\mathscr{C}} = \mathscr{C}'$, and by (i) this means $\mathscr{C}'$ has D/Q as a defect group. By hypothesis, 1 is a defect group of $\mathscr{C}'$, so we conclude $D = Q$. This shows that the mapping is surjective, and (ii) is proved.

Theorem 6.2A Let Q be a normal p-subgroup of G and suppose that $G = QC_G(Q)$. Put $\tilde{G} := G/Q$. Let $\beta: kG \to k\tilde{G}$ be the ring homomorphism induced by the canonical mapping $G \to G/Q = \tilde{G}$ (see the remark above). Then β maps the set $f_1, f_2, \ldots, f_t$ of block idempotents of kG bijectively onto

6.2 BLOCKS WITH NORMAL p-SUBGROUPS

the set of block idempotents of $k\tilde{G}$. Moreover, if the block $f_i(kG)$ has D as a defect group, then the block $\beta(f_i)(k\tilde{G})$ has D/Q as a defect group ($i = 1, 2, \ldots, t$).

Proof Since β is an algebra homomorphism of kG into $k\tilde{G}$, we have the decomposition $1 = \beta(f_1) + \beta(f_2) + \cdots + \beta(f_t)$ into central idempotents of $k\tilde{G}$. [By the remark above, Ker β is nilpotent, so each $\beta(f_i) \neq 0$.] Therefore, to prove the first assertion of the theorem, it remains to show that each $\beta(f_i)$ is a block idempotent. Suppose on the contrary that $\beta(f_1)$, say, is not primitive. Then $\beta(f_1) = e_1 + e_2 + \cdots + e_n$ ($n \geq 2$) as a sum of block idempotents. Let ω_1 and ω_2 be the central characters of $k\tilde{G}$ associated with e_1 and e_2, respectively. Then for $i = 1, 2$ we have $\omega_i(e_i) = 1$ and $\omega_i(e_j) = \omega_i(\beta(f_l)) = 0$ for $j \neq i$ and $l \geq 2$. Thus

$$(\omega_1 \circ \beta)(f_l) = (\omega_2 \circ \beta)(f_l) = \begin{cases} 1 & \text{if } l = 1 \\ 0 & \text{if } l = 2, 3, \ldots, t. \end{cases}$$

This shows that $\omega_1 \circ \beta = \omega_2 \circ \beta$ is the central character of kG associated with the block $f_1(kG)$. Since β is surjective, this shows that $\omega_1 = \omega_2$ and so $e_1 = e_2$ contrary to our assumption. Thus $\beta(f_1)$ (and similarly, $\beta(f_i)$) is a block idempotent; the first part of the theorem is then proved.

Finally suppose that f is a block idempotent of kG. Then by Lemma 4.3B, we can write

$$f = \sum_{i=1}^{s_1} \gamma_i c_i + \sum_{j=s_1+1}^{s} \gamma_j^* c_j,$$

where $c_1, c_2, \ldots, c_{s_1}$ are the class sums of classes $\mathscr{C}_1, \mathscr{C}_2, \ldots, \mathscr{C}_{s_1}$ that have D as a defect group and $c_{s_1+1}, \ldots, c_s$ are the class sums of the classes $\mathscr{C}_{s_1+1}, \ldots, \mathscr{C}_s$ that have a proper subgroup of D as a defect group. Then by Lemma 6.2B(i), $\tilde{\mathscr{C}}_i$ has D/Q as a defect group and $\tilde{\mathscr{C}}_j$ has a proper subgroup of D/Q as a defect group. Hence D/Q is a defect group of $\beta(f)k\tilde{G}$ as asserted.

These results permit us to give a description of the characters that lie in a block whose defect group is contained in the center of G.

Theorem 6.2B (*Reynolds*) Let Q be a subgroup of order p^d lying in the center of G, and let B be a block of kG with Q as a defect group. Then

(i) there is only one irreducible modular character of G lying in B (we shall denote it by ϕ);
(ii) there are exactly p^d irreducible ordinary characters $\xi^1, \xi^2, \ldots, \xi^{p^d}$ of G lying in B. Moreover, if $\lambda^1, \ldots, \lambda^{p^d}$ are the irreducible ordinary characters of Q (all of degree 1 since Q is abelian), then

$$\xi^i(x) = \begin{cases} \lambda^i(z)\phi(y) & \text{when } x = zy \ (z \in Q, y \in G^\circ) \\ 0 & \text{for } x \notin QG^\circ. \end{cases}$$

148   VI   THE MAIN THEOREMS OF BRAUER

Proof We have $G = C_G(Q)$. Let f be the block idempotent for B. Then putting $\tilde{G} := G/Q$, Theorem 6.2A shows that $\tilde{B} := \beta(f)(k\tilde{G})$ is a block of $k\tilde{G}$ with 1 as defect group (so $\tilde{B}$ is a block of defect 0). Therefore by Theorem 4.6A there is only one irreducible modular character $\tilde{\phi}$ and only one irreducible ordinary character $\tilde{\xi}$, say, lying in $\tilde{B}$, and $\tilde{\phi} = \tilde{\xi}$ on $\tilde{G}°$. From these we can define an irreducible modular character ϕ and irreducible ordinary character ξ of G by

$$\xi(x) := \tilde{\xi}(Qx) \quad \text{for all } x \in G,$$

and

$$\phi(x) := \tilde{\phi}(Qx) \quad \text{for all } x \in G°.$$

Note that since $\tilde{\xi} = 0$ on $\tilde{G}\backslash\tilde{G}°$, $\xi = 0$ on $G\backslash QG°$.

We first prove that if ψ is an irreducible modular character of G lying in the block B, then $\psi = \phi$; this will prove (i). Let V be an irreducible kG-module which affords the irreducible character ψ. Since Ker β annihilates V by the remark above, we can define an action of $k\tilde{G}$ on V by $v\beta(a) := va$ for all $a \in kG$. Under this action V is an irreducible $k\tilde{G}$-module. Since V lies in $f(kG)$, $Vf \neq 0$ and so $V\beta(f) \neq 0$. Thus as a $k\tilde{G}$-module, V lies in $\tilde{B} = \beta(f)(k\tilde{G})$. Since $\tilde{\phi}$ is the only irreducible modular character lying in $\tilde{B}$, V as a $k\tilde{G}$-module affords $\tilde{\phi}$, and so V as a kG-module affords the modular character ϕ. Hence $\psi = \phi$ as required.

Now the irreducible ordinary characters, say $\xi^1, \xi^2, \ldots, \xi^t$, lying in B are characters of the form $\xi^i = d_i \phi$ on $G°$ for suitable integers $d_i \geq 1$ (decomposition numbers), and conversely, each irreducible ordinary character of G of this form lies in B (see §4.4). In particular, ξ lies in B since $\xi = \phi$ on $G°$; put $\xi^1 = \xi$. Since K is a splitting field and $Q \subseteq Z(G)$, the representation affording the irreducible character ξ^i will represent each $z \in Q$ as a scalar, say $\lambda^i(z) \cdot 1$, where $\lambda^i(z)$ is a root of unity. Note that λ^i is a character of Q of degree 1. Then since $\xi^i = d_i \phi$ on $G°$ we have

$$\xi^i(zy) = d_i \lambda^i(z)\phi(y) \quad \text{for all } z \in Q \text{ and } y \in G°.$$

In particular, $|\xi^i(x)|^2 = d_i^2 |\xi(x)|^2$ for all $x \in QG°$. Therefore (with $g := |G|$) we have

$$g = g\langle \xi^i, \xi^i \rangle_G \geq \sum_{x \in QG°} |\xi^i(x)|^2 = d_i^2 \sum_{x \in QG°} |\xi(x)|^2$$

$$= d_i^2 g\langle \xi, \xi \rangle_G = gd_i^2$$

because $\xi = 0$ on $G\backslash QG°$. This shows that $d_i = 1$ for all i and (since the first inequality must be an equality) $\xi^i(x) = 0$ for all $x \notin QG°$. Thus for all i we have

$$\xi^i(x) = \begin{cases} \lambda^i(z)\phi(y) & \text{if } x = zy \text{ with } z \in Q, y \in G° \\ 0 & \text{if } x \notin QG°, \end{cases}$$

6.3 THE BRAUER CORRESPONDENCE: THE FIRST MAIN THEOREM

where $\lambda^1, \lambda^2, \ldots, \lambda^t$ are (distinct) characters of degree 1 of Q. Finally by Theorem 4.2C, for each $z \in Q$, $z \neq 1$,

$$0 = \sum_{i=1}^{t} \xi^i(1)\xi^i(z) = \phi(1)^2 \sum_{i=1}^{t} \lambda^i(z).$$

Thus $\sum_{i=1}^{t} \lambda^i(z) = 0$ for all $z \in Q\setminus\{1\}$, and so the inner product $\langle \lambda^1, \lambda^1 + \lambda^2 + \cdots + \lambda^t \rangle_Q = t/|Q|$ and $t \geq p^d = |Q|$. Since Q only has p^d irreducible ordinary characters, $t = p^d$ and $\lambda^1, \lambda^2, \ldots, \lambda^{p^d}$ is the complete set of irreducible characters of Q as asserted.

EXERCISES

1. Under the hypothesis of the Theorem 6.2A, prove that there is a bijection from the set of all irreducible modular characters lying in $\beta(f)(k\tilde{G})$ onto the set of all irreducible modular characters lying in $f(kG)$ given by $\tilde{\phi} \mapsto \phi$, where $\phi(x) := \tilde{\phi}(Qx)$ for all p'-elements $x \in G$.
2. Let $Z(kG)$ be the center of kG and let $\sum_{x \in G} \alpha_x x$ be an idempotent in $Z(kG)$. Show that $\alpha_x = 0$ whenever $x \in G\setminus G^\circ$.

6.3 The Brauer correspondence: The First Main Theorem

Let H be a subgroup of G, b be a block of kH and let D be a subgroup some defect group of b. Let ω be the central character of kH associated with the block b (see §4.1). We recall by Theorem 4.3 that if $\mathscr{C}'$ is a conjugate class of H with class sum c', then $\omega(c') \neq 0$ implies that D is contained in a defect group of $\mathscr{C}'$; in other words, ω vanishes on a class sum c' of kH except when c' corresponds to a class $\mathscr{C}'$ with $\mathscr{C}' \cap C_H(D) \neq \varnothing$. In this situation we can always define a k-linear mapping $\omega^G \colon Z(kG) \to k$ by

(1) $$\omega^G(c_i) := \omega(c_i^*) \quad \text{for } i = 1, 2, \ldots, s,$$

where for each conjugate class $\mathscr{C}_i$ of G with class sum c_i we put $\mathscr{C}_i^* := \mathscr{C}_i \cap H$ and $c_i^* := \sum_{x \in \mathscr{C}_i^*} x$. We have $\omega^G(c_i) = 0$ whenever $\mathscr{C}_i \cap C_H(D) = \varnothing$.

Note 1 Clearly $\mathscr{C}^*$ is a union of conjugate classes in H, so $c_i^* \in Z(kH)$. On the other hand, the class sums $c_1, c_2, \ldots, c_s$ form a k-basis of $Z(kG)$, so ω^G is well defined by (1). Although, in general, ω^G is not a central character of kG, the following lemma gives a condition on H under which ω^G is a central character.

Lemma 6.3 Let H be a subgroup of G and let b be a block of kH. Let ω be the central character of kH associated with b and define ω^G as in (1). If D is a p-subgroup of G such that $DC_G(D) \subseteq H \subseteq N_G(D)$, then the following hold:

(i) For each class $\mathscr{C}'$ of H either $\mathscr{C}' \subseteq C_G(D)$ or $\mathscr{C}' \cap C_G(D) = \varnothing$. Consequently, $\omega^G = \omega \circ \sigma$, where σ is the Brauer homomorphism with respect to D.

(ii) ω^G is a central character of kG, and the block B of kG associated with ω^G has a defect group which contains D.

Remark It follows from Lemma 6.2A that b has a defect group containing D because $D \triangleleft H$.

Proof (i) Since $C_G(D) \triangleleft H$, and $\mathscr{C}'$ is a conjugate class of H, either $\mathscr{C}' \subseteq C_G(D)$ or $\mathscr{C}' \cap C_G(D) = \varnothing$. It now follows at once from (1) and the Brauer homomorphism (see 6.1) that $\omega^G = \omega \circ \sigma$ as asserted.

(ii) Since ω and σ are both k-algebra homomorphisms (Lemma 6.1A), it follows from (i) that ω^G is also a k-algebra homomorphism (from $Z(kG)$ into k), and is therefore a central character of kG.

Let B be the block of kG associated with ω^G. Theorem 4.3 shows that there exists a class $\mathscr{C}$ of p'-elements of G with class sum c such that the defect groups of $\mathscr{C}$ are the defect groups of B and $\omega^G(c) \neq 0$. Then $\omega(\sigma(c)) \neq 0$. This means that some class $\mathscr{C}'$ of p'-elements of H with class sum c' has the property $\mathscr{C}' \subseteq \mathscr{C} \cap H$, and $\omega(c') \neq 0$. By Theorem 4.3 again, each defect group of b is contained in some defect group of $\mathscr{C}'$. Hence by the remark above, D is contained in some defect group of $\mathscr{C}'$. Hence for some $x \in \mathscr{C}' \subseteq \mathscr{C}$ we have $D \subseteq C_H(x) \subseteq C_G(x)$, and so D is contained in some defect group of $\mathscr{C}$, and hence of B.

Definition Under the hypotheses of Lemma 6.3, there exists a function from the set of blocks b of kH with a fixed p-group D as a defect group into the set of blocks B of kG with a defect group containing D. We call this function the *Brauer correspondence* and denote the image of b under the Brauer correspondence by b^G. (See the remark at the end of this section.)

Note 2 Assume that the hypotheses of Lemma 6.3 hold, and let B be a block of kG which has a defect group containing D. Suppose that f is the block idempotent of B and σ is the Brauer homomorphism with respect to D. Then $\sigma(f)$ is an idempotent in $Z(kH)$ by Lemma 6.1B [since $Z(kH) \supseteq Z(kN_G(D)) \cap kH$], so we can write $\sigma(f)$ as a sum of block idempotents of kH. Thus we can enumerate the block idempotents $e_1, e_2, \ldots, e_t$ of kH so that

$$\sigma(f) = e_1 + e_2 + \cdots + e_n \quad \text{and} \quad 1 - \sigma(f) = e_{n+1} + \cdots + e_t.$$

Let ω_i be the central character of kH associated with the block $b_i := e_i(kH)$ for $i = 1, 2, \ldots, t$. Now by Lemma 6.3, $\omega_1^G, \omega_2^G, \ldots, \omega_t^G$ are central characters for kG. For $i = 1, 2, \ldots, n$ we have $\omega_i^G(f) = (\omega_i \circ \sigma)(f) = \omega_i(e_1) + \omega_i(e_2) + \cdots + \omega_i(e_n) = 1$, and so ω_i^G is associated with B; hence $b_1^G = b_2^G = \cdots = b_n^G = B$. On the other hand, for $i = n + 1, \ldots, t$ we have $\omega_i^G(f) =$

6.3 THE BRAUER CORRESPONDENCE: THE FIRST MAIN THEOREM

$(\omega_i \circ \sigma)(f) = \omega_i(1) - \omega_i(e_{n+1}) - \cdots - \omega_i(e_t) = 0$, and so $b_{n+1}^G, b_{n+2}^G, \ldots, b_t^G$ are all different from B (see §4.1).

Theorem 6.3 (*The First Main Theorem*) Let Q be a p-subgroup of a group G and put $N := N_G(Q)$. Then the Brauer correspondence $b \mapsto b^G$ gives a bijection from the set of all blocks of kN with Q as defect group onto the set of all blocks of kG with Q as defect group.

Proof Let X_N (respectively, X_G) denote the set of central characters associated with blocks of kN (respectively, kG) that have Q as a defect group. In view of the definition of the Brauer correspondence and the correspondence between blocks and their associated central characters, it is enough to show that the mapping $\Psi: \omega \to \omega^G$ [defined in (1)] is a bijection from X_N onto X_G. We shall do this in a series of steps. Let σ be the Brauer homomorphism with respect to Q.

Step 1 Let $\mathscr{C}'$ be a class in N with Q as defect group and c' as its class sum. Then there exists a class $\mathscr{C}$ in G with Q as defect group and class sum c such that $\sigma(c) = c'$.

Using Lemma 6.1B(iii), there is a class $\mathscr{C}$ of G with Q as a defect group such that $\mathscr{C}' = \mathscr{C} \cap C_G(Q)$. Then $\sigma(c) = c'$.

Step 2 For each central character ω associated with a block of kN, $\omega \in X_N$ iff $\omega^G \in X_G$. In particular, Ψ maps X_N into X_G.

Let b and b^G be the blocks associated with ω and ω^G, respectively. Suppose $\omega \in X_N$. Then by Theorem 4.3 there is a class $\mathscr{C}'$ of N with Q as defect group such that $\omega(c') \neq 0$ for the class sum c'. By Step 1 there is a class $\mathscr{C}$ of G with Q as defect group, and $\omega^G(c) = \omega(\sigma(c)) = \omega(c') \neq 0$ for the class sum c. Thus Theorem 4.3 again shows that Q is also contained in a defect group of b^G; hence Q itself is a defect group of b^G.

Conversely, using Lemma 6.1B(i) a similar argument shows that if Q is a defect group of b^G, then b has a defect group contained in Q by Theorem 4.3. Since $Q \triangleleft N$, Lemma 6.2A now shows that Q is a defect group of b.

Step 3 Ψ is injective.

Otherwise there exist $\omega_1, \omega_2 \in X_N$ with distinct associated block idempotents e_1 and e_2 of kN such that $\omega_1^G = \omega_2^G$. By Lemma 6.2A we know that $e_1 \in J_Q$, where J_Q is the k-space spanned by all class sums c' corresponding to classes $\mathscr{C}'$ of N that have Q as a defect group. By Step 1, $J_Q \subseteq \operatorname{Im} \sigma$, so $e_1 = \sigma(a)$ for some $a \in Z(kG)$. But then for $i = 1, 2$,

$$\omega_i(e_1) = \omega_i(\sigma(a)) = \omega_i^G(a),$$

so $\omega_1(e_1) = \omega_2(e_1)$. This is impossible since $\omega_1(e_1) = 1$ and $\omega_2(e_1) = 0$. Hence Ψ is injective.

Step 4 Ψ *is surjective.*

Let $\psi \in X_G$ and B be the associated block of kG. Since B has Q as a defect group, it follows from Note 2 above that there is at least one central character ω of kN such that $\omega^G = \psi$. By Step 2, $\omega \in X_N$. This proves Ψ is surjective; and the theorem is proved.

Corollary Under the hypothesis of the theorem suppose that H is a subgroup of G with $H \supseteq N_G(Q)$. Then there is a bijection between the set of blocks of kH that have Q as a defect group, and the blocks of kG that have Q as a defect group.

Proof Both sets correspond bijectively with the set of blocks of kN with Q as a defect group.

Remark The Brauer correspondence has been defined for blocks of kH that have a defect group D satisfying $DC_G(D) \subseteq H \subseteq N_G(D)$. Theorem 6.3 enables us to extend this definition to the case where simply $DC_G(D) \subseteq H$. Indeed, in the latter case, if B is a block of kH with D as defect group, then by Theorem 6.3 there is a unique block b of $kN_H(D)$ with D as defect group such that $b^H = B$. Since $DC_G(D) \subseteq N_H(D) \subseteq N_G(D)$, b^G is already defined as a block of kG with a defect group containing D. We define $B^G := b^G$. It is important to note that the relation (1) is still valid for the central characters ω and ω^G associated with B and B^G, respectively.

EXERCISES

1. Let N be a normal subgroup of G and let $e(kG)$ and $f(kN)$ be blocks of kG and kN, respectively. Show that $e(kG) = (f(kN))^G$ if and only if $ef = e$.

2. Let P be a Sylow p-subgroup of G. Prove that the number of blocks of kG with P as a defect group is equal to the number of conjugate classes of p'-elements with P as a defect group. [*Hint:* Let $H := N_G(P)$ and let β be the natural algebra homomorphism from kH to $k(H/P)$. First show that the number of blocks of kH is equal to $\dim_k(\operatorname{Im} \beta)$ and then show the class sums of the classes of H of p'-elements form a basis of $\operatorname{Im} \beta$.]

6.4 Extension of the First Main Theorem

Let Q be a p-subgroup of the group G and let $N := N_G(Q)$. Then the First Main Theorem (Theorem 6.3) gives a correspondence between the blocks of kG with Q as a defect group and the blocks of kN with Q as a defect group. It is sometimes useful to pursue this reduction one further step, namely down to the blocks of kH where $H := QC_G(Q)$. Since $H \subseteq N$, there is no loss in generality in supposing that we have already carried out the first reduction and so take $G = N$. The reduction to the set of blocks of kH with Q as defect group is no longer one-to-one, but corresponds to a natural equivalence relation on this set.

6.4 EXTENSION OF THE FIRST MAIN THEOREM

Definition Let H be a normal subgroup of the group G. Let e and e' be block idempotents of kH. We say that the block idempotents e and e' [and the corresponding blocks $e(kH)$ and $e'(kH)$] are *G-conjugate* if $e' = y^{-1}ey$ for some $y \in G$.

Note 1 If $\{e_1, e_2, \ldots, e_t\}$ is the set of all block idempotents of kH, then G acts on this set by conjugation (because $H \triangleleft G$). The orbits under this action are the sets of G-conjugate block idempotents of kH. If D is a defect group for $e(kH)$, then $y^{-1}Dy$ is a defect group for $(y^{-1}ey)(kH)$.

EXAMPLE 1 Suppose that with the notation above χ is an irreducible ordinary character of H lying in a block $e(kH)$. Then for any $y \in G$ we can define the character χ^y by $\chi^y(x) := \chi(y^{-1}xy)$ (see §2.2). If χ is afforded by the kH-module V, then χ^y is afforded by the kH-module V^y, which has the same underlying k-space and where the action of H on V^y is defined in terms of its action on V by $v^x := v(yxy^{-1})$. If V is a kH-module which affords χ, then V^y is a kH-module which affords χ^y. Since V lies in $e(kH)$, $Ve \neq 0$, and hence $V^y(y^{-1}ey) \neq 0$. This shows that V^y lies in $(y^{-1}ey)(kH)$. Thus we have shown that G-conjugate characters χ and χ^y lie in G-conjugate blocks $e(kH)$ and $(y^{-1}ey)(kH)$, respectively. Clearly the analogous result holds for an irreducible modular character of H lying in $e(kH)$.

EXAMPLE 2 With the notation above, if $e(kH)$ is the principal block of kH, then the only G-conjugate of $e(kH)$ is $e(kH)$ itself. Indeed, by definition 1_H lies in $e(kH)$ and by Example 1, 1_H^y lies in $y^{-1}ey(kH)$ for each $y \in G$. Since $1_H^y = 1_H$, this shows that $y^{-1}ey(kH) = e(kH)$ for all $y \in G$.

Lemma 6.4 Let Q be a normal p-subgroup of G and put $H := QC_G(Q)$. Let e and e' be block idempotents of kH, $b := e(kH)$ and $b' := e'(kH)$ the corresponding blocks, and ω and ω' the associated central characters of kH. Then the following are equivalent.

(i) b and b' are G-conjugate.
(ii) $\omega = \omega'$ on $Z(kG) \cap Z(kH)$.
(iii) $b^G = (b')^G$.

Proof First assume that (ii) holds; we shall prove (i). Indeed, suppose that $e = e_1, e_2, \ldots, e_n$ are block idempotents of kH so that $e_i(kH)$ ($i = 1, 2, \ldots, n$) are the G-conjugates of b. Then $\{e_1, e_2, \ldots, e_n\}$ is an orbit under conjugation by G, so $a := e_1 + e_2 + \cdots + e_n \in Z(kG) \cap Z(kH)$. Hence $\omega'(a) = \omega(a) = \omega(e_1) + \cdots + \omega(e_n) = 1$. This shows that $\omega'(e_j) \neq 0$ for some j ($1 \leq j \leq n$), and so $\omega'(e_j) = 1$ [since $\omega'(e_j^2) = \omega'(e_j)^2 = \omega'(e_j)$]. Hence ω' is associated with the block $e_j(kH)$. Thus $b' = e_j(kH)$ and hence it is G-conjugate to b.

Next assume that (iii) holds [so $\omega^G = (\omega')^G$]; we shall prove that (ii) holds. First suppose that $\mathscr{C}$ is a conjugate class of H whose class sum c does not lie in $kC_G(Q)$. Since $Q \triangleleft G$, this means that $\mathscr{C} \cap C_G(Q) = \varnothing$, and so Q is contained in no defect group of $\mathscr{C}$. However by Lemma 6.2(iv) Q is contained in all of the defect groups of b and b', and so $\omega(c) = \omega'(c) = 0$ by Theorem 4.3. Since the class sums of H form a basis for $Z(kH)$, and $C_G(Q) \triangleleft H$, this shows that $\omega = \omega' = 0$ on $Z(kH) \backslash kC_G(Q)$. Thus to prove (ii) it remains to show that $\omega = \omega'$ on $Z(kG) \cap Z(kH) \cap kC_G(Q)$. However (iii) and Lemma 6.3 together imply that $\omega \circ \sigma = \omega^G = (\omega')^G = \omega' \circ \sigma$, where σ is the Brauer homomorphism with respect to Q. The definition of σ shows that Im $\sigma = Z(kG) \cap kC_G(Q)$ because $Q \triangleleft G$. Therefore $\omega = \omega'$ on $Z(kG) \cap Z(kH) \cap kC_G(Q)$ as required, and we have shown that (iii) implies (ii).

Finally suppose that (i) holds; we shall prove (iii). Let f be the block idempotent of b^G, and let σ be the Brauer homomorphism with respect to Q. Since $Q \triangleleft G$, the defect groups of b^G contain Q (Lemma 6.2A), and therefore $\sigma(f)$ is a central idempotent of kH (Lemma 6.1B) since Im $\sigma \subseteq Z(kH)$. Write $\sigma(f) = e_1 + e_2 + \cdots + e_n$ as a sum of distinct block idempotents of kH. Using Lemma 6.3 the central character of b^G is $\omega^G = \omega \circ \sigma$, and so $1 = \omega^G(f) = \omega(e_1) + \omega(e_2) + \cdots + \omega(e_n)$. This shows that ω is the central character associated with one of the blocks $e_i(kH)$ ($i = 1, 2, \ldots, n$) [see §4.1], and so $b = e_j(kH)$, say. On the other hand, since b' is G-conjugate to b, $b' = e_l(kH)$ for some l, $1 \leq l \leq n$, and so $\omega'(e_l) = 1$. Then $(\omega')^G(f) = \omega'(\sigma(f)) = \omega'(e_1) + \cdots + \omega'(e_n) = 1$. Hence the central character $(\omega')^G$ is associated with the block $f(kG) = b^G$; this proves that $(b')^G = b^G$.

EXAMPLE 3 With the hypothesis and notation of Lemma 6.4, suppose that b_0 is the principal block of kH. If b is a different block of kH, then $b^G \neq b_0^G$. Indeed $b^G = b_0^G$ implies b and b_0 are G-conjugate by Lemma 6.4, and so $b = b_0$ by Example 2.

Now suppose Q is a normal p-subgroup of G and put $H := QC_G(Q)$; since $Q \triangleleft G$, H is also normal in G. Let σ be the Brauer homomorphism with respect to Q, and let f be a block idempotent of kG. Since $Q \triangleleft G$, the defect groups of $f(kG)$ contains Q by Lemma 6.2A. Since Im $\sigma \subseteq Z(kH)$, Lemma 6.1B then shows that $\sigma(f)$ is a central idempotent of kH and so can be written $\sigma(f) = e_1 + e_2 + \cdots + e_n$ as a sum of block idempotents of kH. We define

(1) $$\Sigma_f := \{e_1(kH), \ldots, e_n(kH)\}.$$

(Note that $\Sigma_f \neq \varnothing$.)

Note 2 If b is a block of kH, then $b \in \Sigma_f$ iff $b^G = f(kG)$. Indeed, if ω is the central character of kH associated with b, then $b \in \Sigma_f$ iff $\omega(e_j) = 1$ for some j ($1 \leq j \leq n$). Since $\omega^G(f) = \omega \circ \sigma(f) = \omega(e_1) + \cdots + \omega(e_n)$, the latter condition is equivalent to $\omega^G(f) = 1$, and that holds iff the block b^G associated with ω^G equals $f(kG)$. In particular, this shows that there is always at least

6.4 EXTENSION OF THE FIRST MAIN THEOREM

one block of kH such that $b^G = f(kG)$. By Lemma 6.4, Σ_f is a complete set of G-conjugate blocks of kH.

Theorem 6.4 *Let Q be a normal p-subgroup of the group G and put $H := QC_G(Q)$. Let l be the largest integer such that $p^l \mid \mid G : H \mid$. Then (with Σ_f defined by (1) above)*

(i) *If $f(kG)$ is a block of kG with Q as (unique) defect group, then Σ_f is a full set of G-conjugate blocks of kH each with Q as defect group, and $\mid \Sigma_f \mid \equiv 0 \pmod{p^l}$;*

(ii) *Conversely, if Ω is a full set of G-conjugate blocks of kH each with Q as defect group, and $\mid \Omega \mid \equiv 0 \pmod{p^l}$, then $\Omega = \Sigma_f$ for some block $f(kG)$ which has Q as defect group.*

Proof The proof will be given in several steps.

Step 1 *If $\mathscr{C}$ is a class in G with Q as defect group, then the class sum c satisfies $\sigma(c) = c$.*

Indeed, the hypothesis implies $Q \subseteq C_G(x)$ for some $x \in \mathscr{C}$. Since Q is normal, this is then true for all $x \in \mathscr{C}$, so $\mathscr{C} \subseteq C_G(Q)$. Hence $\sigma(c) = c$ by the definition of σ.

Step 2 *If f is a block idempotent of kG and $f(kG)$ has Q as defect group, then $\sigma(f) = f$ and Σ_f is a full set of G-conjugate blocks.*

It follows from Lemma 6.2A and Step 1 that $\sigma(f) = f$, and so $f = e_1 + e_2 + \cdots + e_n$. Since $f \in Z(kG)$, $x^{-1}fx = f$ for all $x \in G$, and so G acts on the set $\{e_1, e_2, \ldots, e_n\}$ by conjugation. Suppose $\{e_1, e_2, \ldots, e_m\}$, say, is an orbit under this action. If $m < n$, then $f = (e_1 + e_2 + \cdots + e_m) + (e_{m+1} + \cdots + e_n)$ would give a decomposition of f into a sum of two nonzero central orthogonal idempotents in kG. Since f is primitive, this is impossible. Hence G is transitive on $\{e_1, e_2, \ldots, e_n\}$; that is, Σ_f is a single G-conjugate set of blocks of kH.

Step 3 *Suppose $\mathscr{C}$ is a class in G contained in H and that $\mathscr{C} = \mathscr{C}_1 \cup \cdots \cup \mathscr{C}_m$, where the $\mathscr{C}_i$ are conjugate classes of H. Then $\mathscr{C}$ has Q as defect group iff $\mathscr{C}_1$ (and hence each $\mathscr{C}_i$) has Q as defect group and p^l is the exact power of p in m, where $m := \mid G : N_G(\mathscr{C}_1) \mid$.*

Let $x \in \mathscr{C}_1$. We first show that $N_G(\mathscr{C}_1) = HC_G(x)$. Indeed, if $y \in N_G(\mathscr{C}_1)$, then $yxy^{-1} \in \mathscr{C}_1$, so $yxy^{-1} = zxz^{-1}$ for some $z \in H$. This shows that $y \in zC_G(x) \subseteq HC_G(x)$. Hence $N_G(\mathscr{C}_1) \subseteq HC_G(x)$ and the reverse inequality is trivial. Now $\mathscr{C}$ has Q as defect group iff Q is a Sylow p-subgroup of $C_G(x)$. This is so iff Q is a Sylow p-subgroup of $C_H(x)$ and $p \nmid \mid C_G(x) : C_H(x) \mid$. This will be the case iff Q is a Sylow p-subgroup of $C_H(x)$ and $p \nmid \mid HC_G(x) : H \mid$.

156   VI   THE MAIN THEOREMS OF BRAUER

So $\mathscr{C}$ has Q as a defect group iff $\mathscr{C}_1$ has Q as a defect group and p^l is the exact power of p in $m = |G : N_G(\mathscr{C}_1)|$ (by what we showed above).

Step 4 *If $f(kG)$ is a block with Q as defect group, then Σ_f is a set of blocks $e_i(kH)$ with Q as defect group.*

Let ω_i be the central character associated with $e_i(kH)$. Then $\omega_i(e_j) = \delta_{ij}$, and so from $\sigma(f) = e_1 + e_2 + \cdots + e_n$ we have $\omega_i(\sigma(f)) = 1$. On the other hand, if f' is any other block idempotent of kG, then $ff' = 0$ and so $0 = \omega_i(\sigma(ff')) = \omega_i(\sigma(f)) \cdot \omega_i(\sigma(f'))$, which shows that $\omega_i(\sigma(f')) = 0$. Thus $\omega_i \circ \sigma$ is the central character of kG associated with f. Since $f(kG)$ has Q as defect group, Theorem 4.3 shows that there is a class $\mathscr{C}$ of G with Q as defect group and class sum c such that $(\omega_i \circ \sigma)(c) \ne 0$. With the notation of Step 3, let c'_j be the class sum in kH of $\mathscr{C}_j$. Then by Step 1 we have $\omega_i(c) \ne 0$ so $\omega_i(c'_j) \ne 0$ for some j. Since $\mathscr{C}$ has Q as a defect group, Q is also a defect group of $\mathscr{C}_j$, and so Theorem 4.3 shows that $e_i(kH)$ is a block with a defect group contained in Q. Hence by Lemma 6.2A, Q is the unique defect group of $e_i(kH)$.

Step 5 *If $f(kG)$ is a block with Q as defect group, then $|\Sigma_f| \equiv 0 \pmod{p^l}$.*

With the notation in the proof of Step 4, $0 \ne \omega_1(c) = \omega_1(c'_1) + \cdots + \omega_1(c'_m)$. Let $g := |G|$ and consider the g values of $\omega_1(x^{-1}c'_1 x)$ for $x \in G$. Since ω_1 lies in a set of n G-conjugate central characters (see Note 2 above), we get each of the values $\omega_i(c'_1)$ ($i = 1, 2, \ldots, n$) g/n times. On the other hand, since c'_1 has m conjugates under G, we get each of the values $\omega_1(c'_j)$ ($j = 1, 2, \ldots, n$) g/m times. Now since $\mathscr{C}$ has Q as defect group, Step 3 shows that p^l is the exact power of p in m and so p^l/m is a unit in k. Thus from above

$$0 \ne (p^l/m)\{\omega_1(c'_1) + \omega_1(c'_2) + \cdots + \omega_1(c'_m)\}$$
$$= (p^l/n)\{\omega_1(c'_1) + \omega_2(c'_1) + \cdots + \omega_n(c'_1)\}.$$

Since the values of $\omega_i(c_j)$ are algebraic integers, this shows that $p^l \mid n$, as required.

Step 6 *If Ω is a set of G-conjugate blocks of kH with Q as defect group, and $|\Omega| \equiv 0 \pmod{p^l}$, then $\Omega = \Sigma_f$, where $f(kG)$ is a block with Q as defect group.*

Suppose that $\Omega = \{e_1(kH), \ldots, e_n(kH)\}$. Since Ω is a full set of G-conjugate blocks, $f_0 := e_1 + e_2 + \cdots + e_n$ is a central idempotent of kG. Let $e_1 = \Sigma \gamma_i c'_i$ with $\gamma_i \in k$, where the c'_i are the class sums in kH. Let $x_1, x_2, \ldots, x_h$ be a transversal of H in G; then each of $e_1, e_2, \ldots, e_n$ appears h/n times among the h conjugates $x_j^{-1} e_1 x_j$. Since p^l is the exact power of p in h and $p^l \mid n$, $n/h \in k$ and so

$$f_0 = \frac{n}{h} \sum_{j=1}^{h} x_j^{-1} e_1 x_j = \frac{n}{h} \sum_i \gamma_i \sum_{j=1}^{h} x_j^{-1} c'_i x_j .$$

6.5 GENERALIZED DECOMPOSITION NUMBERS

Since $e_1(kH)$ has Q as defect group, Lemma 6.2A shows that $\gamma_i \neq 0$ implies that the class $\mathscr{C}_i$ of H corresponding to c'_i has Q as defect group. Now suppose that f_0 is written as a k-linear combination of class sums in kG. Suppose that c is a class sum (corresponding to the class $\mathscr{C}$ in G) with nonzero coefficient. This means that for some i, $\gamma_i \neq 0$ and $\sum_{j=1}^{h} x_j^{-1} c'_i x_j \neq 0$, where the class $\mathscr{C}_i$ of H is contained in $\mathscr{C}$. Without loss in generality, suppose that $i = 1$ and $\mathscr{C} = \mathscr{C}_1 \cup \cdots \cup \mathscr{C}_m$. Then each of the class sums $c'_1, c'_2, \ldots, c'_m$ occur h/m times among $x_j^{-1} c'_1 x_j$ ($j = 1, 2, \ldots, h$), and so $0 \neq \sum_{j=1}^{h} x_j^{-1} c'_1 x_j = (h/m)\{c'_1 + c'_2 + \cdots + c'_m\} = (h/m)c$ in kH. Thus h/m is a unit in k and since p^l is the exact power of p in h, this shows that p^l is also the exact power of p in m. As we noted above, $\mathscr{C}_1$ has Q as defect group because $\gamma_1 \neq 0$; therefore $\mathscr{C}$ has Q as defect group by Step 3. Thus we have shown that $f_0 = \sum_i \beta_i c_i$, say, where $\beta_i \in k$ and $\beta_i \neq 0$ only when the class sum c_i corresponds to a class of G with Q as defect group.

Finally write f_0 as a sum of block idempotents in kG, and let f be one of the summands. If ω is the central character associated with f, then $1 = \omega(f) = \omega(f_0)$, and so for some i, $\beta_i \neq 0$ and $\omega(c_i) \neq 0$. Theorem 4.3 now shows that $f(kG)$ is a block with a defect group contained in Q; and so Lemma 6.2A shows that $f(kG)$ has Q itself as defect group. By Step 2, Σ_f is a full set of G-conjugate blocks in kH, and clearly $\Sigma_f \subseteq \Omega$; since Ω is also a set of G-conjugate blocks, this proves that $\Sigma_f = \Omega$ as asserted. This completes the proof of the theorem.

Remark Theorems 6.3 and 6.4 and Theorem 6.2 can be combined to describe a bijective correspondence between the set of all blocks of a group G that have a specified p-subgroup Q as defect group and certain classes of blocks of the group $\tilde{H} := QC_G(Q)/Q$ with 1 as defect group (that is, of defect 0).

EXERCISES

1. Let P be a Sylow p-subgroup of G. Prove that kG has exactly one block if and only if $P \subseteq C_G(P)$.
2. Let G be a group of order $p^n m$, $(m, p) = 1$ with a normal subgroup N of order m. If B is a block of kG, prove that B is a direct sum of isomorphic indecomposable kG-modules.
3. Let G be a group of order $p^n m$, $(p, m) = 1$ with a normal subgroup of order m, and let r be the number of conjugate classes of p'-elements of G. Prove that the Cartan matrix of kG is an $r \times r$ diagonal matrix in which each diagonal entry is a power of p.

6.5 Generalized decomposition numbers: The Second Main Theorem

The First Main Theorem of Brauer describes the Brauer correspondence between certain blocks of a group and blocks of suitable subgroups. The

Second Main Theorem gives a sufficient criterion for two blocks to correspond in this manner. The criterion is in terms of generalized decomposition numbers defined as follows.

Let z be a p-element in the group G, and let $\psi^1, \psi^2, \ldots, \psi^{r'}$ be the irreducible modular characters of the subgroup $H := C_G(z)$. Let $\chi^1, \chi^2, \ldots, \chi^{s'}$ be the irreducible ordinary characters of H. Since $z \in Z(H)$, z is represented by a scalar $\lambda_i \cdot 1$, say in every representation of H affording χ^i; and λ_i is a p-power root of unity since z is a p-element. Then for all p'-elements $y \in H$ we have

$$(1) \qquad \chi^i(zy) = \lambda_i \chi^i(y) = \lambda_i \sum_{j=1}^{r'} d_{ij} \psi^j(y)$$

where the d_{ij} are the decomposition numbers for H. Finally, if $\zeta^1, \zeta^2, \ldots, \zeta^s$ are the irreducible ordinary characters of G, then the restriction ζ_H^i can be written as an integral linear combination of $\chi^1, \chi^2, \ldots, \chi^{s'}$, and so we obtain for all p'-elements $y \in H$

$$\zeta^i(zy) = \sum_{j=1}^{r'} d_{ij}^z \psi^j(y) \qquad (i = 1, 2, \ldots, s),$$

where d_{ij}^z are algebraic integers. These algebraic integers d_{ij}^z are called the *generalized decomposition numbers* of G at z.

Note 1 Since $\psi^1, \psi^2, \ldots, \psi^{r'}$ are linearly independent (Theorem 3.7B), the generalized decomposition numbers at z are uniquely determined for a fixed ordering of $\zeta^1, \zeta^2, \ldots, \zeta^s$ and $\psi^1, \psi^2, \ldots, \psi^{r'}$.

Note 2 When $z = 1$, $d_{ij}^1 = d_{ij}$ are the ordinary decomposition numbers.

Note 3 If for some l and some algebraic integers d'_{lj} we have $\chi^l(zy) = \sum_{j=1}^{r'} d'_{lj} \psi^j(y)$ for all p'-elements $y \in H$, then it follows from (1) and the linear independence of the irreducible modular characters $\psi^1, \psi^2, \ldots, \psi^{r'}$ that $d'_{lj} = \lambda_l d_{lj}$. Hence if some $d'_{lj} \neq 0$ then $d_{lj} \neq 0$, and so χ^l and ψ^j lie in the same block of kH (see §4.4).

Note 4 If z has order 2, then each $\lambda_i = \pm 1$; hence in this case the generalized decomposition numbers are rational integers.

The proof of the Second Main Theorem requires two rather technical lemmas.

Lemma 6.5A Let $P = \langle z \rangle$ be a cyclic group of order $p^d > 1$, and put $Q := \langle z^p \rangle$. Let V be an A-free AQ-module, and W be an A-free AP-module. Then

(i) The A-algebra $\operatorname{End}_{AP}(V^P)$ contains a subalgebra $E \simeq \operatorname{End}_{AQ}(V)$ and

$\operatorname{End}_{AP}(V^p) = E[h_0]$ for some central element h_0 such that $h_0^p \in E$ and $h_0^{p^d} = 1$.

(ii) If V is an indecomposable AQ-module, then V^p is an indecomposable AP-module.

(iii) Suppose there exists an AQ-homomorphism $\lambda: W \to W$ such that $\lambda + z^{-1}\lambda z + \cdots + z^{-p+1}\lambda z^{p-1} = 1$ on W. Then there exists an A-free AQ-module V_0 such that $W \simeq V_0^p$.

Proof (i) For each $h \in \operatorname{End}_{AP}(V^p)$ we have the restriction $h_V \in \operatorname{Hom}_{AQ}(V, V^p)$. Consider the mapping $h \mapsto h_V$ of $\operatorname{End}_{AP}(V^p) \to \operatorname{Hom}_{AQ}(V, V^p)$. Since $V^p = V \otimes 1 + V \otimes z + \cdots + V \otimes z^{p-1}$, and h is an AP-homomorphism, it is clear that h_V completely determines h, and so the mapping $h \mapsto h_V$ is injective. Conversely, given any $h_1 \in \operatorname{Hom}_{AQ}(V, V^p)$ we can extend this to an $h \in \operatorname{End}_{AP}(V^p)$ by defining $(v \otimes z^i)h := vh_1 \otimes z^i$ ($v \in V$, $i = 0, 1, 2, \ldots, p-1$). Thus the mapping $h \mapsto h_V$ is also surjective, and so has inverse ψ, say. Now give the A-module $\operatorname{Hom}_{AQ}(V, V^p)$ an A-algebra structure by defining $h_V g_V := (hg)_V$ for all $h, g \in \operatorname{End}_{AP}(V^p)$. Then we have

(2) $\qquad \psi: \operatorname{Hom}_{AQ}(V, V^p) \simeq \operatorname{End}_{AP}(V^p) \qquad$ as A-algebras.

Now $\operatorname{End}_{AQ}(V)$ can be identified with the subalgebra of $\operatorname{Hom}_{AQ}(V, V^p)$ consisting of the functions with images contained in V. Then $E := \psi(\operatorname{End}_{AQ}(V))$ is a subalgebra of $\operatorname{End}_{AP}(V^p)$ isomorphic to $\operatorname{End}_{AQ}(V)$. Finally, since z centralizes Q, we can define $h_1 \in \operatorname{Hom}_{AQ}(V, V^p)$ by $vh_1 := v \otimes z$, and it is readily verified that h_1 lies in the center of $\operatorname{Hom}_{AQ}(V, V^p)$ and generates the latter as a ring over $\operatorname{End}_{AQ}(V)$. Thus, if we put $h_0 := \psi(h_1)$, then (i) follows from (2).

(ii) If G is any group and U is an A-free AG-module, then we know from Theorem 3.4B that U is indecomposable iff $\bar{U}$ is indecomposable. Since $\bar{U}$ is a kG-module, the latter is true iff $\operatorname{End}_{kG}(\bar{U})/\operatorname{rad} \operatorname{End}_{kG}(\bar{U})$ is a division ring (Theorem 1.7A). Moreover $\operatorname{End}_{kG}(\bar{U}) \simeq \overline{\operatorname{End}_{AG}(U)}$ [see (1) of §3.4]. Thus (ii) will follow at once from (i) provided we can prove the following assertion: If $R (= \bar{E})$ is a k-algebra such that $R/\operatorname{rad} R$ is a division ring, and $S = R[t]$, where t is a central element of S such that $t^{p^d} = 1$, then $S/\operatorname{rad} S$ is a division ring. To prove this we proceed as follows. First, from the binomial theorem, $(t-1)^{p^d} = t^{p^d} - 1 = 0$, and so the central element $t - 1$ generates a nilpotent ideal of S. Hence $t - 1 \in \operatorname{rad} S$ by Theorem 1.3A. Now consider the R-homomorphism $\theta: S \to R/\operatorname{rad} R$, which maps t onto 1. Then θ is surjective, and $\operatorname{Ker} \theta = (t-1)S + \operatorname{rad} R \subseteq \operatorname{rad} S$. Thus $S/\operatorname{rad} S$ is isomorphic to a factor ring of $S/\operatorname{Ker} \theta$; and the latter is isomorphic to $R/\operatorname{rad} R$, which is a division ring. Hence $\operatorname{Ker} \theta = \operatorname{rad} S$ and $S/\operatorname{rad} S$ is also a division ring. This proves (ii).

(iii) By Theorem 5.2B, W is Q-projective and so $W \mid V_0^p$ for some A-free AQ-module V_0. By part (ii), V_0^p is indecomposable and hence $W \simeq V_0^p$ as required.

Lemma 6.5B Let z and y be elements of the group G such that $z \neq 1$ is a

p-element, y is a p'-element, and $zy = yz$. Suppose that W is an A-free $A\langle zy\rangle$-module such that for some $A\langle z^p\rangle$-module U we have $W_{\langle z\rangle} \simeq U^{\langle z\rangle}$. Assume that $W = W_0 \oplus W_1$ as a sum of $A\langle zy\rangle$-submodules. Then the character χ afforded by W_0 satisfies $\chi(zy) = 0$.

Remark Since z and y have relatively prime orders, $\langle zy\rangle = \langle z\rangle \times \langle y\rangle$.

Proof It is clearly enough to prove the result in the case that W_0 is indecomposable. Let T be the representation afforded by W_0, and let ε_1, $\varepsilon_2, \ldots, \varepsilon_m$ be the distinct eigenvalues of $T(y)$. By the hypothesis (see §3.3), each ε_i is an algebraic integer in K and so lies in A. If y has order n, then the minimal polynomial for $T(y)$ divides $X^n - 1 \in K[X]$ and so has no multiple roots. Therefore the minimal polynomial must be $(X - \varepsilon_1)(X - \varepsilon_2) \cdots (X - \varepsilon_m) \in A[X]$ and so we can write

$$W_0 = \operatorname{Ker}(y - \varepsilon_1 \cdot 1) \oplus \cdots \oplus \operatorname{Ker}(y - \varepsilon_m \cdot 1).$$

Since $zy = yz$, the modules on the right hand side are all $A\langle zy\rangle$-modules. But W_0 is assumed to be indecomposable, and so $W_0 = \operatorname{Ker}(y - \varepsilon \cdot 1)$ for some $\varepsilon \in A$; that is, $T(y) = \varepsilon \cdot 1$.

Now write $U = U_1 \oplus U_2 \oplus \cdots \oplus U_t$ as a sum of indecomposable $A\langle z^p\rangle$-submodules. Then by Lemma 6.5A(ii), $U^{\langle z\rangle} = U_1^{\langle z\rangle} \oplus \cdots \oplus U_t^{\langle z\rangle}$ is a sum of indecomposable $A\langle z\rangle$-modules. Since $W_{\langle z\rangle} \simeq U^{\langle z\rangle}$, the Krull–Schmidt theorem for AP-modules (Theorem 3.4B) shows that $(W_0)_{\langle z\rangle} \simeq V^{\langle z\rangle}$ where V is an $A\langle z^p\rangle$-module (a sum of certain U_i). Since U is A-free, W_0 and V are also A-free modules. Thus we may choose an A-basis of W_0 of the form $v_i z^j$ ($i = 1, 2, \ldots, n; j = 0, 1, \ldots, p-1$), where $Av_1 + \cdots + Av_n \simeq V$ as $A\langle z^p\rangle$-modules. If we now calculate the matrix of $T(zy)$ relative to this basis, then all its diagonal entries are 0, and so the character χ afforded by W_0 has the value $\chi(zy) = 0$. The lemma is proved.

Theorem 6.5 (*The Second Main Theorem*) Let z be a p-element in the group G and put $H := C_G(z)$. Let $\zeta^1, \zeta^2, \ldots, \zeta^s$ be the irreducible ordinary characters of G, and $\psi^1, \psi^2, \ldots, \psi^{r}$ be the irreducible modular characters of H. Suppose that for some i and j the generalized decomposition number $d_{ij}^z \neq 0$. If ψ^j lies in the block b of kH, and ζ^i lies in the block B of kG, then $b^G = B$.

Remark Suppose that D is a defect group of b. Then since $\langle z\rangle$ is normal in H, Lemma 6.2A shows that $\langle z\rangle \subseteq D$, and so $C_G(D) \subseteq C_G(z) = H$. Thus $H \supseteq DC_G(D)$ and so the Brauer correspondence $b \mapsto b^G$ is defined (corollary of Theorem 6.3).

Proof Let V be an A-free AG-module which affords the character ζ^i (see Theorem 3.3). Using the Lifting theorem (Theorem 3.4A) we can find a central idempotent e of AG and the central idempotents $f_1, f_2, \ldots, f_n$ of AH

6.5 GENERALIZED DECOMPOSITION NUMBERS

such that $\bar{e}(kG) = B$ and $b_i := \bar{f}_i(kH)$ (for $i = 1, 2, \ldots, n$) are the distinct blocks of kH such that $b_i^G = B$. Put $f_0 = f_1 + f_2 + \cdots + f_n$; f_0 is a central idempotent of AH (possibly $f_0 = 1$). Now $V_H = V(1 - f_0) \oplus Vf_0$, where the summands are A-free AH-modules. Let χ and χ', respectively, be the characters of H afforded by $V(1 - f_0)$ and Vf_0. If $f'(kH)$ is some block of kH distinct from $b_1, b_2, \ldots, b_n$, then $f'\bar{f}_l = 0$ for $l = 1, 2, \ldots, n$, and so $\overline{Vf_0}f' = \overline{V}(\bar{f}_1 + \cdots + \bar{f}_n)f' = 0$. This shows that all indecomposable components of V_H lying in blocks other than $b_1, b_2, \ldots, b_n$ must occur in the summand $V(1 - f_0)$. This means that all irreducible constituents of χ' lie in the blocks $b_1, b_2, \ldots, b_n$. Thus, if we can show that for all p'-elements $y \in H$ we have $\chi(zy) = 0$, then it follows from Note 3 that we can have $d_{ij}^z \neq 0$ only when ψ^j lies in one of the blocks $b_1, b_2, \ldots, b_n$; and that will prove the theorem. Thus it remains to prove that $\chi(zy) = 0$ for all $y \in H°$, the set of p'-elements of H.

Define $\sigma_0: Z(AG) \to Z(AH)$ by

$$\sigma_0\left(\sum_{x \in G} \alpha_x x\right) := \sum_{x \in H} \alpha_x x.$$

This is the analog of the Brauer homomorphism σ of G with respect to $\langle z \rangle$; clearly $\overline{\sigma_0(a)} = \sigma_0(\bar{a})$ for all $a \in Z(AG)$. Note however that σ_0 is only an A-linear mapping, which does not in general preserve the ring structure.

Step 1 There exists $a_0 \in AG$ such that $a_0 z^p = z^p a_0$ and $e - \sigma_0(e) = a_0 + z^{-1}a_0 z + \cdots + z^{-p+1}a_0 z^{p-1}$.

Let $\mathscr{C}_1, \ldots, \mathscr{C}_s$ be the conjugate classes of G with class sums $c_1, c_2, \ldots, c_s$. Define $\mathscr{C}_i^* := \mathscr{C}_i \cap H$ and $c_i^* := \sum_{x \in \mathscr{C}_i^*} x$. Then $\mathscr{C}_i \backslash \mathscr{C}_i^*$ is normalized by $\langle z \rangle$, and so $\langle z \rangle$ acts on this set by conjugation. Since $\langle z \rangle$ is a p-group, the orbits under this action all have lengths which are powers of p; moreover since $H = C_G(z)$, none of the orbits has length 1. Thus we may find a subset S_l of $\mathscr{C}_i \backslash \mathscr{C}_i^*$ which is normalized by $\langle z^p \rangle$ and such that $S_l, z^{-1}S_l z, \ldots, z^{-p+1}S_l z^{p-1}$ is a partition of $\mathscr{C}_i \backslash \mathscr{C}_i^*$. Put $s_l = \sum_{x \in S_l} x$; then $c_l - c_l^* = s_l + z^{-1}s_l z + \cdots + z^{-p+1}s_l z^{p-1}$. Finally, writing $e = \sum_{l=1}^{s} \gamma_l c_l$ (with $\gamma_l \in A$), we have $\sigma_0(e) = \sum_{l=1}^{s} \gamma_l c_l^*$, and so $e - \sigma_0(e)$ has the form described with $a_0 = \sum_{l=1}^{s} \gamma_l s_l$.

Step 2 There exists $a \in AG$ such that $z^p a = az^p$ and $(1 - f_0)e = a + z^{-1}az + \cdots + z^{-p+1}az^{p-1}$.

First note that $\overline{(1 - f_0)\sigma_0(e)} = (1 - \bar{f}_0)\sigma(\bar{e})$, where σ is the Brauer homomorphism with respect to $\langle z \rangle$. By the note before Theorem 6.3, $\sigma(\bar{e}) = \bar{f}_1 + \cdots + \bar{f}_n = \bar{f}_0$, so $(1 - \bar{f}_0)\sigma(\bar{e}) = 0$. Thus $(1 - f_0)\sigma_0(e) \in \pi AH$. Since $\pi^l \in pA$ for suitably large $l \geq 1$ (see Theorem 3.2) and $f_0^2 = f_0$, the binomial theorem shows that $(1 - f_0)\sigma_0(e)^{p^l} \equiv (1 - f_0)^{p^l}\sigma_0(e)^{p^l} \equiv 0 \pmod{pAH}$. Hence, since f_0, e, and $\sigma_0(e)$ all commute and $e^2 = e$, the binomial theorem again shows that $(1 - f_0)e - (1 - f_0)[e - \sigma_0(e)]^{p^l} = pa_2$ for some $a_2 \in AG$. Note that z commutes with $[e - \sigma_0(e)]$, e, and f_0; and so also with a_2. Hence by Step 1, we have

$$(1 - f_0)e = a + z^{-1}az + \cdots + z^{-p+1}az^{p-1} \quad \text{and} \quad z^p a = az^p,$$

where $a = a_0 + (1 - f_0)[e - \sigma_0(e)]^{p^l-1} a_1$.

Step 3 $\chi(zy) = 0$ *for each* $y \in H^\circ$.

Since submodules of A-free modules are A-free, Step 2 and Lemma 6.5A show that there exists an A-free $A\langle z^p \rangle$-module V_0 such that $V(1 - f)_{\langle z \rangle} \simeq V_0^{\langle z \rangle}$. Finally, it now follows from Lemma 6.5B that $\chi(zy) = 0$ for each $y \in H^\circ$.

EXERCISE

Let G be a group and Σ be a set of p-elements of G. Then the set $\{x \in G \mid \text{the } p\text{-part } x_p \text{ of } x \text{ is conjugate to an element of } \Sigma\}$ is called a *p-section* of G. If x and y are elements of G lying in two disjoint p-sections, then show that $\sum_{\zeta \in B} \zeta(x)\overline{\zeta(y)} = 0$ for each block B of kG.

6.6 Principal blocks: The Third Main Theorem

The principal block of kG (the block in which the trivial character 1_G lies) plays an especially important role in applications. This is partly because we know rather more about this block than we do about blocks in general. The Third Main Theorem shows that the Brauer correspondence maps the principal block into the principal block.

Remark The central character ω of kG associated with the principal block B_0 of kG has been described in Theorem 4.2A. If $\sum_{x \in G} \alpha_x x \in Z(kG)$ with $\alpha_x \in k$, then

$$\omega\left(\sum_{x \in G} \alpha_x x\right) = \sum_{x \in G} \alpha_x.$$

Moreover, since the trivial character 1_G has degree 1, it follows from Theorem 4.5A that B_0 is a block of full defect (that is, the Sylow p-subgroups of G are its defect groups).

Lemma 6.6 Let D be a p-subgroup of the group G and let H be a subgroup of G such that $DC_G(D) \subseteq H$. Let B be a block of kH with D as defect group and let B^G be the block of kG defined by the Brauer correspondence.

  (i) If B is the principal block of kH, then B^G is the principal block of kG.

6.6 PRINCIPAL BLOCKS: THE THIRD MAIN THEOREM

(ii) If $D \triangleleft G$ and B^G is the principal block of kG, then B is the principal block of kH.

Proof (i) Let ω and ω^G, respectively, be the central characters associated with the blocks B and B^G, where B is the principal block of kH. Let $\mathscr{C}_1, \mathscr{C}_2, \ldots, \mathscr{C}_s$ be the conjugate classes of G with class sums $c_1, c_2, \ldots, c_s$ and define $\mathscr{C}_i^* := \mathscr{C}_i \cap H$ and $c_i^* := \sum_{x \in \mathscr{C}_i^*} x$. Then by definition (see §6.3), $\omega^G(c_j) = \omega(c_j^*)$ $(j = 1, 2, \ldots, s)$. Since D normalizes H and $C_G(D) \subseteq H$, D acts by conjugation on each set $\mathscr{C}_i \backslash \mathscr{C}_i^*$ and the orbits of this action all have lengths greater than 1. Since D is a p-group, all orbits have lengths equal to a power of p, so we conclude $|\mathscr{C}_i| - |\mathscr{C}_i^*| = |\mathscr{C}_i \backslash \mathscr{C}_i^*| \equiv 0 \pmod{p}$ for each i. Put $h_i := |\mathscr{C}_i|$ and $h_i^* := |\mathscr{C}_i^*|$. Now $\mathscr{C}_i^*$ is a union of conjugate classes of H, and ω is the central character associated with the principal block of kH; so by the remark above, $\omega(c_i^*) = \bar{h}_i^*$. Since $h_i \equiv h_i^* \pmod{p}$ and $\omega^G(c_i) = \omega(c_i^*)$, we conclude that $\omega^G(c_i) = \bar{h}_i$ for $i = 1, 2, \ldots, s$. The remark above now shows that the block B^G associated with ω^G is the principal block of kG.

(ii) Since $D \triangleleft G$, $H_1 := DC_G(D) \triangleleft G$. Since B^G is the principal block of kG, it follows by (i) that $B^G = b_0^G$, where b_0 is the principal block of kH_1. On the other hand by Note 2 of §6.4, there is at least one block b of kH_1 such that $b^H = B$. Then $b_0^G = B^G = (b^H)^G = b^G$, and so by Lemma 6.4 and Example 2 of §6.4 we conclude that b is the principal block b_0 of kH_1. Thus $B = b_0^H$, and so B is the principal block of kH by (i).

Theorem 6.6 (*The Third Main Theorem*) Let Q be a p-subgroup of the group G and let H be a subgroup of G such that $QC_G(Q) \subseteq H$. Let B be a block of kH with Q as defect group. Then B^G is the principal block of kG iff B is the principal block of kH.

Proof If B is the principal block of kH, then B^G is the principal block of kG by Lemma 6.6. Thus suppose that B^G is the principal block of kG; we must show that B is the principal block of kH. Put $N := N_G(Q)$ and $C := QC_G(Q)$. We shall consider three cases.

Case 1 $H \subseteq N$ *and* Q *is a defect group of* B^N. By the First Main Theorem (Theorem 6.3) the mapping $b \mapsto b^G$ is a bijection from the set of blocks of kN with Q as a defect group to the set of blocks of kG with Q as defect group. Thus $(B^N)^G = B^G$ has Q as defect group. Since B^G is the principal block of kG, its defect groups are Sylow p-subgroups of G (see the remark at the beginning of this section), and so Q is a Sylow p-subgroup of G. This implies that Q is a Sylow p-subgroup of N. Let b_0 be the principal block of kN. It has Q as defect group, and so by Lemma 6.6, b_0^G is the principal block of kG. Thus $b_0^G = B^G = (B^N)^G$, and so $b_0 = B^N$ by Theorem 6.3. Since $Q \triangleleft N$, Lemma 6.6 now shows that B is the principal block of kH.

For the second case we define B^* to be the unique block of $k(N \cap H)$ with Q as defect group such that $(B^*)^H = B$ (Theorem 6.3).

Case 2 Q is a defect group of $(B^*)^N$. Since $(B^*)^G = B^G$ is the principal block of kG, we can apply Case 1 to $H \cap N$ in place of H to conclude that B^* is the principal block of $k(N \cap H)$. Thus by Lemma 6.6 (since Q is a defect group of B^*) $B = (B^*)^H$ is the principal block of kH.

Case 3 The general case. Suppose that the theorem is false, and for a given group G choose Q, H, and B as a counterexample with $|Q|$ as large as possible. Define B^* as above, and let D be a defect group of $(B^*)^N$. By Lemma 6.3, $Q \subseteq D$ because $Q \triangleleft N$, and $Q \neq D$ by Case 2. Define $C_1 := DC_G(D) \subseteq DC_G(Q) \subseteq N$, and $N_1 := N_G(D) \cap N \subseteq N$. By Theorem 6.3 there exists a (unique) block b_1 of kN_1 with D as defect group, such that $b_1^N = (B^*)^N$. On the other hand, since $D \triangleleft N_1$, Theorem 6.4 shows that there is at least one block b of kC_1 with D as defect group such that $b^{N_1} = b_1$; then $b^N = b_1^N = (B^*)^N$. In particular, $b^G = (B^*)^G = B^G$, which is the principal block of kG; hence by the choice of the counterexample, b is the principal block of kC_1 (since $|D| > |Q|$). By Lemma 6.6 this shows that $b^N = (B^*)^N$ is the principal block of kN.

Finally, since $Q \triangleleft H \cap N$, Theorem 6.4 shows that there exists a block b_2 of kC such that $b_2^{H \cap N} = B^*$; then $b_2^N = (B^*)^N$ is the principal block. Let b_0 be the principal block of kC. If D_0 is a defect group of b_0, then $Q \subseteq D_0$ by Lemma 6.3 because $Q \triangleleft C$. Therefore $D_0 C_G(D_0) \subseteq D_0 C_G(Q) \subseteq C$, and so by Lemma 6.6, b_0^N is the principal block of kN. This shows that $b_0^N = b_2^N$ and so $b_2 = b_0$ by Example 3 of §6.4. Thus $B = (B^*)^H = b_2^H = b_0^H$ is the principal block of kH by Lemma 6.6. This contradicts the choice of counterexample, so the theorem is proved.

6.7 The characters in the principal block

For a group G we defined $O_p(G)$ and $O_{p'}(G)$ as the maximal normal p-subgroup and the maximal normal p'-subgroup of G, respectively. Since the product of two normal p-subgroups (respectively, two normal p'-subgroups) is again a normal p-subgroup (p'-subgroup), $O_p(G)$ and $O_{p'}(G)$ are uniquely determined. We define $O_{p', p}(G)$ as the normal subgroup given by $O_{p', p}(G)/O_{p'}(G) = O_p(G/O_{p'}(G))$.

Note G has a normal p-complement iff $G/O_{p'}(G)$ is a p-group and hence iff $O_{p', p}(G) = G$.

These subgroups can be characterized in the following way.

Theorem 6.7A Let B_0 be the principal block of kG. Then

6.7 THE CHARACTERS IN THE PRINCIPAL BLOCK

(i) $x \in O_{p'}(G)$ iff $x \in \operatorname{Ker} \zeta$ for each irreducible ordinary character ζ lying in B_0.

(ii) $x \in O_{p', p}(G)$ iff $x \in \operatorname{Ker} \phi$ for each irreducible modular character ϕ lying in B_0.

Proof (i) Let $\zeta^1, \zeta^2, \ldots, \zeta^t$ be the irreducible ordinary characters lying in B_0 and put $H := \bigcap_{i=1}^{t} \operatorname{Ker} \zeta^i$. Then $y \in H$ implies $\zeta^i(y) = \zeta^i(1)$ for each i; and so $\sum_{i=1}^{t} \zeta^i(y)\zeta^i(1) = \sum_{i=1}^{t} \zeta^i(1)^2 > 0$. Hence Theorem 4.2C shows that each $y \in H$ is a p'-element; therefore the normal subgroup H of G is contained in $O_{p'}(G)$. It remains to show that $O_{p'}(G) \subseteq H$.

Put $n := |O_{p'}(G)|$. Since $p \nmid n$, n is a unit in A, and so

$$f := \frac{1}{n} \sum_{x \in O_{p'}(G)} x \in AG.$$

Clearly $f \in Z(AG)$ [because $O_{p'}(G) \triangleleft G$] and $f^2 = f$. Write $f = e_1 + e_2 + \cdots + e_n$ as a sum of primitive idempotents in $Z(AG)$. By Theorem 3.4A, $\bar{e}_1, \bar{e}_2, \ldots, \bar{e}_n$ are block idempotents of kG. If V_0 is a trivial A-free AG-module, then $vf = v$ for all $v \in V_0$, and so $\bar{V}_0 \bar{e}_j \neq 0$ for some j; hence the principal block B_0 (in which V_0 lies) is equal to the block $\bar{e}_j(kG)$. Now let V_i ($i = 1, 2, \ldots, t$) be an A-free AG-module which affords the character ζ^i (Theorem 3.3). Since ζ^i lies in $B_0 = \bar{e}_j(kG)$, we have $\bar{v}f = \bar{v}\bar{e}_j = \bar{v}$ for all $v \in V_i$. On the other hand, for each $x \in O_{p'}(G)$, $xf = f$, and so $\bar{v} = \overline{vf} = \overline{(vx)f} = \bar{v}x$; so $x \in \operatorname{Ker} \bar{V}_i$ for each i. Finally, we know that $\operatorname{Ker} \bar{V}_i / \operatorname{Ker} V_i$ is always a p-group by Example 2 of §3.3, and so $O_{p'}(G) \subseteq \operatorname{Ker} \bar{V}_i$ implies that $O_{p'}(G) \subseteq \operatorname{Ker} V_i$ as required.

(ii) Let ϕ be an irreducible modular character lying in B_0. Then there exists an indecomposable A-free AG-module V lying in B_0 such that ϕ is afforded by some irreducible kG-constituent of $\bar{V}$. Let ζ be the character afforded by V. Since the irreducible constituents of $V \otimes_A K$ all lie in B_0, (i) shows that $O_{p'}(G) \subseteq \operatorname{Ker} \zeta$ so V may be considered as an $A[G/O_{p'}(G)]$-module. Since k has characteristic p, any irreducible representation of a group $\tilde{G}$ over k contains in its kernel the normal p-subgroup $O_p(\tilde{G})$ (see the corollary of Theorem 2.2A). In particular, $\operatorname{Ker} \phi \supseteq O_{p', p}(G)$. Thus if $L := \bigcap \operatorname{Ker} \phi$ as ϕ ranges over the irreducible modular character lying in B_0, then $L \supseteq O_{p', p}(G)$. It remains to show that we have equality.

If we did not have equality, then the definition of $O_p(G)$ shows that there exists a p'-element $x \in L$, $x \notin O_{p', p}(G)$. Then $\phi(x) = \phi(1)$ for all x, so we conclude that $\zeta(x) = \zeta(1)$ for each irreducible ordinary character lying in B_0 [see Theorem 4.4(iii)]. But this implies that $x \in O_{p'}(G) \subseteq O_{p', p}(G)$ by (i), contrary to the choice of x. Thus $L = O_{p', p}(G)$ and (ii) is proved.

Corollary (i) The trivial character is the only irreducible modular character in the principal block of kG iff G has a normal p-complement.

(ii) The trivial character is the only irreducible ordinary character in the principal block of kG iff G is a p'-group.

EXAMPLE Let z be a p-element in the group G such that $H := C_G(z)$ has a normal p-complement. If ζ is an irreducible ordinary character lying in the principal block of kG, then $\zeta(zy) = \zeta(z)$ for all p'-elements $y \in H$. Indeed, by the corollary of Theorem 6.7A, the trivial character ψ, say, is the only irreducible modular character in the principal block B_0 of kH. If B is any block of kH with defect group D, say, then $D \supseteq \langle z \rangle$ by Lemma 6.2A, so $DC_G(D) \subseteq DC_G(z) = H$. Therefore by Theorem 6.6, B_0 is the only block B of kH such that B^G is the principal block of kG. We can apply Theorem 6.5 to conclude that there exists an algebraic integer d such that $\zeta(zy) = d\psi(y)$ for all p'-elements $y \in H$. But ψ is the trivial character, so $\zeta(zy) = d\psi(1) = \zeta(z)$ for all p'-elements $y \in H$.

We shall need the following classical theorem of Burnside in the next chapter. The usual proof is based on ideas from transfer theory, but we take this opportunity to use some of the results we have just developed to give an alternative proof.

Lemma 6.7 Let P be a Sylow p-subgroup of G and let S_1 and S_2 be normal subsets of P. [In particular, we may have $S_i = \{z_i\}$, $i = 1, 2$, where z_i lies in the center $Z(P)$ of P.] If S_1 is conjugate to S_2 in G, then S_1 is conjugate to S_2 in $N_G(P)$.

Proof Suppose $S_1 = y^{-1} S_2 y$ for some $y \in G$. Then both P and $y^{-1}Py$ are contained in $N_G(S_1)$. Hence by the Sylow theorems, $x^{-1}Px = y^{-1}Py$ for some $x \in N_G(S_1)$. Then $xy^{-1} \in N_G(P)$ and $(xy^{-1})^{-1} S_1 (xy^{-1}) = S_2$.

Remark Let P be a Sylow p-subgroup of G and let H be a normal subgroup of G. Put $\tilde{G} := G/H$. If $C_G(P) = N_G(P)$, then $C_{\tilde{G}}(HP/H) = N_{\tilde{G}}(HP/H)$. Indeed, we have $C_{\tilde{G}}(HP/H) \subseteq N_{\tilde{G}}(HP/H) = N_G(P)H/H = C_G(P)H/H \subseteq C_{\tilde{G}}(HP/H)$, from which the assertion follows.

Theorem 6.7B Let G be a group with a Sylow p-subgroup P such that $C_G(P) = N_G(P)$ (so in particular, P is abelian). Then G has a normal p-complement.

Proof We proceed by induction on $|G|$.

Case 1 The center $Z(G)$ contains a p-subgroup $Q \neq 1$. By the remark above, G/Q satisfies the hypothesis of the theorem. By induction G/Q has a normal p-complement L_0/Q. Since L_0/Q is a p'-group, the Schur–Zassenhaus theorem shows that L_0 has a p-complement L; and $L \triangleleft L_0$ because $Q \subseteq Z(G)$. Then $L = O_{p'}(L_0)$ is a characteristic subgroup of the normal subgroup L_0, and so $L \triangleleft G$. Since L is a p'-group and G/L is a p-group, L is a normal p-complement of G.

6.7 THE CHARACTERS IN THE PRINCIPAL BLOCK

Case 2 The center $Z(G)$ is a p'-group. In this case, for each p-element $z \neq 1$ of G, $C_G(z) \neq G$ and $C_G(z)$ contains a Sylow p-subgroup of G, so it satisfies the hypothesis of the theorem. By induction $C_G(z)$ has a normal p-complement. Since the theorem is trivial when G is a p'-group, by Corollary (ii) of Theorem 6.7A we may assume that the principal block B_0 has more than one irreducible ordinary character. Let $\zeta \neq 1_G$ be any irreducible ordinary character of degree d lying in B_0. By the example above, this shows that $\zeta(zy) = \zeta(z)$ for each p-element $z \neq 1$ in G and each $y \in C_G(z)^\circ$. Note that $C_G(z)^\circ$ is the normal p'-complement of $C_G(z)$, and since $P \subseteq C_G(z)$, $|C_G(z)^\circ| = |C_G(z) : P|$. Now each element of G can be written uniquely as a product zy of commuting elements (z a p-element and y a p'-element) (see §1.5). Moreover each p-element $z \neq 1$ is conjugate in G to one (and, by Lemma 6.7, only one) element of P. Then $\zeta - d \cdot 1_G$ is a class function which vanishes on 1 and so

$$|G|\langle \zeta - d \cdot 1_G, \zeta - d \cdot 1_G \rangle_G = \sum_{z \in P} |G : C_G(z)| \sum_{y \in C_G(z)^\circ} |\zeta(zy) - d|^2$$

$$\geq \sum_{\substack{z \in P \\ z \neq 1}} |G : C_G(z)| \sum_{y \in C_G(z)^\circ} |\zeta(z) - d|^2$$

$$= \sum_{\substack{z \in P \\ z \neq 1}} |G : C_G(z)| \, |C_G(z) : P| \, |\zeta(z) - d|^2$$

$$= |G : P| \sum_{z \in P} |\zeta(z) - d|^2$$

$$= |G|\langle \zeta_P - d1_P, \zeta_P - d1_P \rangle_P.$$

The left-hand side equals

$$|G|\{\langle \zeta, \zeta \rangle + d^2 \langle 1_G, 1_G \rangle - 2d \langle \zeta, 1_G \rangle\} = (d^2 + 1)|G|$$

and the right-hand side is at least $|G|(d^2 + 1)$. Hence we must have equality throughout. In particular, this shows that $|\zeta(y) - d| = 0$ for all $y \in C_G(1)^\circ = G^\circ$, so $\zeta(y) = d$ for all y in G°. Since this is true for each ordinary irreducible character lying in B_0, it follows that the trivial character is the only irreducible modular character in B_0. Hence by Corollary (i) of Theorem 6.7A, G has a normal p-complement.

Corollary If G is a group with a cyclic Sylow 2-subgroup P, then G has a normal 2-complement.

Proof We shall prove that $N_G(P) = C_G(P)$. Put $P = \langle x \rangle$. If x has order 2^n, then the automorphism group Aut P has order 2^{n-1}. Indeed for each integer m, $1 \leq m \leq 2^n$, relatively prime to 2^n there is an automorphism defined by $x^i \mapsto x^{im}$ ($i = 0, 1, \ldots, 2^n - 1$). On the other hand, $N_G(P)$ acts on P

by conjugation, and the kernel of this action is $C_G(P)$. This gives a group homomorphism $N_G(P) \to \text{Aut } P$ with kernel $C_G(P)$. Since P is abelian, $P \subseteq C_G(P)$ and so $N_G(P)/C_G(P)$ has odd order. Since Aut P is even, this shows that $N_G(P) = C_G(P)$ as required.

EXERCISES

1. Let G be a group of order $p^n m$, $(m, p) = 1$, and let N be a normal subgroup of order m. Put $\tilde{G} := G/N$. If P is any p-subgroup of G, prove that $C_{\tilde{G}}(NP/N) = C_G(P)N/N$.
2. Let G be a group and let p be the smallest prime factor of $|G|$. If the Sylow p-subgroups of G are cyclic, prove that G has a normal p-complement.
3. Let G be a group of order $p^n m$, $(m, p) = 1$, and let N be a normal subgroup of order m. Let f be a block idempotent of kN and let $S := \{x \in G \mid x^{-1}fx = f\}$. If $x_1 = 1, x_2, \ldots, x_t$ is a right transversal of S in G, show that $\sum_{i=1}^{t} x_i^{-1} f x_i$ is a block idempotent of kG.

6.8 Notes and comments

Several of the results of this chapter can be proved by ring theoretic arguments (without any recourse to character theory). For an expository work in this line, see Michler [1]. The First Main Theorem was proved in Brauer [8] and a somewhat different (although not quite complete) proof was given by Osima [1]. Both proofs use the theory of ordinary and modular characters. The proof given here is due to Rosenberg [1]. For other proofs of the First Main Theorem see Brauer [10] and Scott [1]. For the relation between the blocks of normal subgroups and the blocks of groups see Fong [1], Passman [2], and Reynolds [1]; in particular, Theorem 6.2B is due to Reynolds [1].

There are several proofs of the Second Main Theorem. The original proof of this theorem uses a combination of arithmetical and norm-theoretic arguments. Iizuka [1] simplified some of Brauer's lemmas. The proof of Nagao [1] is based on results of Green [1,2] concerning P-projective modules. An arithmetical proof based on the choice of a suitable order in the group ring was given by Dade [1]. The present proof (based on Nagao [1]) was given by Dade [5]. Also see Scott [1].

The Third Main Theorem was proved by Brauer [9] and the present proof closely resembles Brauer's proof. For a group theoretic proof of the result that the modular kernel of a block of a finite group G is p-nilpotent, see Michler [2]. Theorem 6.7B is a well-known result of Burnside, but our proof is believed to be new. For a proof by transfer theory see Hall [1] and for a character theoretic proof see Feit [1].

CHAPTER **VII**

Fusion of 2-Groups

An important problem in the theory of finite groups is the classification of simple groups. Feit and Thompson [2] have proved that the orders of nonabelian simple groups are even, and so the simplicity of a nonabelian group depends on its Sylow 2-subgroups. Two conjugate classes of a subgroup H of a group G are *fused* in G if they are contained in a single conjugate class of G. In the present chapter we shall prove three main results (Theorem 7.3B, Theorems 7.4A and 7.4B, and Theorem 7.5), each of which can be interpreted as saying that certain kinds of fusion can never occur when H is a Sylow 2-subgroup with specified structure and G is a simple group. Alternatively, they can be read as giving sufficient conditions for the group G to be nonsimple.

7.1 Further results on generalized decomposition numbers

Let $z \neq 1$ be a p-element of G and put $H := C_G(z)$. Let $\phi^1, \phi^2, \ldots, \phi^r$ be the irreducible modular characters of H (so r is the number of classes of p'-elements of H). Then $\phi^1, \phi^2, \ldots, \phi^r$ are linearly independent elements of $\text{Class}_K(H°)$ (from Theorem 3.7B) and so form a **Z**-basis of the free **Z**-module that they generate and which we shall denote by $\text{Char}(H°)$. Since $\text{rank}(\text{Char}(H°)) = r$, any **Z**-basis of $\text{Char}(H°)$, say $\psi^1, \psi^2, \ldots, \psi^r$, has r

elements and for some invertible $r \times r$ matrix $[a_{ij}]$ with integer entries we have

(1) $$\phi^i = \sum_{j=1}^{r} a_{ij} \psi^j \quad \text{on } H° \quad (i = 1, 2, \ldots, r).$$

Now relative to this new **Z**-basis of Char($H°$) we can define generalizations of the concept of "decomposition numbers," "Cartan matrix," and "generalized decomposition numbers." We proceed to do this.

(a) Let $\chi^1, \chi^2, \ldots, \chi^s$ be the irreducible ordinary characters of H. Then the *decomposition numbers* (with respect to $\psi^1, \psi^2, \ldots, \psi^r$) are the (uniquely determined) integers d_{ij} such that

(2) $$\chi^i = \sum_{j=1}^{r} d_{ij} \psi^j \quad \text{on } H° \quad (i = 1, 2, \ldots, s).$$

These decomposition numbers can be calculated from the ordinary decomposition numbers using (1).

(b) The *Cartan matrix* (relative to $\psi^1, \psi^2, \ldots, \psi^r$) is defined to be the $r \times r$ matrix $\Gamma_H = \Delta_H^T \Delta_H$, where $\Delta_H := [d_{ij}]$ is the matrix of decomposition numbers given in (a) (compare with Theorem 3.7A).

(c) Let $\zeta^1, \zeta^2, \ldots, \zeta^n$ be the irreducible ordinary characters of G. Then the *generalized decomposition numbers* at z (relative to $\psi^1, \psi^2, \ldots, \psi^r$) are the (uniquely determined) algebraic integers d_{ij}^z satisfying

(3) $$\zeta^i(zy) = \sum_{j=1}^{r} d_{ij}^z \psi^j(y) \quad \text{for all } y \in H° \quad (i = 1, 2, \ldots, n).$$

These generalized decomposition numbers can be calculated from the ordinary generalized decomposition numbers (see §6.5) using (1). In particular, suppose that B is a block of kG and the characters ζ^i are ordered so that $\zeta^1, \zeta^2, \ldots, \zeta^t$ are the irreducible ordinary characters lying in B. Suppose further that the irreducible modular characters of H are ordered so that $\phi^1, \phi^2, \ldots, \phi^m$ are those characters that lie in blocks b with the property $b^G = B$; and that the ψ^i are chosen so that $\psi^1, \psi^2, \ldots, \psi^m$ form a **Z**-basis of the submodule of Char($H°$) generated by $\phi^1, \phi^2, \ldots, \phi^m$. Then it follows from Theorem 6.5 (The Second Main Theorem) that $d_{ij}^z = 0$ for all $i = t+1, \ldots, n$ and $j = 1, 2, \ldots, m$.

Note 1 Suppose that $z \neq 1$ is a p-element in G and $w := xzx^{-1}$ is a conjugate of z. Then $C_G(w) = xC_G(z)x^{-1}$. Since ζ^i is a class function, it follows from (3) that

$$\zeta^i(wy) = \zeta^i(zx^{-1}yx) = \sum_{j=1}^{r} d_{ij}^z \psi^j(x^{-1}yx)$$

7.1 FURTHER RESULTS ON GENERALIZED DECOMPOSITION NUMBERS

for all $y \in C_G(w)^\circ$ and $i = 1, 2, \ldots, n$. Hence relative to the conjugate **Z**-basis $(\psi^j)^x$ ($j = 1, 2, \ldots, r$) of $\text{Char}(C_H(w)^\circ)$ we have the same generalized decomposition numbers at w as we do at z.

Note 2 Using (1) and the definition of Γ_H, it follows by a direct calculation from Theorem 3.7B that $\langle \psi^i, \psi^j \rangle_{H^\circ} = c'_{ij}$, where $\Gamma_H^{-1} = [c'_{ij}]$. (Of course this is not the same Cartan matrix as the one referred to in Theorem 3.7A.)

Theorem 7.1 With the notation above, if $\Gamma_H = [c_{ij}]$, then

(4) $$\sum_{l=1}^{t} d_{li}^z \overline{d_{lj}^z} = c_{ji} \quad \text{for} \quad i = 1, 2, \ldots, m \text{ and all } j$$

(for this section, the bar denotes complex conjugate). Moreover, if w is a nontrivial p-element of G not conjugate to z, and d_{ij}^w are the generalized decomposition numbers at w with respect to some **Z**-basis of $\text{Char}(C_G(w)^\circ)$, then

(5) $$\sum_{l=1}^{t} d_{li}^z \overline{d_{lj}^w} = 0 \quad \text{for all} \quad i = 1, 2, \ldots, m \text{ and all } j.$$

Remark The case w is conjugate but not equal to z is covered in the Note 1 above.

Proof Let $\mathscr{C}_1, \mathscr{C}_2, \ldots, \mathscr{C}_r$ be the classes of p'-elements in H, and for $i = 1, 2, \ldots, n$, define $\zeta_j^i := \zeta^i(zy)$ when $y \in \mathscr{C}_j$ (the value of ζ_j^i is independent of the choice of y because $z \in Z(H)$). Similarly write $\psi_j^i = \psi^i(y)$ when $y \in \mathscr{C}_j$. Consider the $n \times r$ matrix $X := [\zeta_j^i]$, the $r \times r$ matrix $U := [\psi_j^i]$ and the $n \times r$ matrix $\Delta := [d_{ij}^z]$; $\Gamma_H = [c_{ji}]$ is an $r \times r$ matrix. By definition we have $X = \Delta U$. Now $\sum_{l=1}^{n} \zeta_i^l \overline{\zeta_j^l} = n_j \delta_{ij}$, where $n_j := |C_G(zy)|$ for $y \in \mathscr{C}_j$. Note that $C_G(zy) = C_G(y) \cap C_G(z) = C_H(y)$ because z and y are, respectively, the p-part and the p'-part of zy; therefore $n_j = |C_H(y)|$ for $y \in \mathscr{C}_j$. This shows that the complex conjugate transpose X^* of X satisfies $X^*X = [n_i \delta_{ij}]$. On the other hand, with $g := |G|$ and $h_i := g/n_i$ it follows from Note 2 that

$$\sum_{l=1}^{r} h_l \psi_l^i \overline{\psi_l^j} = g \langle \psi^i, \psi^j \rangle = c'_{ij} g \quad \text{where} \quad \Gamma_H^{-1} = [c'_{ij}].$$

Thus $U[n_i^{-1} \delta_{ij}]U^* = \Gamma_H^{-1}$. Since $\psi^1, \psi^2, \ldots, \psi^r$ are linearly independent, U is invertible, and so

$$\Gamma_H = (U^*)^{-1}[n_i \delta_{ij}]U^{-1} = (U^*)^{-1}X^*XU^{-1} = \Delta^*\Delta.$$

Hence $\sum_l \overline{d_{li}^z} d_{lj}^z = c_{ij}$ for all i and j. However if $i \leq m$, then the Second Main Theorem shows that $d_{li}^z = 0$ for $l = t+1, \ldots, n$ [see (c) above]. Hence (4) is proved.

Consider (5). Since z and w are not conjugate, the two elements zy and wx

are not conjugate for any $y \in C_G(z)^\circ$ and $x \in C_G(w)^\circ$. Therefore by Theorem 2.2A we have

(6) $$\sum_{l=1}^{n} \zeta^l(zy)\overline{\zeta^l(wx)} = \sum_{l=1}^{n} \zeta^l(zy)\zeta^l((wx)^{-1}) = 0$$

for all $y \in C_G(z)^\circ$, $x \in C_G(w)^\circ$. Let $\xi^1, \xi^2, \ldots, \xi^{r'}$ be any basis of $\mathrm{Char}(C_G(w)^\circ)$. Then substituting in (6) the values of $\zeta^l(zy)$ and $\zeta^l(wx)$ in terms of the generalized decomposition numbers we obtain

$$\sum_{i=1}^{r} \sum_{j=1}^{r'} \left(\sum_{l=1}^{n} d_{li}^z \, \overline{d_{lj}^w} \right) \psi^i(y)\xi^j(x) = 0$$

for all $y \in C_G(z)^\circ$ and all $x \in C_G(w)^\circ$. The linear independence of $\psi^1, \psi^2, \ldots, \psi^r$ and $\xi^1, \xi^2, \ldots, \xi^{r'}$ now show that

$$\sum_{l=1}^{n} d_{li}^z \, \overline{d_{lj}^w} = 0 \quad \text{for all } i \text{ and } j.$$

Finally using the fact that $d_{li}^z = 0$ for $i = 1, 2, \ldots, m$ and $l = t + 1, \ldots, n$ [see (c) above], we have (5).

Corollary Let z be a nontrivial p-element in G and put $H := C_G(z)$. Let B_0 be the principal block of kG and b_0 be the principal block of kH. Let $\zeta^1, \zeta^2, \ldots, \zeta^t$ and $\chi^1, \chi^2, \ldots, \chi^{t'}$ be the irreducible ordinary characters lying in B_0 and b_0, respectively. Then for all $y \in H^\circ$,

(i) $\sum_{l=1}^{t} |\zeta^l(zy)|^2 = \sum_{l=1}^{t'} |\chi^l(y)|^2$

(ii) $\sum_{l=1}^{t} \zeta^l(zy)\overline{\zeta^l(wy)} = 0$

whenever w is a p-element not conjugate to z in G.

Proof (i) If b is any block of kH with defect group Q, say, then $\langle z \rangle \subseteq Q$ by Lemma 6.2A because $\langle z \rangle \triangleleft H$. Hence $H \supseteq QC_G(Q)$, and so by the Third Main Theorem (Theorem 6.6), $b^G = B_0$ iff $b = b_0$. Choose $\psi^1, \psi^2, \ldots, \psi^m$ as those characters lying in the block b_0. Then it follows from (3) and the theorem that for all $y \in H^\circ$

$$\sum_{l=1}^{t} |\zeta^l(zy)|^2 = \sum_{i=1}^{m} \sum_{j=1}^{m} \left(\sum_{l=1}^{t} d_{li}^z \, \overline{d_{lj}^z} \right) \psi^i(y)\overline{\psi^j(y)}$$

$$= \sum_{i=1}^{m} \sum_{j=1}^{m} c_{ji} \psi^i(y)\overline{\psi^j(y)}$$

since $d_{li}^z = 0$ for $l = 1, 2, \ldots, t$ when $i > m$. On the other hand, we can write

7.1 FURTHER RESULTS ON GENERALIZED DECOMPOSITION NUMBERS

the χ^l in terms of the ψ^j using the decomposition numbers d_{ij} [Eq. (2)]. Since $d_{li} = 0$ for $l = t' + 1, \ldots, n$ and $i = 1, 2, \ldots, m$ by ordering of the ψ^j, the relation $[c_{ij}] = \Gamma_H = [d_{ij}]^T[d_{ij}]$ shows that $c_{ji} = \sum_{l=1}^{t'} d_{lj} d_{li}$. Hence for all $y \in H°$, we have

$$\sum_{l=1}^{t'} |\chi^l(y)|^2 = \sum_{i=1}^{m} \sum_{j=1}^{m} \left(\sum_{l=1}^{t'} d_{li} d_{lj} \right) \psi^i(y) \overline{\psi^j(y)}$$

$$= \sum_{i=1}^{m} \sum_{j=1}^{m} c_{ji} \psi^i(y) \overline{\psi^j(y)}.$$

This proves (i).

(ii) Let $\xi^1, \xi^2, \ldots, \xi^{r'}$ be the irreducible modular characters of $C_G(w)$. Then as above

$$\sum_{l=1}^{t} \zeta^l(zy)\overline{\zeta^l(wy)} = \sum_{i=1}^{m} \sum_{j=1}^{m} \left(\sum_{l=1}^{t} d_{li}^z \overline{d_{lj}^w} \right) \psi^i(y) \overline{\xi^j(y)} = 0$$

by (5) whenever w is a p-element not conjugate to z.

EXAMPLE   Suppose that the hypotheses of the corollary hold and suppose that $H = C_G(z)$ has a normal p-complement. In this case, for each $y \in H°$, we have

(7) $$\sum_{l=1}^{t} |\zeta^l(zy)|^2 = |P|,$$

where P is a Sylow p-subgroup of H. Indeed, since H has a normal p-complement, $H/O_{p'}(H) \simeq P$. Let χ be an irreducible ordinary character of H. If χ lies in the principal block b_0, then $\operatorname{Ker} \chi \supseteq O_{p'}(H)$ by Theorem 6.7A. Conversely, since $O_{p'}(H) = H°$ in this case, Theorem 4.2B shows that if $\operatorname{Ker} \chi \supseteq O_{p'}(H)$, then χ lies in b_0. Thus the characters $\chi^1, \chi^2, \ldots, \chi^{t'}$ lying in b_0 are precisely the irreducible characters of H whose kernels contain $O_{p'}(H)$. These characters correspond in a natural way with the set of all irreducible ordinary characters of $P \simeq H/O_{p'}(H)$. Hence by Theorem 2.2A we have for each $y \in H°$

$$\sum_{l=1}^{t'} |\chi^l(y)|^2 = \sum_{l=1}^{t'} |\chi^l(1)|^2 = |P|.$$

Thus (7) now follows from the corollary.

EXERCISES

1. Let B be a block of kG and $\zeta^1, \zeta^2, \ldots, \zeta^t$ be the irreducible ordinary characters lying in B. If $z \neq 1$ is a p-element of G, let d_{ij}^z be the ordinary generalized decomposition numbers at z. For each p-element w of G, show

that $\sum_{l=1}^{t} d_{li}^{z} \overline{d_{lj}^{w}}$ is a rational integer divisible by p. [*Hint:* For any $y \in C_G(z)$ and $x \in C_G(w)$, show that $\sum_{l=1}^{t} \zeta^l(zy)\overline{\zeta^l(wx)}$ belongs to πA.]

2. Let B be a block of defect 0 and let ζ^i be the (unique) irreducible ordinary character lying in B. Show that the generalized decomposition number $d_{ij}^z = 0$ for all j and all p-elements $z \neq 1$ in G.

7.2 Some technical lemmas

In the remaining theorems in this chapter we shall be interested in the Sylow 2-subgroups of the group and consequently be applying our results on modular representations in the case where the characteristic of k is $p = 2$. The prime 2 plays a special role in many group theoretic investigations, partly because the existence of involutions (elements of order 2) in a group enables one to analyze the structure of the group more easily. The lemmas of the present section deal with results of this type; they will find application in the succeeding sections.

Lemma 7.2A ($p = 2$) Let z be a 2-element in the group G, let $\mathscr{C}_1$, $\mathscr{C}_2$ be two classes of involutions in G such that no element in $\mathscr{C}_1 \mathscr{C}_2$ has z as its 2-part, and let $\zeta^1, \zeta^2, \ldots, \zeta^t$ be the irreducible ordinary characters lying in a block B of kG. Let $\psi^1, \psi^2, \ldots, \psi^r$ be a Z-basis of $\mathrm{Char}(C_G(z)^\circ)$ such that $\psi^1, \psi^2, \ldots, \psi^m$ is a Z-basis of the submodule generated by the modular characters of $C_G(z)$ that lie in blocks b with the property $b^G = B$. Let d_{ij}^z be the corresponding generalized decomposition numbers at z. Then for each $x \in \mathscr{C}_1$, $y \in \mathscr{C}_2$ we have

(1) $$\sum_{i=1}^{t} d_{ij}^z \zeta^i(x)\zeta^i(y)/\zeta^i(1) = 0 \quad \text{for } j = 1, 2, \ldots, m$$

(2) $$\sum_{i=1}^{t} \zeta^i(z)\zeta^i(x)\zeta^i(y)/\zeta^i(1) = 0.$$

Proof Let $\mathscr{C}_1, \mathscr{C}_2, \ldots, \mathscr{C}_s$ be the classes of G and $c_1, c_2, \ldots, c_s$ the corresponding class sums in KG; we have $x \in \mathscr{C}_1$, $y \in \mathscr{C}_2$. Then we can write

(3) $$c_1 c_2 = \sum_{i=1}^{s} v_i c_i \quad (\text{with } v_i \in \mathbf{Z}).$$

If T is a representation of G over K which affords the irreducible character ζ, then $T(c_1)T(c_2) = \sum_{j=1}^{s} v_j T(c_j)$ by (3). However $T(c_j)$ is a scalar because K is a splitting field, and so $T(c_i) = \lambda_i \cdot 1$ gives $\lambda_i \zeta(1) = h_i \zeta_i$ where $h_j := |\mathscr{C}_j|$ and ζ_j is the value of ζ on $\mathscr{C}_j$. Thus we have

(4) $$h_1 h_2 \zeta_1 \zeta_2 = \zeta(1) \sum_{j=1}^{s} v_j h_j \zeta_j.$$

7.2 SOME TECHNICAL LEMMAS

If w is a $2'$-element in $C_G(z)$, then $zw \notin \mathscr{C}_1 \mathscr{C}_2$ by hypothesis, and so $(zw)^{-1} \notin \mathscr{C}_2^{-1}\mathscr{C}_1^{-1} = \mathscr{C}_2\mathscr{C}_1$ (since $\mathscr{C}_1$, $\mathscr{C}_2$ consist of involutions). Suppose that $zw \in \mathscr{C}_l$ and put $\mathscr{C}_{l*} := \mathscr{C}_l^{-1}$. Then since $\mathscr{C}_{l*} \cap \mathscr{C}_1\mathscr{C}_2 = \varnothing$, (3) shows that $v_{l*} = 0$. On the other hand if $\zeta^1, \zeta^2, \ldots, \zeta^s$ are all the irreducible ordinary characters of G, then (4) together with the orthogonality relations (Theorem 2.2A) shows that

$$h_1 h_2 \cdot \sum_{i=1}^s \zeta_1^i \zeta_2^i \zeta_l^i / \zeta^i(1) = v_{l*} |G| = 0.$$

Writing $\zeta_l^i = \zeta^i(zw) = \sum_{j=1}^r d_{ij}^z \psi^j(w)$, this gives

$$\sum_{j=1}^r \left\{ \sum_{i=1}^s d_{ij}^z \zeta_1^i \zeta_2^i / \zeta^i(1) \right\} \psi^j(w) = 0$$

for each $2'$-element w in $C_G(z)$. The linear independence of $\psi^1, \psi^2, \ldots, \psi^r$ now shows that

$$\sum_{i=1}^s d_{ij}^z \zeta_1^i \zeta_2^i / \zeta^i(1) = 0$$

for each j. From the ordering of the ψ^j it follows from the Second Main Theorem that $d_{ij}^z = 0$ for $i = t+1, \ldots, s$ and $j = 1, 2, \ldots, m$ (see (c) of §7.1). Therefore for $j = 1, 2, \ldots, m$ we have

$$\sum_{i=1}^t d_{ij}^z \zeta_1^i \zeta_2^i / \zeta^i(1) = 0.$$

This proves (1). Equation (2) now follows because $\zeta^i(z) = \sum_{j=1}^s d_{ij}^z \psi^j(1) = \sum_{j=1}^m d_{ij}^z \psi^j(1)$ for $i = 1, 2, \ldots, t$.

Remark If z is a 2-element of the group G and $\mathscr{C}_1$ and $\mathscr{C}_2$ are classes of involutions in G such that for each $x \in \mathscr{C}_1$, $x^{-1}zx \neq z^{-1}$, then no element in $\mathscr{C}_1 \mathscr{C}_2$ has z as its 2-part; so in this case (1) and (2) hold. Indeed, suppose on the contrary that $x \in \mathscr{C}_1$, $y \in \mathscr{C}_2$ and $xy = zw$ for some $2'$-element $w \in C_G(z)$. Then $x^{-1}zwx = yx = y^{-1}x^{-1} = (xy)^{-1} = (zw)^{-1} = w^{-1}z^{-1}$. But equating the 2-parts then gives $x^{-1}zx = z^{-1}$, contrary to the hypothesis.

Let $\zeta^1, \zeta^2, \ldots, \zeta^s$ be the irreducible ordinary characters of G and suppose that $\zeta^1, \zeta^2, \ldots, \zeta^t$ are the characters lying in the principal block B_0. Let Q be a p-subgroup of G. Then for each $\alpha \in \text{Char}(Q)$, the induced generalized character $\alpha^G \in \text{Char}(G)$; suppose $\alpha^G = \sum_{i=1}^s a_i \zeta^i$ (where each $a_i \in \mathbb{Z}$). We define $\hat{a}(\alpha) := [a_1, a_2, \ldots, a_t]$, the t-tuple of coefficients corresponding to the characters $\zeta^1, \zeta^2, \ldots, \zeta^t$ lying in B_0. If $\beta \in \text{Char}(Q)$ and $\hat{a}(\beta) = [b_1, b_2, \ldots, b_t]$, then we define the *dot product* as usual by $\hat{a}(\alpha) \cdot \hat{a}(\beta) = a_1 b_1 + \cdots + a_t b_t$.

Note By the Frobenius reciprocity (Theorem 2.5A) we have

$$a_i = \langle \alpha^G, \zeta^i \rangle_G = \langle \alpha, \zeta_Q^i \rangle_Q.$$

Hence if $\chi^1, \chi^2, \ldots, \chi^{s'}$ are the irreducible ordinary characters of Q, then for $i = 1, 2, \ldots, t$

$$\zeta_Q^i = \sum_{j=1}^{s'} a_{ij}\chi^j, \quad \text{where } \hat{a}(\chi^j) = [a_{1j}, a_{2j}, \ldots, a_{tj}].$$

Lemma 7.2B $(p = 2)$ Let Q be a 2-subgroup of the group G and let S be a subset of Q. Let H be a subgroup of G such that $Q \subseteq H \subseteq N_G(Q)$. Let α, $\beta \in \text{Char}(Q)$ such that $\alpha(x) = 0$ for all x in $Q \backslash S$ and let $\zeta^1, \zeta^2, \ldots, \zeta^t$ be the irreducible ordinary characters of G lying in the principal block of kG. Suppose that

(i) if $x \in S$ and $y \in Q$ are conjugate in G, then they are conjugate in H; and

(ii) there is a constant γ such that $\gamma |C_H(x)| = \sum_{i=1}^{t} |\zeta^i(x)|^2$ for all $x \in S$.

Then

$$\hat{a}(\alpha) \cdot \hat{a}(\beta) = \gamma \left\langle \alpha, \sum_{x \in H} \beta^x \right\rangle_Q \bigg/ |Q|.$$

Proof By the note above

$$\hat{a}(\alpha) \cdot \hat{a}(\beta) = \sum_{i=1}^{t} \langle \alpha, \zeta_Q^i \rangle_Q \langle \beta, \zeta_Q^i \rangle_Q$$

$$= |Q|^{-2} \sum_{x \in Q} \sum_{y \in Q} \alpha(x)\beta(y^{-1}) \sum_{i=1}^{t} \zeta^i(x)\zeta^i(y^{-1}).$$

Now

$$\sum_{i=1}^{t} \zeta^i(x)\zeta^i(y^{-1}) = \begin{cases} \sum_{i=1}^{t} |\zeta^i(x)|^2 & \text{if } y \text{ is conjugate to } x \\ 0 & \text{otherwise} \end{cases}$$

by the corollary of Theorem 7.1. Therefore taking into account (i) and (ii) we conclude that

$$\hat{a}(\alpha) \cdot \hat{a}(\beta) = |Q|^{-2} \sum_{x \in S} \alpha(x)\gamma |C_H(x)| \sum_{\substack{y \in Q \\ y \text{ conjugate to } x}} \beta(y^{-1})$$

$$= \gamma |Q|^{-2} \sum_{x \in S} \alpha(x) \sum_{z \in H} \beta(z^{-1}x^{-1}z)$$

$$= \gamma |Q|^{-1} \left\langle \alpha, \sum_{z \in H} \beta^z \right\rangle_Q.$$

Theorem 7.2 $(p = 2)$ Let Q be a 2-subgroup of the group G, u an involution in G, and S a subset of Q such that for each $x \in G$ conjugate to an

7.2 SOME TECHNICAL LEMMAS

element in S, $u^{-1}xu \neq x^{-1}$. Let $\alpha \in \text{Char}(Q)$ with $\alpha(x) = 0$ for all $x \in Q\backslash S$. Finally, let $\zeta^1, \zeta^2, \ldots, \zeta^t$ be the irreducible ordinary characters in the principal block of kG with $\zeta^1 = 1_G$. Then

$$\text{(5)} \qquad \sum_{i=1}^{t} a_i(\alpha)\zeta^i(u)^2/\zeta^i(1) = 0$$

$$\text{(6)} \qquad \sum_{i=1}^{t} a_i(\alpha)\zeta^i(1) = 0$$

$$\text{(7)} \qquad \sum_{i=1}^{t} a_i(\alpha)\zeta^i(u) = 0,$$

where $\hat{a}(\alpha) := [a_1(\alpha), a_2(\alpha), \ldots, a_t(\alpha)]$.

Proof By the remark following Lemma 7.2A we see that the hypotheses of Lemma 7.2A are satisfied in our case with u in place of x and y, and any element of S^{-1} in place of z. Thus by (2)

$$\text{(8)} \qquad \sum_{i=1}^{t} \zeta^i(x^{-1})\zeta^i(u)^2/\zeta^i(1) = 0$$

for all $x \in S$. Since $a_i(\alpha) = \langle \alpha^G, \zeta^i \rangle_G = \langle \alpha, \zeta_Q^i \rangle_Q = |Q|^{-1} \sum_{x \in Q} \alpha(x)\zeta^i(x^{-1})$ and $\alpha(x) = 0$ on $Q\backslash S$, (5) follows at once from (8). On the other hand since $u^{-1}uu = u^{-1}$, u is not conjugate in G to any $x \in S$ by the hypothesis on S. Therefore by the corollary of Theorem 7.1, $\sum_{i=1}^{t} \zeta^i(1)\zeta^i(x) = 0$ and $\sum_{i=1}^{t} \zeta^i(u)\zeta^i(x) = 0$ for all $x \in S$, since 1, x, and u are mutually nonconjugate elements of G. Again using the expression for $a_i(\alpha)$ given above, (6) and (7) follow at once.

Corollary If $t > 1$ and $\langle \alpha, 1_Q \rangle_Q \neq 0$, then either there exist at least two $a_i(\alpha) > 0$ and at least two $a_i(\alpha) < 0$, or else G has a proper normal subgroup containing $O_{2'}(G)$.

Proof Since $\langle \alpha, 1_Q \rangle \neq 0$, not all $a_i(\alpha) = 0$. Suppose that the characters are ordered so that $a_i(\alpha) \neq 0$ for $i \leq m$, but $a_i(\alpha) = 0$ for $i = m+1, \ldots, t$. By (6) not all $a_i(\alpha)$ have the same sign. Suppose that the first alternative does not hold. Taking $-\alpha$ in place of α if necessary and reordering the ζ^i we may assume that $a_i(\alpha) > 0$ for $i = 1, 2, \ldots, m-1$ and $a_m(\alpha) < 0$. Then

$$\sum_{1 \leq i < j \leq m-1} \frac{a_i(\alpha)a_j(\alpha)}{\zeta^i(1)\zeta^j(1)} [\zeta^i(u)\zeta^j(1) - \zeta^i(1)\zeta^j(u)]^2$$

$$= \left\{ \sum_{i=1}^{m-1} \frac{a_i(\alpha)}{\zeta^i(1)} \zeta^i(u)^2 \right\} \left\{ \sum_{i=1}^{m-1} \frac{a_i(\alpha)}{\zeta^i(1)} \zeta^i(1)^2 \right\}$$

$$- \left\{ \sum_{i=1}^{m-1} \frac{a_i(\alpha)}{\zeta^i(1)} \zeta^i(u)\zeta^i(1) \right\}^2,$$

which equals 0 using (5), (6), and (7). This implies that $\zeta^i(u)\zeta^j(1) = \zeta^i(1)\zeta^j(u)$ for all $i, j = 1, 2, \ldots, m - 1$ and hence [using (6) and (7)] for $i = m$ as well. In particular, taking ζ^j as the principal character 1_G, we get $\zeta^i(u) = \zeta^i(1)$ for all i. Since Ker $\zeta^i \supseteq O_{2'}(G)$ for all i (by Theorem 6.7A), the corollary follows.

7.3 Groups with Sylow 2-subgroups of type $(2^m, 2^m)$

In the present section we investigate the structure of a group G which has a Sylow 2-subgroup P of the type $(2^m, 2^m)$, (that is, P is isomorphic to a direct product of two cyclic groups each of order 2^m). In particular, when $m > 1$, Theorem 7.3 shows that a group with such a Sylow 2-subgroup cannot be simple.

Remark In dealing with involutions the following fact is often useful. If u is an involution in the group G, and ζ is an irreducible ordinary character of G, then $\zeta(u)$ is an integer and $\zeta(u) \equiv \zeta(1) \pmod 2$.

[*Proof* If T is a representation which affords ζ, then the eigenvalues of $T(u)$ are 1 and -1, and so $\zeta(u)$ is an integer. The second assertion follows from the observation $\langle \zeta_{\langle u \rangle}, 1_{\langle u \rangle} \rangle = \frac{1}{2}(\zeta(1) + \zeta(u))$.]

Lemma 7.3A Let P be a Sylow 2-subgroup of the group G and suppose P is of type $(2^m, 2^m)$. Then

(i) Aut $P/O_2(\text{Aut } P) \simeq GL(2, 2)$ and hence $|N_G(P) : C_G(P)| = 1$ or 3.
(ii) If $N_G(P) \neq C_G(P)$ and $z \neq 1$ lies in P, then $C_G(z) \cap N_G(P) = C_G(P)$. Hence [by (i)] z has 3 conjugates in $N_G(P)$.

Proof (i) Choose $x, y \in P$ of order 2^m such that $P = \langle x \rangle \times \langle y \rangle$. Then each $\alpha \in \text{Aut } P$ is determined by its values $x^\alpha = x^{m_{11}} y^{m_{12}}$ and $y^\alpha = x^{m_{21}} y^{m_{22}}$ (where the m_{ij} are integers with $0 \leq m_{ij} < 2^m$). It is readily verified that the mapping

$$\alpha \mapsto \begin{bmatrix} m_{11} & m_{12} \\ m_{21} & m_{22} \end{bmatrix} \pmod{2^m}$$

defines an isomorphism of Aut P onto $GL(2, \mathbf{Z}/(2^m))$. On the other hand, the canonical mapping (mod $2\mathbf{Z}/(2^m)$) gives a homomorphism of $GL(2, \mathbf{Z}/(2^m))$ onto $GL(2, 2)$, and the elements of the kernel have the form

$$\begin{bmatrix} 1 + 2n_{11} & 2n_{12} \\ 2n_{21} & 1 + 2n_{22} \end{bmatrix} \pmod{2^m}.$$

Counting the elements we see that this kernel has order $(2^{m-1})^4$, so the kernel lies in $O_2(GL(2, \mathbf{Z}/(2^m)))$. Since $GL(2, 2)$ is nonabelian of order 6, it has

7.3 GROUPS WITH SYLOW 2-SUBGROUPS OF TYPE $(2^m, 2^m)$

no nontrivial normal 2-subgroup; hence we conclude

$$\text{Aut } P/O_2(\text{Aut } P) \simeq GL(2, \mathbf{Z}/(2^m))/O_2(GL(2, \mathbf{Z}/(2^m))) \simeq GL(2, 2).$$

Finally, consider the action of $N_G(P)$ on P defined by conjugation. Each $u \in N_G(P)$ acts as an automorphism of P and the kernel of the action of $N_G(P)$ is $C_G(P)$. Since $P \subseteq C_G(P)$, this shows that $N_G(P)/C_G(P)$ is isomorphic to a $2'$-subgroup of Aut P. But 3 is the largest $2'$-factor of $|\text{Aut } P|$; therefore $|N_G(P)/C_G(P)| = 1$ or 3.

(ii) Consider the automorphism $\alpha: z \mapsto u^{-1}zu$ of P (for fixed $u \in N_G(P) \setminus C_G(P)$). With the notation above, α corresponds to a matrix

$$\begin{bmatrix} m_{11} & m_{12} \\ m_{21} & m_{22} \end{bmatrix} \quad (\text{mod } 2^m)$$

of order 3 in $GL(2, \mathbf{Z}/(2^m))$. If $z^\alpha = z$ for some $z \neq 1$ in P, then $w^\alpha = w$ for some involution w in P (take w as a power of z). But the involutions of P are $x^{2^{m-1}}$, $y^{2^{m-1}}$, and $(xy)^{2^{m-1}}$, so we have $m_{11} \equiv 1$, $m_{21} \equiv 0$ (mod 2); $m_{12} \equiv 0$, $m_{22} \equiv 1$ (mod 2); or $m_{11} + m_{12} \equiv 1$, $m_{21} + m_{22} \equiv 1$ (mod 2); in the respective cases. However, an obvious calculation shows that in all these cases

$$\begin{bmatrix} m_{11} & m_{12} \\ m_{21} & m_{22} \end{bmatrix}^2 \in O_2(GL(2, \mathbf{Z}/(2^m))),$$

and so $\alpha^2 \in O_2(\text{Aut } P)$. This is impossible since α has order 3; thus α fixes no nontrivial element of P. Part (ii) now follows.

Lemma 7.3B ($p = 2$) Let P be a sylow 2-subgroup of the group G and suppose that P is of type $(2^m, 2^m)$. Let $\zeta^1 = 1_G$, $\zeta^2, \ldots, \zeta^t$ be the irreducible ordinary characters of G lying in the principal block B_0. Then

(i) For all $z \neq 1$ in P and each $y \in C_G(z)^\circ$ we have

$$C_G(z) \cap N_G(P) = C_G(P) \quad \text{and} \quad \sum_{i=1}^{t} |\zeta^i(zy)| = |P|.$$

(ii) Either G has a normal 2-complement or else G has only one class of involutions.

(iii) If G has only one class of involutions and $m = 1$, then $t = 4$ and for some $\varepsilon = 1$ or -1 and a suitable ordering of $\zeta^2, \zeta^3, \zeta^4$ we have

(a) $\zeta^2(z) = 1$, $\zeta^3(z) = \varepsilon$, $\zeta^4(z) = -1$, and $\zeta^i(z) \equiv \zeta^i(1) \pmod 4$ for each i;
(b) $1 + \zeta^2 + \varepsilon\zeta^3 - \zeta^4 = 0$ on G°;
(c) $\zeta^i(zy) = \zeta^i(z)$ for $i = 1, 2, 3, 4$ and $y \in C_G(z)^\circ$;
(d) The restrictions of $\zeta^1, \zeta^2, \varepsilon\zeta^3$ to G° form a **Z**-basis for the submodule

of Char($G°$) generated by the modular characters lying in B_0, and the corresponding decomposition matrix and Cartan matrix, respectively, are

$$\begin{bmatrix} 1 & 0 & 0 \\ 0 & 1 & 0 \\ 0 & 0 & \varepsilon \\ 1 & 1 & 1 \end{bmatrix} \quad \text{and} \quad \begin{bmatrix} 2 & 1 & 1 \\ 1 & 2 & 1 \\ 1 & 1 & 2 \end{bmatrix}$$

Proof (i) Put $H := C_G(z)$. It follows from Lemma 7.3A that $N_H(P) = N_G(P) \cap H = C_H(P)$. In particular, this shows that H has a normal 2-complement (Theorem 6.7B). Thus the final assertion of (i) follows from the example of §7.1.

(ii) If $C_G(P) = N_G(P)$, then G has a normal 2-complement by Theorem 6.7B. On the other hand if $N_G(P) \neq C_G(P)$, then Lemma 7.3A shows that the three involutions that P contains lie in a single conjugate class of $N_G(P)$. Hence by the Sylow theorems, G has a single class of involutions.

(iii) (a) If $m = 1$ then $|P| = 4$. Let $z \neq 1$ lie in P. Then by the remark above, $\zeta^i(z)$ is an integer and $\zeta^i(z) \equiv \zeta^i(1)$ (mod 2). Since the principal block of G has defect 2, it follows from Theorem 4.5A that $2 \nmid \zeta^i(1)$ ($i = 1, 2, \ldots, t$), and so $\zeta^i(z)$ is an odd integer. In particular, $\zeta^i(z)$ is a nonzero integer for $i = 1, 2, \ldots, t$, and so (i) (applied with $y = 1$) shows that $t = 4$ and $|\zeta^i(z)| = 1$ for each i; hence $\zeta^1(z) = 1_G(z) = 1$ and $\zeta^i(z) = \varepsilon_i$ ($i = 2, 3, 4$) for suitable $\varepsilon_i = \pm 1$. Since the nontrivial elements of P are all conjugate in G by hypothesis, $\langle \zeta^i_P, 1_P \rangle = \frac{1}{4}\{\zeta^i(1) + 3\varepsilon_i\}$ for $i = 1, 2, 3, 4$, which shows that $\zeta^i(1) \equiv \varepsilon_i$ (mod 4). Finally, from Theorem 4.2C we have

$$0 = \sum_{i=1}^{4} \zeta^i(z)\zeta^i(1) = 1 + \varepsilon_2 \zeta^2(1) + \varepsilon_3 \zeta^3(1) + \varepsilon_4 \zeta^4(1).$$

So with a suitable ordering we may take $\varepsilon_2 = 1$ and $\varepsilon_4 = -1$; put $\varepsilon_3 = \varepsilon$ and then (a) is proved.

(b) Again Theorem 4.2C shows that for all $y \in G°$,

$$0 = \sum_{i=1}^{4} \zeta^i(z)\zeta^i(y) = 1 + \zeta^2(y) + \varepsilon\zeta^3(y) - \zeta^4(y).$$

(c) Since z lies in the center of $H = C_G(z)$, $\langle z \rangle \triangleleft H$. Then $H/\langle z \rangle$ has a (cyclic) Sylow 2-subgroup of order 2, so it has a normal 2-complement $L_0/\langle z \rangle$ by the Corollary of Theorem 6.7B. Again, $\langle z \rangle$ is a Sylow 2-subgroup of L_0, so L_0 has a normal 2-complement of H. Now (c) follows from the Example of §6.7.

(d) By Theorem 4.4, the **Z**-submodule of Char($H°$) spanned by the modular characters lying in B_0 is the same as that spanned by the restriction of $\zeta^1, \zeta^2, \zeta^3$, and ζ^4 to $G°$. By (b) this module is spanned by the restriction of ζ^1, ζ^2, and ζ^3 to $G°$. We claim that ζ^1, ζ^2, and ζ^3 are linearly independent on $G°$.

7.3 GROUPS WITH SYLOW 2-SUBGROUPS OF TYPE $(2^m, 2^m)$ 181

Otherwise there exist $a_i \in K$ not all zero such that $a_1\zeta^1 + a_2\zeta^2 + a_3\zeta^3 = 0$ on G°. Define $a_0 := a_1\zeta^1(z) + a_2\zeta^2(z) + a_3\zeta^3(z)$. Then since each element of $G\backslash G^\circ$ is conjugate to an element of the form zy ($y \in H^\circ$), it follows from (a), (b), and (c) that $a_0(\zeta^1 + \zeta^2 + \varepsilon\zeta^3 - \zeta^4) = 4(a_1\zeta^1 + a_2\zeta^2 + a_3\zeta^3)$ on G. Since this contradicts the linear independence of $\zeta^1, \zeta^2, \zeta^3$, and ζ^4 on G, we conclude that ζ^1, ζ^2, and ζ^3 are linearly independent on G°, and so that ζ^1, ζ^2, and $\varepsilon\zeta^3$ form a $\mathbf{Z}$-basis of the $\mathbf{Z}$-module generated by the modular characters lying in B_0. With respect to the basis $\{\zeta^1, \zeta^2, \varepsilon\zeta^3\}$, the decomposition matrix and the Cartan matrix have the required forms.

Theorem 7.3 Let G be a group with a Sylow 2-subgroup P of type $(2^m, 2^m)$ with $m > 1$. If $O_{2'}(G) = 1$, then $P = C_G(P) \triangleleft G$ and $|G:P| = 1$ or 3.

Proof We shall proceed by induction on $|G|$. By Lemma 7.3A, $|N_G(P)/C_G(P)| = 1$ or 3. If $|N_G(P)| = |C_G(P)|$, then G has a normal 2-complement by Theorem 6.7B; and since $O_{2'}(G) = 1$, that implies that $G = P$. Thus we may suppose that $|N_G(P)/C_G(P)| = 3$. Since P is abelian of order 2^{2m}, it has 2^{2m} irreducible ordinary characters ψ_i ($i = 0, 1, 2, \ldots, 2^{2m} - 1$) each of degree 1. (We use lower indices to simplify the later notation.) The group $N_G(P)$ acts on this set of characters by $\psi_i^x(z) := \psi_i(x^{-1}zx)$ for all $z \in P$ and $x \in N_G(P)$. Since the orbits of P under conjugation by $N_G(P)$ all have length 3 except the trivial orbit $\{1\}$ (Lemma 7.3A), the same is true of the orbits in the set $\{\psi_i \mid i = 0, 1, \ldots, 2^{2m} - 1\}$. Thus we may choose the notation so that $\psi_0 = 1_P$, and $\psi_1, \ldots, \psi_r$ ($r := \frac{1}{3}(2^{2m} - 1)$) are representatives of the orbits of length 3 under the action of $N_G(P)$. The major step is to calculate the values of $\hat{a}(\psi_j - \psi_0)$ for the characters ψ_j of P.

We shall apply Lemma 7.2B to the case where $Q = P$, $G = H := N_G(P)$ and $S = P\backslash\{1\}$. Since $P \triangleleft H$, Lemma 7.3B shows that for all $x \in S$, $C_H(x) = C_H(P)$, and $\sum_{i=1}^{t} |\chi^i(x)|^2 = |P|$, where $\chi^1, \chi^2, \ldots, \chi^t$ are the irreducible ordinary characters lying in the principal 2-block of H. Thus condition (iii) of Lemma 7.2B is satisfied with $\gamma = |P|/|C_H(P)| = 3|P|/|H|$. On the other hand, condition (ii) is trivially satisfied and condition (i) holds for $\alpha = \psi_i - \psi_0$ and $\beta = \psi_j$. Thus we conclude from Lemma 7.2B that

(1) $\quad \hat{a}(\psi_i - \psi_0) \cdot \hat{a}(\psi_j) = 3\left\langle \psi_i - \psi_0, \sum_{x \in H} \psi_j^x \right\rangle_P \Big/ |H|$

$\quad = \begin{cases} \delta_{ij} & \text{for } i, j = 1, 2, \ldots, r \\ -3 & \text{for } j = 0, i \neq 0, \end{cases}$

because of the way that H acts on the characters of P. This yields the identities

(2) $\quad \hat{a}(\psi_i - \psi_0) \cdot \hat{a}(\psi_j - \psi_0) = 3 + \delta_{ij} \quad$ for $i, j = 1, 2, \ldots, r$

(3) $\quad \hat{a}(\psi_i - \psi_r) \cdot \hat{a}(\psi_j - \psi_r) = 1 + \delta_{ij} \quad$ for $i, j = 1, 2, \ldots, r - 1$

(4) $\quad\hat{a}(\psi_0) \cdot \hat{a}(\psi_i - \psi_j) = 0 \qquad$ for $\quad i, j = 1, 2, \ldots, r$

(5) $\quad\hat{a}(\psi_r - \psi_0) \cdot \hat{a}(\psi_j - \psi_r) = -1 \qquad$ for $\quad j = 1, 2, \ldots, r - 1$

using the linearity of the function $\hat{a}$, and the commutativity and distributivity of the dot product.

The entries of the vectors $\hat{a}(\psi_i)$ are all nonnegative integers by definition. Therefore it follows from (3) that for $i = 1, 2, \ldots, r - 1$, $\hat{a}(\psi_i - \psi_r)$ has only two nonzero entries, each equal to ± 1; and (4) shows that these have opposite signs. Furthermore, (3) also shows that for two different $\hat{a}(\psi_i - \psi_r)$, $\hat{a}(\psi_j - \psi_r)$, two nonzero entries (with the same sign) occur at the same place. Thus choosing the order of the characters $\psi_1, \psi_2, \ldots, \psi_{r-1}$ and the order of the characters $\zeta^1, \zeta^2, \ldots, \zeta^t$ in the principal 2-block of G suitably, we get the table

$$\hat{a}(\psi_1 - \psi_r) = (\varepsilon_0, 0, \ldots, 0, -\varepsilon_0, 0, \ldots, 0)$$
$$\hat{a}(\psi_2 - \psi_r) = (0, \varepsilon_0, \ldots, 0, -\varepsilon_0, 0, \ldots, 0)$$
$$\cdots\cdots\cdots\cdots\cdots\cdots\cdots\cdots\cdots\cdots\cdots\cdots$$
$$\hat{a}(\psi_{r-1} - \psi_r) = (0, 0, \ldots, \varepsilon_0, -\varepsilon_0, 0, \ldots, 0)$$

for some $\varepsilon_0 = 1$ or -1. It then follows from (5) that

$$\hat{a}(\psi_r - \psi_0) = (a_1, a_1, \ldots, a_1, a_1 + \varepsilon_0, a_{r+1}, \ldots, a_t)$$

for some integers a_i; and by (2) we must have

(6) $\qquad (r - 1)a_1^2 + (a_1 + \varepsilon_0)^2 + \sum_{i=r+1}^{t} a_i^2 = 4.$

By definition we know that the entry in $\hat{a}(\psi_i)$ corresponding to the principal character ζ^j is 1 if $i = 0$ and 0 otherwise (see the note of §7.2). Thus $\zeta^j = 1_G$ for some $j > r$ and then $a_j = -1$. On the other hand, $r = \frac{1}{3}(2^{2m} - 1) \geq 5$ since $m > 1$. Therefore it follows from (6) that $a_1 = 0$ and (reordering if necessary) $a_{r+1} = -\varepsilon_1$, $a_{r+1} = -\varepsilon_2$, and $a_{r+3} = -\varepsilon_3$ for suitable $\varepsilon_i = 1$ or -1, whilst $a_j = 0$ for all $j > r + 3$. Substituting these values and calculating $\hat{a}(\psi_i - \psi_0) = \hat{a}(\psi_i - \psi_r) + \hat{a}(\psi_r - \psi_0)$, we obtain

$$\hat{a}(\psi_1 - \psi_0) = (\varepsilon_0, 0, \ldots, 0, 0, \varepsilon_1, \varepsilon_2, \varepsilon_3, 0, \ldots, 0)$$
$$\hat{a}(\psi_2 - \psi_0) = (0, \varepsilon_0, \ldots, 0, 0, \varepsilon_1, \varepsilon_2, \varepsilon_3, 0, \ldots, 0)$$
$$\cdots\cdots\cdots\cdots\cdots\cdots\cdots\cdots\cdots\cdots\cdots\cdots$$
$$\hat{a}(\psi_{r-1} - \psi_0) = (0, 0, \ldots, \varepsilon_0, 0, \varepsilon_1, \varepsilon_2, \varepsilon_3, 0, \ldots, 0)$$
$$\hat{a}(\psi_r - \psi_0) = (0, 0, \ldots, 0, \varepsilon_0, \varepsilon_1, \varepsilon_2, \varepsilon_3, 0, \ldots, 0).$$

7.3 GROUPS WITH SYLOW 2-SUBGROUPS OF TYPE $(2^m, 2^m)$

Finally, if ψ_i is any character of P ($i = 0, 1, \ldots, 2^{2m} - 1$), then $\psi_i^x = \psi_j$ for some $x \in H$ and some $j \leq r$. Then

$$\langle \psi_i - \psi_0, \zeta_P^l \rangle = \langle \psi_i^x - \psi_0, (\zeta_P^l)^x \rangle$$
$$= \langle \psi_i^x - \psi_0, \zeta_P^l \rangle = \langle \psi_j - \psi_0, \zeta_P^l \rangle,$$

for $l = 1, 2, \ldots, t$ (since ζ_P is constant on classes of G). Thus by the definition of $\hat{a}$ we have proved that for $i = 1, 2,$ and 3,

$$\langle \psi_j, \zeta_P^{r+i} \rangle - \langle \psi_0, \zeta_P^{r+i} \rangle = \langle \psi_j - \psi_0, \zeta_P^{r+i} \rangle = \varepsilon_i$$

for $j = 1, 2, \ldots, 2^{2m} - 1$; and so $\zeta_P^{r+i} = -\varepsilon_i \psi_0 + d_i \sum_{j=0}^{2^{2m}-1} \psi_j$ for some integer d_i. However, $\sum_{j=0}^{2^{2m}-1} \psi_j = (1_1)^P$ (see the example of §2.5A) and so takes the value 0 except at 1. This shows that for all $z \neq 1$ in P

(7) $$\zeta^{r+i}(z) = -\varepsilon_i \quad (i = 1, 2, 3).$$

In particular, $\langle \zeta_P^{r+i}, 1_P \rangle = \{\zeta^{r+i}(1) - (2^{2m} - 1)\varepsilon_i\}/2^{2m}$ and so

(8) $$\zeta^{r+i}(1) \equiv -\varepsilon_i \pmod{2^{2m}} \quad \text{for } i = 1, 2, 3.$$

As the next step we choose any character $\psi_j \neq \psi_0$ with $|P : \text{Ker } \psi_j| = 2$. Then the values of ψ_j are all ± 1 and every element of order $< 2^m$ in P lies in Ker ψ_j. Let us define S as the set of all elements of order 2^m, and u as any involution in P. If x is of order 2^m in G, and $u^{-1}\langle x \rangle u = \langle x \rangle$, then u and x lie in the same Sylow 2-subgroup of G and so $u^{-1}xu = x$. This shows that the hypotheses of Theorem 7.2 hold with $Q = P$ and $\alpha = \psi_j - \psi_0$. Since $\hat{a}(\psi_j - \psi_0)$ has only four nonzero entries, Theorem 7.2 together with (7) shows that

(9) $$\varepsilon_0 \zeta^j(u)^2/z_j + \varepsilon_1/z_{r+1} + \varepsilon_2/z_{r+2} + \varepsilon_3/z_{r+3} = 0$$

(10) $$\varepsilon_0 z_j + \varepsilon_1 z_{r+1} + \varepsilon_2 z_{r+2} + \varepsilon_3 z_{r+3} = 0$$

(11) $$\varepsilon_0 \zeta^j(u) - 3 = 0$$

where $z_j := \zeta^j(1)$.

Since $m > 1$, (8) implies that $z_{r+i} \equiv -\varepsilon_i \pmod{16}$ for $i = 1, 2, 3$; in particular, either $z_{r+i} = 1$ or $z_{r+i} \geq 15$. From (9) we obtain

$$\varepsilon_0 \zeta^j(u)^2 \equiv 3z_j \pmod{16}$$

and so (11) shows that $z_j \equiv \pm 3 \pmod{16}$. We saw above (in the calculation of $\hat{a}(\psi_r - \psi_0)$) that one of the characters ζ^{r+i} ($i = 1, 2, 3$) is the principal character of G; say $\zeta^{r+1} = 1_G$, and then $\varepsilon_1 = -z_{r+1} = -1$. On the other hand, since $|\zeta^j(u)| = 3$ by (11), these conditions on z_j, z_{r+1}, z_{r+2}, and z_{r+3} together with (9) easily show that at least one of z_{r+2} and z_{r+3} is also 1. Suppose $z_{r+2} = 1$; then $\varepsilon_{r+2} = -z_{r+2} = -1$.

Finally, from (7) we conclude $\zeta^{r+2}(z) = -\varepsilon_{r+2} = 1$ for all $z \in P\backslash\{1\}$; so P is contained in the kernel L of ζ^{r+2}. Since $O_2(G) = 1$, and $O_{2'}(L)$ is a characteristic subgroup of L, $O_{2'}(L) = 1$. Since P is a Sylow 2-subgroup of L and $L \neq G$, the induction hypothesis shows that $P \triangleleft L$; hence $P \triangleleft G$ and so $C_G(P) \triangleleft G$. Moreover, $C_G(P)$ has a normal 2-complement M by Theorem 6.7B. Since $C_G(P) \triangleleft G$, M is normal in G; since $O_{2'}(G) = 1$, this means $M = 1$ and $P = C_G(P)$. By Lemma 7.3A, $|G : C_G(P)| = 1$ or 3 and so the theorem is proved.

Remark The above theorem shows that when $m > 1$ there are no simple groups which have a Sylow 2-subgroup of type $(2^m, 2^m)$. In the case $m = 1$ this is no longer true. Gorenstein and Walter [1] have shown that the simple groups with Sylow 2-subgroups of type $(2, 2)$ are precisely the projective special linear groups $PSL(2, q)$, where q is a prime such that $q \equiv 3$ or $5 \pmod 8$.

EXERCISES

1. Let G be a group with a Sylow 2-subgroup Q of type $(2, 2)$ and let u be an involution of G. If G has no 2-complement, prove that

$$(d_2 + \varepsilon_2)(d_3 + \varepsilon_3)(d_4 + \varepsilon_4)g |C_G(Q)|^2 = 8d_2 d_3 d_4 |C_G(u)|^2,$$

where $1, d_2, d_3, d_4$ are the degrees of the irreducible ordinary characters in the principal 2-block and $\varepsilon_i = \pm 1$.

2. Let G be a group with Sylow 2-subgroup Q of type $(2, 2)$ and let $Q = \langle u, v \rangle$. If G has a normal 2-complement M, then prove Wielandt's fixed point formula

$$|M| |C_M(Q)|^2 = |C_M(u)| |C_M(v)| |C_M(uv)|.$$

7.4 Groups with quaternion Sylow 2-subgroups

A *generalized quaternion group* of order 2^{n+1} ($n \geq 2$) is a group P which can be generated by two elements z and w such that z has order 2^n, $w^2 = z^{2^{n-1}}$ and $w^{-1}zw = z^{-1}$. The following properties of generalized quaternion groups are known (see Hall [1]).

(a) Any generalized quaternion group has a normal cyclic subgroup of index 2 (generated by z), and posesses only one involution (namely $z^{2^{n-1}} = w^2$).

(b) Any two generalized quaternion groups of the same order are isomorphic and any homomorphic image of a quaternion group is either of order less than or equal to 4 or is also generalized quaternion.

(c) A 2-group possessing only one involution is either cyclic or a generalized quaternion group; consequently any subgroup of a quaternion group is either cyclic or generalized quaternion.

The generalized quaternion group of order 8 is known as *the* quaternion group. It has the property (which we shall need) of being a nonabelian group in which each subgroup is normal. The generalized quaternion groups of order greater than 8 do not have this property. The theorems of this section investigate the structure of groups that have generalized quaternion Sylow 2-subgroups; in particular such groups are never simple. The case of the quaternion group of order 8 is treated first, and then used to give an inductive proof of the general result.

Theorem 7.4A $(p = 2)$ Let G be a group with a Sylow 2-subgroup P which is a quaternion group of order 8. If $O_{2'}(G) = 1$, then $|Z(G)| = 2$.

Proof By hypothesis, P is generated by two elements z and w of order 4 with $z^2 = w^2$ and $z^{-1}wz = w^{-1}$. Then $u = z^2$ is the unique involution in P. We shall suppose that the theorem is false and take G as a counterexample of minimal order, and obtain a contradiction in a series of steps.

Let $\zeta^1, \zeta^2, \ldots, \zeta^t$ be the irreducible ordinary characters of G lying in the principal 2-block.

Step 1 All elements of order 4 in P are conjugate in $N_G(P)$.

The elements of order 4 in P are $z, z^{-1}, w, w^{-1}, zw,$ and $(zw)^{-1}$ (the two other elements are 1 and u). Since $z^{-1}wz = w^{-1}$, $w^{-1}zw = z^{-1}$, and $z^{-1}(zw)z = (zw)^{-1}$, there are at most 3 classes of elements of order 4 in $N_G(P)$. If the elements of order 4 are not all conjugate in $N_G(P)$, then one of the sets $\{z, z^{-1}\}, \{w, w^{-1}\},$ and $\{zw, (zw)^{-1}\}$ is not conjugate to either of the other. Then symmetry of P allows us to suppose that $\{z, z^{-1}\}$, say, is not conjugate to either $\{w, w^{-1}\}$ or $\{zw, (zw)^{-1}\}$ in $N_G(P)$. By Lemma 6.7, $\{z, z^{-1}\}$ is not conjugate to either $\{w, w^{-1}\}$ or $\{zw, (zw)^{-1}\}$ in G. Thus we can partition P into two disjoint sets $P_1 := \{1, u, z, z^{-1}\}$ and $P_{-1} := \{w, w^{-1}, zw, (zw)^{-1}\}$ such that no element of P_1 is conjugate in G to an element of P_{-1}. This permits us to define a class function $\theta: G \to K$ by $\theta(x) := 1$ if the 2-part of x is conjugate to an element in P_1 and $\theta(x) := -1$ if the 2-part of x is conjugate to an element in P_{-1}. We shall apply Theorem 2.6B to prove that θ is an irreducible character of G. Let E be a subgroup of G of the form $E = Q \times S$, where Q is 2-group and S is a $2'$-subgroup. It is readily verified that $\theta_E \in \text{Char}(E)$. Since every elementary subgroup of G is nilpotent and hence conjugate to a subgroup of this form, it follows that $\theta_E \in \text{Char}(E)$ for every elementary subgroup of G. Moreover, $\langle \theta, \theta \rangle_G = 1$ because $|\theta(x)|^2 = 1$ for all $x \in G$, and $\theta(1) = 1$. Therefore by Theorem 2.6B, θ is an irreducible character of G, and clearly $|G/\text{Ker } \theta| = 2$. Since the subgroups of index 2 in

P are cyclic, Ker θ has a cyclic Sylow 2-subgroup and therefore it has a normal 2-complement L by the corollary of Theorem 6.7B. Since L is characteristic in Ker θ, $L \triangleleft G$. But $O_{2'}(G) = 1$, so $L = 1$. Hence $G = P$ contrary to the hypothesis that G is a counterexample.

Step 2 For each i, $\zeta^i(z)$ is an integer congruent to $\zeta^i(u)$ and $\zeta^i(1)$ (mod 2).

Let T be a representation of G which affords ζ^i. Since z has order 4, the eigenvalues of $T(z)$ are the fourth roots of unity, namely 1, -1, $\sqrt{-1}$ and $-\sqrt{-1}$, with multiplicities m_1, m_2, m_3, and m_4, say. Then $\zeta^i(z) = m_1 - m_2 + (m_3 - m_4)\sqrt{-1}$, $\zeta^i(u) = m_1 + m_2 - m_3 - m_4$, and $\zeta^i(1) = m_1 + m_2 + m_3 + m_4$. Since z is conjugate z^{-1} (even in P), $\zeta^i(z) = \zeta^i(z^{-1})$, and so $m_3 - m_4 = 0$. The assertion now follows immediately.

Step 3 Analysis at z.

Since $z \notin Z(P)$, $\langle z \rangle$ is the Sylow 2-subgroup of $C_G(z)$. Therefore by the corollary of Theorem 6.7B, $C_G(z)$ has a normal 2-complement. The example of §7.1 now shows that $\sum_{i=1}^{t} |\zeta^i(z)|^2 = |\langle z \rangle| = 4$. Since $\zeta^1(z) = 1_G(z) = 1$, it follows (using Step 2) that $\zeta^i(z) = \pm 1$ for four of the characters ζ^i [and for these $\zeta^i(1)$ is odd] whilst $\zeta^i(z) = 0$ for the remaining characters (and for these $\zeta^i(1)$ is even).

Step 4 Analysis at u.

Put $H := C_G(u)$ and $\tilde{H} := H/\langle u \rangle$. Then there is a bijection from the set of $2'$-elements of H onto the set of $2'$-elements of $\tilde{H}$ given by $x \mapsto x\langle u \rangle$; and $\tilde{P} := P/\langle u \rangle$ is a Sylow 2-subgroup of H and is of type $(2, 2)$. Since $N_G(P) \subseteq C_G(u)$, Step 1 implies that all involutions in $\tilde{P}$ are conjugate in $\tilde{H}$. Thus we can apply Lemma 7.3B(iii) to $\tilde{H}$ and conclude that the submodule of Char($\tilde{H}°$) corresponding to the principal block of $k\tilde{H}$ has a $\mathbf{Z}$-basis consisting of three generalized modular characters $\tilde{\psi}^1 = 1_{\tilde{H}}$, $\tilde{\psi}^2$, and $\tilde{\psi}^3$. The associated Cartan matrix has the form

$$\begin{bmatrix} 2 & 1 & 1 \\ 1 & 2 & 1 \\ 1 & 1 & 2 \end{bmatrix}.$$

Now since $\langle u \rangle$ is a normal 2-subgroup of H, every irreducible representation of H over k has $\langle u \rangle$ in its kernel (corollary to Theorem 2.2A). Thus there is a one-to-one correspondence between the irreducible modular characters of H and those of $\tilde{H}$; namely, to each irreducible modular character θ of H there is the irreducible modular character $\tilde{\theta}$ of $\tilde{H}$ given by $\tilde{\theta}(x\langle u \rangle) = \theta(x)$ for all $x \in H$. In particular, the submodule of Char($H°$) corresponding to the principal block on kH has a $\mathbf{Z}$-basis $\psi^1 = 1_H$, ψ^2, and ψ^3 of generalized modular characters of H related to $\tilde{\psi}^1$, $\tilde{\psi}^2$, and $\tilde{\psi}^3$ in the same way. Since $|H| = 2|\tilde{H}|$,

7.4 GROUPS WITH QUATERNION SYLOW 2-SUBGROUPS

we have
$$\langle \psi^i, \psi^j \rangle_{H^\circ} = \tfrac{1}{2} \langle \tilde{\psi}^i, \tilde{\psi}^j \rangle_{\tilde{H}^\circ} \quad \text{for} \quad i, j = 1, 2, 3.$$
Thus, by Note 2 of §7.1, the Cartan matrix associated with ψ^1, ψ^2, and ψ^3 is twice the Cartan matrix above, namely
$$\begin{bmatrix} 4 & 2 & 2 \\ 2 & 4 & 2 \\ 2 & 2 & 4 \end{bmatrix}.$$

Finally, since $\tilde{P}$ is a defect group of the principal block of $k\tilde{H}$, the latter has defect 2; and so the degrees of the modular irreducible characters in this block have odd degrees (Theorem 4.5B). From the correspondence between the modular characters of H and the modular characters of $\tilde{H}$, we conclude that the degrees of the irreducible modular characters lying in the principal block of kH are also odd.

Step 5 Decomposition relative to ψ^1, ψ^2, ψ^3.

By Theorem 6.6 the principal block b of H is the only block with the property that b^G is the principal block of G. Therefore by Theorem 6.5, there exist algebraic integers a_{ij} such that $\zeta^i(uy) = \sum_{j=1}^{3} a_{ij} \psi^j(y)$ ($i = 1, 2, \ldots, t$) for all $y \in H^\circ$ because ψ^1, ψ^2, ψ^3 is a **Z**-basis of b. From §6.5 (Note 4) we know that the (ordinary) generalized decomposition numbers at u are rational integers. Since the irreducible modular characters are **Z**-linear combinations of ψ^1, ψ^2, ψ^3 (by the definition of a **Z**-basis of a block), it follows that $a_{ij} \in \mathbf{Z}$. By Theorem 7.1 these integers satisfy the following properties:

(1) $$\sum_{l=1}^{t} a_{li}^2 = 4 \quad \text{for} \quad i = 1, 2, 3$$

(2) $$\sum_{l=1}^{t} a_{li} a_{lj} = 2 \quad \text{for} \quad i \neq j$$

and by Step 2,

(3) $$\sum_{i=1}^{3} a_{li} \equiv \zeta^l(1) \pmod{2}$$

since the modular characters in b have odd degrees (by Step 4). Moreover the orthogonality condition (Theorem 4.2C) shows that
$$0 = \sum_{i=1}^{t} \zeta^i(1)\zeta^i(uy) = \sum_{j=1}^{3} \left\{ \sum_{i=1}^{t} a_{ij} \zeta^i(1) \right\} \psi^j(y)$$
for all $y \in H^\circ$. Hence by the linear independence of ψ^1, ψ^2, ψ^3 we conclude

(4) $$\sum_{i=1}^{t} a_{ij} \zeta^i(1) = 0 \quad \text{for} \quad j = 1, 2, 3.$$

We shall now determine the form of the matrix $[a_{ij}]$. First of all, no row of $[a_{ij}]$ is 0. For suppose $a_{i1} = a_{i2} = a_{i3} = 0$. Then $\zeta^i(u) = 0$, and $\zeta^i(1)$ is even by (3). Step 3 now shows that $\zeta^i(z) = 0$ and so $\langle \zeta_P^i, 1_P \rangle = \zeta^i(1)/8$. Hence $8 \mid \zeta^i(1)$ and so ζ^i lies in a block of defect 0 by theorem 4.5A. This is impossible since the principal block of G has P as a defect group (see §4.5).

Next we observe that no entry $a_{ij} = \pm 2$. For otherwise this would be the only nonzero entry in its column [by (1)] and this contradicts (4). Hence the entries of $[a_{ij}]$ are 0, 1, and -1, and each column has 4 nonzero entries by (1). We also note that in any row the nonzero entries have the same sign. Otherwise (2) implies that the nonzero entries of two columns would be $\pm(1, -1)$, $\pm(1, 1)$, $\pm(1, 1)$ and $\pm(1, 1)$; and once again this would contradict (4).

Let m_1, m_2, and m_3 denote the number of rows of $[a_{ij}]$ with 1, 2, and 3 nonzero entries, respectively. Since by Step 3 exactly four $\zeta^l(1)$ are odd, it follows from (1), (2), and (3) that

(5) $\quad m_1 + 2m_2 + 3m_3 = 12, \quad m_2 + 3m_3 = 6, \quad$ and $\quad m_1 + 2m_3 = 5.$

These equations yield $m_1 = 3$, $m_2 = 3$, and $m_3 = 1$. As we saw above, all rows of $[a_{ij}]$ are nonzero, so $[a_{ij}]$ has 7 rows. It is easily seen [using (1) and (2)] that to within a permutation of the rows, $[a_{ij}]$ has the form given on the left in the accompanying table (with each $\varepsilon_i = \pm 1$). The first row corresponds to the trivial character ζ^1 (and $\psi^1 = 1_H$). On the right in the table we have introduced symbols for the values of the characters at 1, u, and z. In the last column (each $\delta_i = \pm 1$) we have used (3) and Step 3. Note that this last column also gives the generalized decomposition numbers of ζ^i at z with respect to the basis $1_{C_G(z)}$ [since $C_G(z)$ has a normal 2-complement, the basis has a single element].

	ψ^1	ψ^2	ψ^3	$\zeta^i(1)$	$\zeta^i(u)/\zeta^i(1)$	$\zeta^i(z)$
ζ^1	1	0	0	$d_1 = 1$	$\lambda_1 = 1$	1
ζ^2	ε_2	ε_2	ε_2	d_2	λ_2	δ_2
ζ^3	ε_3	ε_3	0	d_3	λ_3	0
ζ^4	ε_4	0	ε_4	d_4	λ_4	0
ζ^5	0	ε_5	ε_5	d_5	λ_5	0
ζ^6	0	ε_6	0	d_6	λ_6	δ_6
ζ^7	0	0	ε_7	d_7	λ_7	δ_7

Now applying Theorem 7.1 we obtain from these two sets of generalized decomposition numbers the relations $1 + \varepsilon_2 \delta_2 = \varepsilon_2 \delta_2 + \varepsilon_6 \delta_6 = \varepsilon_2 \delta_2 + \varepsilon_7 \delta_7 = 0$. These show that

(6) $\quad\quad \delta_2 = -\varepsilon_2, \quad\quad \delta_6 = \varepsilon_6, \quad$ and $\quad \delta_7 = \varepsilon_7.$

7.4 GROUPS WITH QUATERNION SYLOW 2-SUBGROUPS

Step 6 u lies in the kernel of ζ^2.

Suppose that x and y are involutions in G, and that xy is a 2-element. Since $x^{-1}(xy)x = yx = (xy)^{-1}$, we see that $\langle x, y \rangle = \langle x, xy \rangle$ is a 2-subgroup of G. Since the Sylow 2-subgroups of G are quaternion and each contains a single involution, this implies $x = y$. Hence xy cannot be equal to a nontrivial 2-element of G. This shows that we can apply Lemma 7.2A in the two cases where we take $\mathscr{C}_1 = \mathscr{C}_2$ as the unique class of involutions of G and take u or z in place of z. The equation (1) of Lemma 7.2A then yields

$$\sum_{i=1}^{7} a_{ij} \zeta^i(u)^2/\zeta^i(1) = 0 \qquad (j = 1, 2, 3)$$

and

$$1 + \delta_2 \zeta^2(u)^2/\zeta^2(1) + \delta_6 \zeta^6(u)^2/\zeta^6(1) + \delta_7 \zeta^7(u)^2/\zeta^7(1) = 0.$$

We write $\zeta^i(u)^2/\zeta^i(1) = \lambda_i^2 \, d_i$ with the notation above. Now we take the last four equations, subtract the middle two from the sum of the first and last, use (6) and the known values of a_{ij}, and we get

(7) $$1 - \varepsilon_2 \lambda_2^2 \, d_2 - \varepsilon_5 \lambda_5^2 \, d_5 = 0$$

(after we have divided through by a factor 2). Similarly from (4) we have

$$\sum_{i=1}^{7} a_{ij} \, d_i = 0 \qquad (j = 1, 2, 3)$$

and the analogous equation $1 + \delta_2 \, d_2 + \delta_6 \, d_6 + \delta_7 \, d_7 = 0$. Therefore we derive in a similar way

(8) $$1 - \varepsilon_2 \, d_2 - \varepsilon_5 \, d_5 = 0.$$

Finally, from the decomposition table, $\zeta^2(u) = \varepsilon_2(1 + \psi^2(1) + \psi^3(1))$ and $\zeta^5(u) = \varepsilon_6(\psi^2(1) + \psi^3(1))$. Hence we get

(9) $$1 - \varepsilon_2 \lambda_2 \, d_2 + \varepsilon_5 \lambda_5 \, d_5 = 0.$$

Eliminating d_5 from (7) and (9) and from (8) and (9) we get

$$(\lambda_5 + 1) - \varepsilon_2 \lambda_2(\lambda_2 + \lambda_5) \, d_2 = (\lambda_5 + 1) - \varepsilon_2(\lambda_2 + \lambda_5) \, d_2 = 0.$$

Together these imply $\lambda_2 = 1$. Hence $\zeta^2(u) = \zeta^2(1)$, and so $u \in \text{Ker } \zeta^2$.

Step 7 Conclusion.

Let $N = \text{Ker } \zeta^2$. If $P \subseteq N$, then $Z(N) = \langle u \rangle$ since $N \neq G$ and G is a minimal counterexample to the theorem, and so $\langle u \rangle$ is characteristic in N. On the other hand if $P \nsubseteq N$, then the Sylow 2-subgroup $P \cap N$ of N is cyclic of order 2 or 4. By the corollary of Theorem 6.7B, N has a normal 2-

complement L, say. Since $L = O_{2'}(N) \subseteq O_{2'}(G) = 1$, this means $N = P \cap N$; and again $\langle u \rangle$ is a characteristic subgroup of N. Hence in either case, $\langle u \rangle \triangleleft G$. Since $\langle u \rangle$ has order 2 this means $\langle u \rangle \subseteq Z(G)$. But $\langle u \rangle = Z(P) \supseteq Z(G)$ because $O_{2'}(G) = 1$, so $Z(G) = \langle u \rangle$. This contradicts the choice of G and so the theorem is proved.

Theorem 7.4B Let G be a group whose Sylow 2-subgroups are generalized quaternion groups (of order greater than or equal to 16). If $O_{2'}(G) = 1$, then $|Z(G)| = 2$.

Proof Let P be a Sylow 2-subgroup of order $2^n (n \geq 4)$ of G. Then we can find z, w in P such that $P = \langle z, w \rangle$, z has order 2^{n-1}, $w^2 = z^{2^{n-2}}$, and $w^{-1}zw = z^{-1}$; $Z(P) = \langle u \rangle$ where $u := w^2$ has order 2. Put $Q = \langle z \rangle$ and let $S = Q \backslash \langle z^4 \rangle$ so S consists of all elements of order greater than or equal to 2^{n-2} $(n \geq 4)$ in Q. Let $\zeta^1, \zeta^2, \ldots, \zeta^t$ be the irreducible ordinary characters of G lying in the principal 2-block. Suppose the theorem does not hold and let G be a minimal counterexample; we shall obtain a contradiction using Theorem 7.4A.

Step 1 The hypotheses (ii) and (iii) of Lemma 7.2B hold with $H = P$.

First suppose $x \in S$ and $v^{-1}xv \in Q$ for some $v \in G$. Suppose $v^{-1}xv \neq x$; then we claim that $v^{-1}xv = x^{-1}$. Indeed, since Q is a cylcic 2-group and x and $v^{-1}xv$ have the same order, say 2^d, therefore $v^{-1}xv = x^h$ for some odd integer $h > 1$. For any integer m we have $v^{-m}xv^m = x^{h^m}$. Since h is odd, we know from elementary number theory that $h^{2^{d-1}} \equiv 1 \pmod{2^d}$ (since 2^{d-1} is the value of the Euler phi-function at 2^d); thus $v^{2^{d-1}} \in C_G(x)$. This shows that the 2'-part of v commutes with x. Hence without loss in generality we may assume that v is a 2-element. But then $\langle v, x \rangle$ is a 2-group (with $\langle x \rangle$ as a normal subgroup) and so contained in a Sylow 2-subgroup P_1, say. But in P_1 the element x of order greater than or equal to 4 has $|P_1 : C_{P_1}(x)| = 2$ because P_1 is a generalized quaternion group. Hence x can only be conjugate to x or x^{-1} in P_1; thus $v^{-1}xv = x^{-1}$ as asserted. This shows that hypothesis (ii) of Lemma 7.2B holds.

Secondly, for all $x \in S$, Q is a cyclic Sylow 2-subgroup of $C_G(x)$. Hence $C_G(x)$ has a normal 2-complement by the corollary of Theorem 6.7B, and so by the example of §7.1 we have

$$\sum_{i=1}^{t} |\zeta^i(x)|^2 = |Q| = 2^{n-1}.$$

Since $|C_P(x)| = |Q| = 2^{n-1}$, hypothesis (iii) of Lemma 7.2B holds with $\gamma = 1$.

Step 2 Application of Lemma 7.2B and Theorem 7.2.

Let $\chi_0 = 1_Q$ and let χ be the irreducible ordinary character of $Q = \langle z \rangle$

7.5 GLAUBERMAN'S Z^*-THEOREM

defined by $\chi(z) = \sqrt{-1}$. Then $\alpha = \chi - \chi_0$ is identically 0 on $Q \backslash S$, and so by Lemma 7.2B and Step 1

$$\hat{a}(\alpha) \cdot \hat{a}(\alpha) = \left\langle \alpha, \sum_{x \in P} \alpha^x \right\rangle \bigg/ |Q|$$
$$= \langle \chi - \chi_0, \chi - \chi_0 + \chi^w - \chi_0 \rangle = 3$$

since

$$(\chi - \chi_0)^x = \begin{cases} \chi - \chi_0 & \text{if } x \in Q \\ \chi^w - \chi_0 & \text{if } x \in P \backslash Q. \end{cases}$$

Since the entries of $\hat{a}(\alpha)$ are nonzero integers, this shows that it has exactly three nonzero entries, each ± 1. Next, the hypothesis of Theorem 7.2 is satisfied. Indeed if a 2-element x has order greater than or equal to 4 in G then $u^{-1}xu \in \langle x \rangle$ implies that $\langle u, x \rangle$ is a 2-group; and then $u^{-1}xu = x$ since the involution u is in the center of each Sylow 2-subgroup in which it lies. Hence the corollary of Theorem 7.2 shows that G has a proper normal subgroup N, say, containing u. Since $O_{2'}(G) = 1$ we have $O_{2'}(N) = 1$.

Step 3 Conclusion.

Since the subgroups of a generalized quaternion group are either cyclic or generalized quaternion (see the beginning of this section), the Sylow 2-subgroups of N are either cyclic, quaternion (of order 8), or generalized quaternion (of order greater than or equal to 16). In the second and third cases (using Theorem 7.4A and the minimality of G, respectively) we see that $|Z(N)| = 2$. But $Z(N) \triangleleft G$, and since it has order 2, this means $Z(P) = \langle u \rangle = Z(N) \subseteq Z(G)$; thus $Z(G) = \langle u \rangle$, contrary to the assumption on G. On the other hand, in the first case N has a normal 2-complement by the corollary of Theorem 6.7B; and since $O_{2'}(N) = 1$ by Step 2, this means N is a 2-group. As a subgroup of a generalized quaternion group, N has a unique involution, namely, u. Then $\langle u \rangle \subseteq Z(G)$, and since $\langle u \rangle = Z(P)$ this means $Z(G) = \langle u \rangle$ again. Thus in all cases we obtain a contradiction to the hypothesis that G is a counterexample. This proves the theorem.

7.5 Glauberman's Z^*-theorem

The theorems of the previous section give sufficient criteria for a group to have a nontrivial center. These have been generalized by Glauberman in what is referred to as the "Z^*-theorem," which we shall consider now.

Definition Let G be a group. Then $Z^*(G)$ is defined as the subgroup of G such that $Z^*(G)/O_{2'}(G) = Z(G/O_{2'}(G))$. In particular, if $O_{2'}(G) = 1$, then $Z^*(G) = Z(G)$.

EXAMPLE 1   If the Sylow 2-subgroups of G are generalized quaternion (of order greater than or equal to 8), then each involution of G lies in $Z^*(G)$. This follows at once from the theorems of §7.4 applied to $G/O_{2'}(G)$.

EXAMPLE 2   If the Sylow 2-subgroups of G are cyclic, then $G = Z^*(G)$. Indeed in this case, if P is a Sylow 2-subgroup of G, then $G = PO_{2'}(G)$ by the corollary of Theorem 6.7B; so $Z(G/O_{2'}(G)) = G/O_{2'}(G)$.

We shall use the following elementary group theoretic lemma in the proof of Glauberman's theorem (Theorem 7.5).

Lemma 7.5   Let P be a Sylow 2-subgroup of the group G, and let u and v be involutions in G with $u \in P$.

(i) If uv has order n, then $\langle u, v \rangle$ is a subgroup of order $2n$ with $u^{-1}(uv)u = v^{-1}(uv)v = (uv)^{-1}$ ("dihedral group of order $2n$"). Moreover, if n is odd, then u is conjugate to v in $\langle u, v \rangle$.

(ii) $x^{-1}u^{-1}xu$ is a $2'$-element for each $x \in G$ iff u is not conjugate in G to any other element of P.

Proof (i) Since u and v have order 2, we have $u^{-1}(uv)u = v^{-1}u^{-1} = (uv)^{-1}$ and similarly $v^{-1}(uv)v = (uv)^{-1}$. Since $\langle u, v \rangle = \langle u, uv \rangle$, this proves the first assertion. The second assertion follows at once since, if n is odd, $\langle u \rangle$ and $\langle v \rangle$ are Sylow 2-subgroups of $\langle u, v \rangle$.

(ii) First suppose that $x^{-1}u^{-1}xu$ is a $2'$-element for all $x \in G$. If for some $x \in G$ we have $x^{-1}ux \in P$, then $x^{-1}u^{-1}xu \in P$ is a $2'$-element and so $x^{-1}u^{-1}xu = 1$. This shows that the only conjugate $x^{-1}ux$ of u lying in P is u itself. Conversely, suppose that the only conjugate of u in G lying in P is u itself. Consider $v := x^{-1}ux$ for $x \in G$. Then for any integer $m \geq 1$, it follows from (i) that $y := (vu)^{-m}v(vu)^m u = (vu)^{-m}(vu)^{-m}vu = (vu)^{-2m+1}$. Now suppose that for some $x \in G$, $x^{-1}u^{-1}xu = vu$ is not a $2'$-element. Then we can choose the integer m so that y is a nontrivial 2-element (lying in $\langle u, v \rangle$). Then $\langle u, y \rangle$ is a 2-subgroup and so contained in a conjugate $z^{-1}Pz$ of P. But then zuz^{-1} and $z(yu)z^{-1} = z(vu)^{-m}x^{-1}ux(vu)^m z^{-1}$ both lie in P and are conjugate to u in G. By hypothesis this means that they are both equal to u; hence $u = yu$ and so $y = 1$. This is contrary to the choice of m. Hence we have shown that $x^{-1}u^{-1}xu$ is always a $2'$-element in this case, and the lemma is proved.

Theorem 7.5   (Z^*-theorem)   Let u be an involution in the group G. Then $u \in Z^*(G)$ iff $x^{-1}u^{-1}xu$ is a $2'$-element for each $x \in G$.

Proof   If $u \in Z^*(G)$, then for each $x \in G$ we have $x^{-1}u^{-1}xu \in O_{2'}(G)$ because $Z^*(G)/O_{2'}(G)$ is the center of $G/O_{2'}(G)$; thus $x^{-1}u^{-1}xu$ has odd order. This proves the theorem in one direction. We now turn to the proof in the other direction. Suppose that the assertion is false and take G as a minimal

7.5 GLAUBERMAN'S Z^*-THEOREM

counterexample. Then for all $x \in G$, $x^{-1}u^{-1}xu$ is a $2'$-element, but $u \notin Z^*(G)$. We shall proceed by a series of steps to obtain a contradiction.

Step 1 $O_{2'}(N) = 1$ *for each normal subgroup N of G.*

Otherwise, we shall have $O_{2'}(G) \neq 1$ since $O_{2'}(N)$ is normal in G. But then $G/O_{2'}(G)$ would be counterexample of smaller order.

Step 2 $Z(G)$ *contains no involutions.*

Suppose v is an involution in $Z(G)$; clearly $v \neq u$. Put $\tilde{G} := G/\langle v \rangle$ and define $L \triangleleft G$ by $L/\langle v \rangle = O_{2'}(\tilde{G})$. Then $\langle v \rangle$ is a Sylow 2-subgroup of L and so by the corollary of Theorem 6.7B, L has a normal 2-complement. But $O_{2'}(L) = 1$ by Step 1, so we have $L = \langle v \rangle$. Hence $Z^*(\tilde{G}) = Z(\tilde{G})$. Now since G is a minimal counterexample to the theorem, it follows easily that $u\langle v \rangle \in Z^*(\tilde{G}) = Z(\tilde{G})$, and so $\langle u, v \rangle \triangleleft G$. But then for each $x \in G$, $x^{-1}u^{-1}xu$ has odd order and lies in the 2-subgroup $\langle u, v \rangle$; hence $x^{-1}u^{-1}xu = 1$. This implies that $u \in Z(G) \subseteq Z^*(G)$, contrary to our choice of G.

Step 3 *Let P be a Sylow 2-subgroup containing u. Then $u \in Z(P)$ and P contains an involution v different from u. Moreover, v is not conjugate to u in G.*

For each $x \in P$, $x^{-1}u^{-1}xu \in P$ and $x^{-1}u^{-1}xu$ is a $2'$-element by hypothesis; hence $x^{-1}u^{-1}xu = 1$. This proves that $u \in Z(P)$. If u were the only involution in P, then P would be cyclic or the generalized quaternion group. Since $O_{2'}(G) = 1$, the corollary of Theorem 6.7B and Theorems 7.4A and 7.4B then show that $u \in Z(G) = Z^*(G)$, contrary to hypothesis. Hence P contains at least one other involution v, say. Finally, any conjugate $x^{-1}ux$ of u in G lying in P has the property that $u^{-1}x^{-1}ux \in P$ but $u^{-1}x^{-1}ux$ is a $2'$-element by hypothesis. This implies that $u^{-1}x^{-1}ux = 1$; and so each conjugate $x^{-1}ux$ of u lying in P equals u. Since $v \in P$ and $v \neq u$, we conclude that v is not conjugate in G to u.

Step 4 *Characters in the principal 2-block of G.*

Since G is not a $2'$-group we can choose an irreducible ordinary character ζ in the principal 2-block of G with $\zeta \neq 1_G$ (by the corollary of Theorem 6.7A). Suppose v is an involution in the Sylow 2-subgroup P with $v \neq u$ (see Step 3). Then for each z in G conjugate to v we shall show that

(1) $$\zeta(uv) = \zeta(uz).$$

Consider $y := uz$. If y were a $2'$-element, then the involutions u and z are conjugate in $\langle u, z \rangle$ by Lemma 7.5. This implies that u and v are conjugate, which is contrary to Step 3. Thus y has an even order $2n$, say. Put $w := y^n$. Then w is an involution. Put $H := C_G(w)$. Since $u^{-1}yu = z^{-1}yz = y^{-1}$ and $w = y^n = w^{-1}$, we have $u, z \in C_G(w)$.

We shall now apply the Second and Third Main Theorems to H. First of all, if B is any 2-block of H with Q, say, as a defect group, then $\langle w \rangle \subseteq Q$ by Lemma 6.2A because $\langle w \rangle \triangleleft H$. Thus $QC_G(Q) \subseteq QC_G(w) = H$, and so Theorem 6.6 shows that B^G is the principal 2-block of G exactly when B is the principal 2-block of H. Let $\phi^1, \phi^2, \ldots, \phi^m$ be the irreducible modular characters of H lying in the principal 2-block of H. Then Theorem 6.5 shows that

$$\zeta(wx) = \sum_{j=1}^{m} d_j^w \phi^j(x) \quad \text{for all } x \in H^\circ, \tag{2}$$

where d_j^w $(j = 1, 2, \ldots, m)$ are the generalized decomposition numbers at w with respect to ζ.

Since G is a minimal counterexample to the theorem, and $H \neq G$ by Step 2, $u \in Z^*(H)$. In particular, $y^2 = (uz)^2 = u^{-1}z^{-1}uz \in O_{2'}(H)$, and so the order n of y^2 is odd. Therefore $y^{n+1} \in \langle y^2 \rangle \subseteq O_{2'}(H)$ and so y^{n+1} is a $2'$-element of H. Applying (2) with $x = y^{n+1} = wy$, we get

$$\zeta(y) = \zeta(wy^{n+1}) = \sum_{j=1}^{m} d_j^w \phi^j(y^{n+1}) = \sum_{j=1}^{m} d_j^w d^j \phi^j(1)$$

by Theorem 6.7A(ii), since $y^{n+1} \in O_{2'}(H)$. With (2) this shows that

$$\zeta(uz) = \zeta(w) \tag{3}$$

by the definition of y.

Next we note that since $u \in C_G(w)$, $(uw)^2 = 1$. On the other hand, $uw = uy \cdot y^{n-1} \in z \langle y^2 \rangle \subseteq zO_{2'}(H)$. Therefore uw is an involution and $(uw)z$ is a $2'$-element. By Lemma 7.5 this shows that uw is conjugate to z and hence conjugate to v in G. Choose $x \in G$ such that $x^{-1}vx = uw$.

Finally we have $x^{-1}ux \in x^{-1}C_G(v)x = C_G(uw)$ by Step 3. Since $x^{-1}u^{-1}xu$ has odd order by hypothesis on G, we conclude from Lemma 7.5 that there exists $s \in \langle x^{-1}ux, u \rangle \subseteq C_G(uw)$ such that $s^{-1}x^{-1}uxs = u$. Then $(xs)^{-1}v(xs) = s^{-1}(uw)s = uw = (xs)^{-1}u(xs)w$, and so $uv = (xs)w(xs)^{-1}$. Since ζ is a class function, this latter result together with (3) proves (1).

Step 5 *If v is an involution in P different from u, and ζ is an irreducible ordinary character from the principal 2-block of G such that $\zeta \neq 1_G$ and $\zeta(v) \neq 0$, then $\zeta(u) = -\zeta(1)$.*

Let $\mathscr{C}_1$ and $\mathscr{C}_2$ be the conjugate classes of G containing u and v, respectively, and let c_1 and c_2 be the corresponding class sums. Let T be a representation of G affording ζ. Since K is a splitting field, $T(c_1)$ and $T(c_2)$ are scalars; taking traces we obtain $T(c_1) = (h_1 \zeta(u)/\zeta(1)) \cdot 1$ and $T(c_2) = (h_2 \zeta(v)/\zeta(1)) \cdot 1$ where $h_i := |\mathscr{C}_i|$. Now each product $xy (x \in \mathscr{C}_1, y \in \mathscr{C}_2)$ is

7.5 GLAUBERMAN'S Z^*-THEOREM

conjugate to an element of the form $uz (z \in \mathscr{C}_2)$, and so by Step 4 we have $\zeta(xy) = \zeta(uv)$. Thus taking traces in the equation

$$T(c_1)T(c_2) = \sum T(xy) \quad \text{(summed over } x \in \mathscr{C}_1, y \in \mathscr{C}_2\text{)}$$

we obtain $h_1 h_2 \zeta(u)\zeta(v)/\zeta(1) = h_1 h_2 \zeta(uv)$. Therefore

(4) $$\zeta(u)\zeta(v) = \zeta(1)\zeta(uv).$$

Similarly, replacing v by the involution uv (see Step 3) we obtain

(5) $$\zeta(u)\zeta(uv) = \zeta(1)\zeta(v).$$

By hypothesis $\zeta(v) \neq 0$, so (4) and (5) imply that $\zeta(u)^2 = \zeta(1)^2$; that is $\zeta(u) = \pm \zeta(1)$.

It remains to show that $\zeta(u) \neq \zeta(1)$. Since $\zeta \neq 1_G$, $N := \operatorname{Ker} \zeta$ is a proper normal subgroup of G, and $\zeta(u) = \zeta(1)$ implies $u \in N$. By Step 1, $O_{2'}(G) = 1$ and so $Z^*(N) = Z(N)$ and $Z(N)$ is a 2-group. Since G is a minimal counterexample, and $N \neq G$, we conclude that $u \in Z(N)$. But $Z(N)$ is characteristic in N and hence normal in G. Therefore for each $x \in G$, $x^{-1}u^{-1}xu \in Z(N)$, and $x^{-1}u^{-1}xu$ is a $2'$-element by hypothesis on G. Since $Z(N)$ is a 2-group, this shows that $x^{-1}u^{-1}xu = 1$ for all $x \in G$, and so $u \in Z(G) \subseteq Z^*(G)$, contrary to hypothesis. This shows that $\zeta(u) \neq \zeta(1)$, and so $\zeta(u) = -\zeta(1)$.

Step 6 The orthogonality relations.

Let $\zeta^1 = 1_G, \zeta^2, \ldots, \zeta^t$ be the irreducible ordinary characters of G that belong to the principal 2-block of G. With the notation in Step 5 we have $\zeta^i(u)\zeta^i(v) = -\zeta^i(1)\zeta^i(v)$ for all $i \geq 2$. However since v is not conjugate to u (Step 3), the corollary of Theorem 7.1 together with Theorem 4.2C gives a contradiction:

$$0 = \sum_{i=1}^t \zeta^i(u)\zeta^i(v) = -\sum_{i=1}^t \zeta^i(1)\zeta^i(v) + 2\zeta^1(1)\zeta^1(v)$$

$$= 2\zeta^1(1)\zeta^1(v) = 2.$$

Thus we have arrived at a contradiction to the assumption that G was a counterexample to the theorem. This completes the proof of the theorem.

EXERCISE

Let G be a nonabelian simple group, P a Sylow 2-subgroup of G, and u an involution in G. Then prove that there exists $x \in G$ such that $x^{-1}ux \in P$ but $x^{-1}ux \neq u$.

7.6 Notes and comments

The general problem of fusions of 2-groups is far from complete solution. Usually there are several possibilities for the fusion of a 2-group (and in general a p-group) and this makes the problem more difficult and complicated. In addition to the groups of type $(2^m, 2^m)$ and quaternion groups considered here, the problems of fusions of a dihedral group, quasi-dihedral group, and wreathed Sylow 2-subgroups have been completely solved. The proofs of Theorems 7.4A and 7.4B were outlined by Brauer and Suzuki [1], and detailed proofs appeared in Suzuki [1] and in Brauer [9, II]. We have closely followed Brauer in the proofs of Theorems 7.3, 7.4A, and 7.4B. Theorems 7.4A and 7.4B can also be proved by using ordinary characters; see Feit [1] and Glauberman [2].

The problem of fusions of a dihedral group was solved by Gorenstein and Walter [2] in a series of three papers, and the cases of quasi-dihedral group and wreathed 2-group were solved in two long papers by Alperin et al. [1, 2]. The necessary character theory was developed by Brauer [9, 12]. For further extensions of these results see Harada [1] and Gorenstein and Harada [1, 2]. Theorem 7.5 and a generalization (proved purely by group theoretical argument) appeared in Glauberman [1]. For another kind of generalization see Goldschmidt [1].

For some general problems of fusion of 2-groups, see Alperin [1] and Goldschmidt [2].

CHAPTER VIII

Blocks with Cyclic Defect Groups

The earliest attempt to determine and describe the irreducible ordinary and modular characters lying in a block was made by Brauer [1], who successfully analyzed blocks of defect 0 and 1. Although Brauer's methods did not extend to other cases, Thompson [1] later gave a new proof of these results which did generalize in certain special cases. By exploiting Thompson's technique, Dade [2] extended Brauer's results to all blocks with cyclic defect groups. We shall not prove here the full theorem of Dade, but using his methods we shall give the analysis in an especially important case, namely, for blocks of defect 1 for a group whose order is divisible by the first power of p only (Theorem 8.5). A number of theorems dealing more generally with blocks with cyclic defect groups lead up to this result. In the final two sections, we apply these results to prove theorems of Brauer, Feit, and Thompson on the existence of normal Sylow p-subgroups in linear groups.

8.1 Extending characters from normal subgroups

In the following sections we shall need on several occasions a method of extending characters from a normal subgroup H of a group G to functions that are characters on G. There are several results of this kind known but here we only deal with the simplest ones.

Recall that if θ is an ordinary (or a modular) character of a normal subgroup H of G, then for each $y \in G$ we define θ^y as an ordinary (respectively, modular) character of H by $\theta^y(x) := \theta(y^{-1}xy)$ for all x in H (respectively, H°). Our first result deals with extension of a character θ from a normal subgroup H of G up to a character of G. Clearly a necessary condition for this to be possible is that $\theta^y = \theta$ for all $y \in G$. In some cases this is also sufficient.

Theorem 8.1 Let H be a normal subgroup of the group G such that G/H is a cyclic group of order h.

(i) Let χ be an irreducible ordinary character of H such that $\chi^y = \chi$ for all $y \in G$. Then there exists an irreducible ordinary character ζ of G such that $\zeta_H = \chi$.

(ii) Let ψ be an irreducible modular character of H such that $\psi^y = \psi$ for all $y \in G$, and let μ be an ordinary character of degree 1 of G with $H = \text{Ker } \mu$. If $p \nmid h$, then there exists an irreducible modular character ϕ of G such that $\phi_H = \psi$ and $\psi^G = \phi + \mu\phi + \cdots + \mu^{h-1}\phi$, where each $\mu^i\phi$ is an irreducible modular character of G (μ^i denotes the ith power of μ).

Proof First let T be any irreducible matrix representation of H over an algebraically closed field K_0. For each $y \in G$ define the representation T^y of H by $T^y(x) := T(y^{-1}xy)$ and suppose that for all $y \in G$, T^y is equivalent to T. Then there exists an invertible matrix a_y such that $T^y(x) = a_y^{-1}T(x)a_y$ for all $x \in H$. In particular, if we choose y so that Hy generates G/H, and write a for a_y, then we get $a^{-h}T(x)a^h = T^{y^h}(x) = T(y^{-h}xy^h) = T(y^{-h})T(x)T(y^h)$ since $y^h \in H$. Thus $T(y^h)a^{-h}$ commutes with $T(x)$ for all $x \in H$. Since T is absolutely irreducible, $T(y^h)a^{-h} = \alpha_0 1$ for some scalar $\alpha_0 \in K_0$. Since K_0 is algebraically closed we can find $\alpha \in K_0$ such that $\alpha^h = \alpha_0$ and then $T(y^h) = (\alpha a)^h$. Put $a_0 = \alpha a$, and define T_0 on G by $T_0(xy^i) := T(x)a_0^i$ for all $x \in H$, $i = 0, 1, \ldots, h - 1$. We claim that T_0 is a representation of G over K_0.

Indeed, if $x,x' \in H$ and $0 \leq i, j < h$, then

$$T_0(xy^i)T_0(x'y^j) = T(x)a_0^i T(x')a_0^{-i}a_0^{i+j}$$
$$= T(x)a^iT(x')a^{-i}a_0^{i+j}$$
$$= T(x)T(y^ix'y^{-i})a_0^{i+j}$$
$$= T(xy^ix'y^{-i})a_0^{i+j}.$$

Since $a_0^{i+j} = T(y^h)a_0^{i+j-h}$ if $h \leq i + j < 2h$, we see that

$$T_0(xy^i)T_0(x'y^j) = T_0(xy^ix'y^j)$$

as required. Note that the restriction of the representation T_0 of G to the subgroup H equals the original representation T of H; in particular, this shows that T_0 is also irreducible. We now consider the situations described in (i) and (ii).

8.2 BLOCKS WITH NORMAL CYCLIC DEFECT GROUPS

(i) In this case take K_0 as a field of characteristic 0 and let T be a representation affording χ. Since $\chi^y = \chi$ for all $y \in G$ by hypothesis, and T^y affords χ^y, we have T^y equivalent to T for all $y \in G$ (Corollary 1 of Theorem 2.3). Thus we obtain an irreducible representation T_0 of G extending T and the character ζ afforded by T_0 satisfies $\zeta_H = \chi$.

(ii) In this case take K_0 as a field of characteristic p, and let T be a representation of H over K_0 which affords the irreducible modular character ψ. Since $\psi^y = \psi$ for all $y \in G$ by hypothesis, the (Frobenius) characters of T^y and T are equal, and so T^y is equivalent to T. Hence we again obtain an irreducible representation T_0 of G over K_0 extending T. The irreducible modular character ϕ afforded by T_0 has the property $\phi_H = \psi$. Define $\check{\psi}$ on G by $\check{\psi} = \psi$ on H and $\check{\psi} = 0$ on $G\backslash H$. Then

$$\psi^G(z) := \sum_{i=0}^{h-1} \check{\psi}(y^{-i}zy^i) \qquad \text{for all} \quad z \in G,$$

where y is chosen such that Hy generates G/H. Hence

$$\psi^G(z) = \begin{cases} h\psi(z) & \text{if } z \in H \\ 0 & \text{otherwise.} \end{cases}$$

On the other hand, for $i = 0, 1, \ldots, h-1$, $\mu^i\phi$ is the irreducible modular character of G afforded by the representation T_i defined by $T_i(z) := T_0(z)\mu(z)^i$. Since $\mathrm{Ker}\,\mu = H$, $\mu(z)$ is a nontrivial hth root of unity for all $z \notin H$. Therefore

$$\sum_{i=0}^{h-1} \mu(z)^i = \begin{cases} h & \text{if } z \in H \\ 0 & \text{if } z \notin H. \end{cases}$$

This shows that

$$\sum_{i=0}^{h-1} (\mu^i\phi)(z) = \left(\sum_{i=0}^{h-1} \mu(z)^i\right)\phi(z) = \begin{cases} h\phi(z) & \text{if } z \in H \\ 0 & \text{otherwise.} \end{cases}$$

Hence $\psi^G = \sum_{i=0}^{h-1} \mu^i\phi$ as asserted.

8.2 Blocks with normal cyclic defect groups

Throughout the remainder of this chapter we shall use the notation first introduced in §3.3. Thus G will be a group of order g, A will be a p-adic algebra which is an integral domain of characteristic 0 containing a primitive gth root of unity, and K will be the field of quotients of A. Also πA is the unique maximal ideal of A and $k = A/\pi A$ is the residue field of characteristic p. Both K and k are splitting fields for G and all its quotient groups as well as subgroups.

Let Q be a cyclic normal p-subgroup of G. As we know from Lemma 6.2A, Q is contained in each defect group of a block B of kG. In the present section we shall consider the case where Q is the (unique) defect group of B. The theorem below describes the set of irreducible ordinary and modular characters that lie in such a block. The proof requires the following lemma.

Lemma 8.2 Let Q be a normal p-subgroup of the group G and put $H := QC_G(Q)$. Let b be a block of kH. If χ is an irreducible ordinary character of H lying in b, then each irreducible constituent of χ^G lies in b^G.

Proof Let $\mathscr{C}_1, \mathscr{C}_2, \ldots, \mathscr{C}_s$ be the conjugate classes of G with class sums $c_1, c_2, \ldots, c_s$ in kG and put $h_i = |\mathscr{C}_i|$, $\mathscr{C}_i^* = \mathscr{C}_i \cap H$ and $c_i^* := \sum_{x \in \mathscr{C}_i^*} x$ ($i = 1, 2, \ldots, s$). If ω is the central character of kH associated with the block b, then (see §6.3) the central character of kG associated with the block b^G is ω^G given by

(1) $$\omega^G(c_i) := \omega(c_i^*) \qquad (i = 1, 2, \ldots, s).$$

If $\mathscr{C}'$ is a class of H with class sum c' and $h' := |\mathscr{C}'|$, then by Theorem 4.2B we have

$$\omega(c') = \overline{h'\chi(y)/\chi(1)} \quad \text{(where the bar denotes reduction modulo } \pi \text{)}$$
$$= \overline{\sum_{x \in \mathscr{C}'} \chi(y)/\chi(1)} \quad \text{with} \quad y \in \mathscr{C}'.$$

Since $H \triangleleft G$, we have for each i either $\mathscr{C}_i^* = \mathscr{C}_i$ or $\mathscr{C}_i^* = \varnothing$. Therefore

(2) $$\omega^G(c_i) = \omega(c_i^*) = \begin{cases} \overline{\sum_{x \in \mathscr{C}_i} \chi(y)/\chi(1)} & \text{if } \mathscr{C}_i \subseteq H \\ 0 & \text{otherwise} \end{cases}$$

for $i = 1, 2, \ldots, s$.

Now suppose that ζ is an irreducible constituent of χ^G. Then (see Theorem 4.2A) the central character ψ associated with the block of kG in which ζ lies is given by $\psi(c_i) = \overline{h_i \zeta_i/\zeta(1)}$ ($i = 1, 2, \ldots, s$), where ζ_i is the value of ζ on $\mathscr{C}_i$. We have to prove that $\psi = \omega^G$. First of all, by the Frobenius reciprocity theorem (Theorem 2.5A) χ is an irreducible constituent of ζ_H. Moreover by Clifford's theorem (Theorem 2.2A) all irreducible constituents of ζ_H are of the form χ^y for some $y \in G$ (since $H \triangleleft G$). Since $\langle \chi^y, \zeta_H \rangle_H = \langle \chi, \zeta_H^{y^{-1}} \rangle_H = \langle \chi, \zeta_H \rangle_H$, it follows that each of the different conjugates of χ occurs with the same multiplicity in ζ_H, say m. Let $\chi^1 = \chi, \chi^2, \ldots, \chi^n$ be the distinct conjugates of χ. Then $\zeta_H = m\{\chi^1 + \cdots + \chi^n\}$ and so for all $x \in H$ we have

$$\zeta(x) = m \sum_{i=1}^{n} \chi^i(x) = \frac{mn}{g} \sum_{y \in G} \chi(y^{-1}xy).$$

8.2 BLOCKS WITH NORMAL CYCLIC DEFECT GROUPS

Thus if $\mathscr{C}_i \subseteq H$ and $z \in \mathscr{C}_i$, then $\zeta_i = \zeta(z)$ and $\zeta(1) = mn\chi(1)$, so

(3)
$$h_i\zeta_i/\zeta(1) = h_i \sum_{y \in G} \chi(y^{-1}zy)/g\chi(1)$$
$$= \sum_{x \in \mathscr{C}_i} \chi(x)/\chi(1).$$

Together with (2) this shows that

(4) $\qquad \omega^G(c_i) = \psi(c_i) \qquad$ whenever $\quad \mathscr{C}_i \subseteq H$.

Now suppose $\mathscr{C}_i \not\subseteq H$. Let $P = O_p(H)$; then $P \triangleleft G$ because it is characteristic in H. Moreover $Q \subseteq P$ because $Q \triangleleft H$. Let $z \in \mathscr{C}_i$. Since $\mathscr{C}_i \not\subseteq H = QC_G(Q)$, therefore $z \notin C_G(Q)$ and so Q (and therefore P) is not contained in $C_G(z)$. This shows that the defect groups of $\mathscr{C}_i$ do not contain P. Therefore by Lemma 6.2A, $c_i \in \mathrm{rad}\, Z(kG)$ and so c_i lies in the kernel of every central character of kG. In particular,

(5) $\qquad \omega^G(c_i) = 0 = \psi(c_i) \qquad$ whenever $\quad \mathscr{C}_i \not\subseteq H$.

The equations (4) and (5) prove the lemma. ∎

NOTATION During the remainder of this chapter we shall be examining the following situation. The group G will contain a cyclic p-subgroup Q and we put $C := C_G(Q) \supseteq Q$ and $N := N_G(Q) \supseteq C$. Let B be a block of kG with Q as a defect group. Then by the First Main Theorem (Theorem 6.3) there is a unique block B_1 of kN with Q as a defect group such that $B = B_1^G$. By Theorem 6.4 the blocks of kC that correspond to the block B_1 of kN are all N-conjugate; fix b as one of these blocks of kC. Then Q is the (unique) defect group of b, $b^G = B$, and the N-conjugates of b are the only blocks of kC that correspond to B under the Brauer correspondence. Write $b = e(kC)$, where e is a central primitive idempotent of kC, and define $F := \{x \in N \mid x^{-1}ex = e\}$ as the stabilizer of the block b under N. We shall put $q := |F : C|$.

Let $|Q| = p^d$. Then the automorphism group $\mathrm{Aut}\, Q$ has order $p^{d-1}(p-1)$. Since N/C is isomorphic to a subgroup of $\mathrm{Aut}\, Q$, this shows that $|N : C|$ divides $p^{d-1}(p-1)$; and since $q \mid |N : C|$ we conclude that $q \mid p^{d-1}(p-1)$. On the other hand, b has $|N : F|$ N-conjugates. Therefore if p^l is the highest power of p dividing $|N : C|$, then Theorem 6.4 shows that $p^l \mid |N : F|$. In particular, this shows that $q = |N : C|/|N : F|$ is relatively prime to p, and so we must have $q \mid p - 1$. Note that F/C is a cyclic group because the subgroup of order $p - 1$ in $\mathrm{Aut}\, Q$ is cyclic (see Hall [1], §6.2).

We next note that if $y \in F \setminus C$, then $C_Q(y) = 1$. Indeed, choose z as a generator of the cyclic group Q. Then $y^{-1}zy = z^m$ for some integer m with $1 \leq m < p^d$. Since $|F/C|$ divides $p - 1$, $y^{p-1} \in C$. Therefore $z = y^{-(p-1)}zy^{p-1} = z^{m^{p-1}}$; and so $m^{p-1} \equiv 1 \pmod{p^d}$. On the other hand, if $z^i \in C_Q(y)$, then $z^i = y^{-1}z^iy = z^{mi}$; and so $mi \equiv i \pmod{p^d}$. If $z^i \neq 1$, then we would have $m \equiv 1 \pmod{p}$. Since $y \notin C$, $m \neq 1$ and so we would have $m = 1 + p^j l$ for some integers j and l with $1 \leq j < d$ and $p \nmid l$. But then the binomial theorem shows that

$$m \equiv mm^{p-1} = (1 + p^j l)^p = 1 + p^{j+1} l + \cdots \equiv 1 \pmod{p^{j+1}},$$

which contradicts the condition $p \nmid l$. Hence we have proved $C_Q(y) = 1$ for all $y \in F \backslash C$. This implies that $Q = \{x^{-1} y^{-1} xy \mid x \in Q\}$ for each $y \in F \backslash C$ since the elements $x^{-1} y^{-1} xy$ ($x \in Q$) must all be different.

The cyclic group Q has $p^d - 1$ irreducible ordinary characters (all of degree 1) different from 1_Q. The group F acts by conjugation on this set of characters, and no character $\chi \neq 1_Q$ is fixed by any $y \in F \backslash C$; indeed $\chi^y = \chi$ implies $1 = \chi^y(x) \chi(x)^{-1} = \chi(y^{-1} xyx^{-1})$, so from what we have just proved, $\text{Ker } \chi = Q$. Thus each orbit of F on the set of nontrivial irreducible characters of Q has length $|F : C| = q$; and there are $(p^d - 1)/q$ orbits. We shall choose a set Λ of representative characters, one from each orbit. Then $|\Lambda| = (p^d - 1)/q$ and each nontrivial irreducible character of Q is conjugate under F to exactly one character from Λ.

With this notation we can now state the first theorem on blocks with cyclic defect groups.

Theorem 8.2 With the notation above suppose that Q is a normal p-subgroup of G (so the block B has Q as its unique defect group). Let b be a block of kC such that $b^G = B$, and using the notation of Theorem 6.2B let ϕ and $\xi_1, \xi_2, \ldots, \xi_{p^d}$ be the irreducible modular character and the irreducible ordinary characters, respectively, lying in b (to simplify later notation we use lower indices). Then

  (i) There are exactly q irreducible modular characters $\psi_1, \psi_2, \ldots, \psi_q$ lying in B. They all have the same degree and satisfy $\psi_1 + \psi_2 + \cdots + \psi_q = \phi^G$.
  (ii) There are exactly $q + (p^d - 1)/q$ irreducible ordinary characters lying in B. These can be divided into two classes $\chi_1, \chi_2, \ldots, \chi_q$ and χ_λ ($\lambda \in \Lambda$). The χ_i ($i = 1, 2, \ldots, q$) are precisely those characters whose kernels contain Q; their degrees are equal and $\chi_1 + \chi_2 + \cdots + \chi_q = \xi_1^G$. The χ_λ all have the same degree and each χ_λ has the form ξ_j^G for exactly q values of j ($2 \leq j \leq p^d$).
  (iii) The q principal indecomposable modular characters lying in B are given by $\chi_i + \sum_{\lambda \in \Lambda} \chi_\lambda$ (restricted to $G°$) for $i = 1, 2, \ldots, q$.
  (iv) For each $\lambda \in \Lambda$, $\chi_\lambda = \sum_{i=1}^{q} \chi_i$ on $G°$.

Remark Suppose ϕ has degree m. Then $\deg \psi_i = |N : F| m = \deg \chi_i$ ($i = 1, 2, \ldots, q$), whilst $\deg \chi_\lambda = |N : C| m = q \deg \chi_i$ ($\lambda \in \Lambda$).

Proof With the notation above, $G = N$. Let $\alpha_1 = 1_Q, \ldots, \alpha_{p^d}$ be the

8.2 BLOCKS WITH NORMAL CYCLIC DEFECT GROUPS

irreducible ordinary characters of the cyclic group Q. Consider the characters that lie in the block b of kC. Since $Q \subseteq Z(C)$ and b has Q as defect group, Theorem 6.2B shows that exactly one irreducible modular character, say ϕ, and p^d irreducible ordinary characters, say $\xi_1, \ldots, \xi_{p^d}$, lie in b where for each i

(1) $\qquad \xi_i(y) = \begin{cases} \alpha_i(z)\phi(x) & \text{if } y = zx \ (z \in Q, x \in C°) \\ 0 & \text{otherwise.} \end{cases}$

Now $b^G = B$ and $C = C_G(Q) \triangleleft G$, so all irreducible constituents of ξ_i^G lie in B by Lemma 8.2. Conversely, suppose that χ is an irreducible ordinary character of G lying in B. Let ξ be an irreducible constituent of χ_H, lying in a block b_1, say, of kC. By the Frobenius reciprocity theorem, χ is an irreducible constituent of ξ^G, and so $b_1^G = B$ by Lemma 8.2. Since $G = N$, it follows from our remarks above that b_1 is a G-conjugate of b. Thus by Example 1 of §6.4, $\xi = \xi_i^y$ for some i and some $y \in G$. Then $\xi^G = \xi_i^G$ and χ is an irreducible constituent of ξ_i^G. Hence we have shown that the irreducible ordinary characters of G in b^G are precisely the irreducible constituents of ξ_i^G ($i = 1, 2, \ldots, p^d$). We now turn to the proof of the particular assertions of the theorem.

(i) Since $\xi_i = \phi$ on $C°$ for each i, by (1), the irreducible modular characters lying in B are precisely the irreducible constituents of ϕ^G. By definition of F, $\phi^y = \phi$ for all $y \in F$. Therefore Theorem 8.1 shows that there is an irreducible modular character ψ of F with $\psi_C = \phi$ and a character μ of degree 1 of F with Ker $\mu = C$ such that

(2) $\qquad\qquad\qquad \phi^F = \psi + \mu\psi + \cdots + \mu^{q-1}\psi$

and hence $\phi^G = \psi^G + (\mu\psi)^G + \cdots + (\mu^{q-1}\psi)^G$. The characters $\psi^G, (\mu\psi)^G, \ldots, (\mu^{q-1}\psi)^G$ are distinct since they take different values on p'-elements of $C°$. To prove (i) it remains to show that each $(\mu^i\psi)^G$ is irreducible. Let V_i be an irreducible kF-module which affords $\mu^i\psi$. Then $V_i | (V_i^G)_F$ by definition of V_i^G. Let W be an indecomposable summand of V_i^G such that $V_i | W_F$. Since $(V_i)_C$ is irreducible and $(V_i)_C | W_C$, $\dim_k W \geq |G:F| \dim_k V_i$ because there are $|G:F|$ distinct conjugates of V_i. But $\dim_k W \leq \dim_k V_i^G$, so that $W = V_i^G$. If U is an irreducible kG-submodule of V_i^G, then U_C must contain a conjugate of V_i and so contain all conjugates. But V_i^G is a direct sum of $|G:F|$ conjugates of V_i, and therefore $U = V_i^G$ is irreducible. Thus the irreducible modular characters of G lying in B are $\psi_i := (\mu^i\psi)^G$; $i = 0, 1, 2, \ldots, q-1$.

(ii) As we noted in our remarks above, the nontrivial irreducible ordinary characters of Q are permuted under the action of F in orbits of length q whilst the trivial character is left fixed. Since ϕ is left fixed by F, (1) shows that ξ_1 is fixed under the action of F but the other ξ_i ($i = 2, 3, \ldots, p^d$) are permuted in orbits of length q. First consider ξ_1.

It follows from (1) that

$$\xi_1^G(y) = \begin{cases} \phi^G(x) & \text{if } y = zx \ (z \in Q, x \in G^\circ) \\ 0 & \text{otherwise.} \end{cases}$$

We can choose an A-free AG-module W that affords ξ_1^G (see Theorem 3.3). Then reduction modulo π gives a kG-module $\bar{W}$ of G which affords ϕ^G as its modular character. From the proof of (i), ϕ^G (and hence $\bar{W}$) has q irreducible constituents that are all inequivalent. This shows that W (and hence ξ_1^G) has at most q irreducible constituents that must be inequivalent. Now $\phi^y = \phi$ exactly when $y \in F$, and so the multiplicity of ξ_1 as an irreducible constituent in $(\xi_1^G)_C$ is $|F:C|$. Therefore

$$\langle \xi_1^G, \xi_1^G \rangle_G = \langle \xi_1, (\xi_1^G)_C \rangle_C = |F:C| = q.$$

Hence ξ_1^G is a sum of q distinct irreducible constituents, which we denote by $\chi_1, \chi_2, \ldots, \chi_q$. Moreover, these irreducible constituents correspond with those of ϕ^G, so $\chi_i = \psi_i$ on G° for $i = 1, 2, \ldots, q$; and $\operatorname{Ker} \chi_i \supseteq \operatorname{Ker} \xi_1^G \supseteq Q$.

Now suppose $i > 1$. Then ξ_i has q conjugates, say $\xi_i, \xi_{i+1}, \ldots, \xi_{i+q-1}$ under F. This means that ξ_i^F is irreducible and

$$\xi_i^F(x) = \begin{cases} \xi_i(x) + \cdots + \xi_{i+q-1}(x) & \text{if } x \in C \\ 0 & \text{otherwise.} \end{cases}$$

Now $\langle \xi_i^G, \xi_i^G \rangle_G = \langle \xi_i, (\xi_i^G)_C \rangle_C = 1$ because $\xi_i^y \neq \xi_i$ when $y \notin C$; so ξ_i^G is also irreducible. Since $(\xi_i^G)_F = |G:F| \xi_i^F$, and ξ_i^G is 0 outside F, it follows that $\xi_i^G = \xi_j^G$ iff ξ_i is F-conjugate to ξ_j; moreover, since $Q \not\subseteq \operatorname{Ker} \xi_i^G$, no ξ_i^G ($i \geq 2$) is equal to any of $\chi_1, \chi_2, \ldots, \chi_q$. Thus we obtain $(p^d - 1)/q = |\Lambda|$ further irreducible characters χ_λ ($\lambda \in \Lambda$) corresponding to the orbits of characters under the action of F, and $Q \not\subseteq \operatorname{Ker} \chi_\lambda$. Since all irreducible ordinary characters of G in B are constituents of ξ_i^G ($i = 1, 2, \ldots, p^d$), this proves (ii).

(iii) Since $\xi_i = \phi$ on C° and $C \triangleleft G$, we have $\xi_i^G = \phi^G$ on G°. Thus we have

(3) $$\chi_\lambda = \psi_1 + \cdots + \psi_q \quad \text{on} \quad G^\circ$$

for all $\lambda \in \Lambda$. As we noted in the proof of (ii) we also have

(4) $$\chi_i = \psi_i \quad \text{on} \quad G^\circ$$

for $i = 1, 2, \ldots, q$. Equations (3) and (4) describe the decomposition numbers for B and then (iii) follows from Note 4 in §3.7.

(iv) This follows at once from (3) and (4).

EXAMPLE Suppose under the hypothesis of the theorem that $B = B_0$ is the principal block of kN ($N = G$). Then by the Third Main Theorem (Theorem 6.6), the principal block of b of kC is the only block of kC such

8.3 GROUPS WITH CYCLIC SYLOW p-SUBGROUPS

that $b^N = B_0$. Clearly in this case ϕ is the trivial modular character, so $\phi(1) = 1$ and $F = N$. Hence there are $n := |N:C|$ characters $\chi_1, \chi_2, \ldots, \chi_n$ and $(p^d - 1)/n$ characters χ_λ ($\lambda \in \Lambda$) lying in B_0. In this case, it is clear that $\xi_i = \phi = 1$ on C°. Therefore $\xi_i^N = n$ on C° for each i and $C^\circ \subseteq \text{Ker } \xi_i^N$. This shows that

$$\chi_1 = \cdots = \chi_n = 1 \quad \text{on } C^\circ \quad \text{and} \quad \chi_\lambda = n \quad \text{on } C^\circ \quad \text{for all } \lambda \in \Lambda.$$

EXERCISE

With the hypothesis of Theorem 8.2, calculate the decomposition matrix and Cartan matrix of the block B.

8.3 Groups with cyclic Sylow p-subgroups

In the present section we shall give a description of all indecomposable kG-modules in the case G is a group with a cyclic Sylow p-subgroup. It turns out that (up to isomorphism) there are only a finite set of such modules. In contrast, in the exercises that follow this section, it will be seen that if G does not have a cyclic Sylow p-subgroup, there are always indecomposable kG-modules of arbitrarily large dimension.

Lemma 8.3 Let P be a cyclic group of order p^d. Then up to isomorphism there are exactly p^d indecomposable kP-modules. These have dimensions 1, 2, ..., p^d, respectively, and each is isomorphic to a homomorphic image of the principal indecomposable kP-module kP.

Proof Each kP-module V may be considered as a module over the polynomial ring $k[X]$ if we define $vX := vz$ for some fixed generator z of the group P. Since $k[X]$ is a principal ideal domain, every $k[X]$-module is a direct sum of cyclic $k[X]$-modules. Thus, if V is an indecomposable $k[X]$-module, then V has the form $v_0 k[X]$ for some $v_0 \in V$. Then $f(X) \mapsto v_0 f(X) = v_0 f(z)$ is a homomorphism of the $k[X]$-module $k[X]$ onto V with kernel I, say. Since k has characteristic p, $v_0(z-1)^{p^d} = v_0(z^{p^d} - 1) = 0$ and so $(X-1)^{p^d} \in I$. Since $k[X]$ is a principal ideal domain, this means $I = (X-1)^m k[X]$ for some m, $1 \le m \le p^d$; and $V \simeq k[X]/(X-1)^m k[X]$. Conversely, if $1 \le m \le p^d$, then $V_m := k[X]/(X-1)^m k[X]$ is an indecomposable $k[X]$-module of dimension m over k on which $X^{p^d} - 1 = (X-1)^{p^d}$ acts as 0; hence V_m may be considered as a kP-module with $vz := vX$. Thus, each indecomposable kP-module is isomorphic to exactly one of V_m ($m = 1, \ldots, p^d$). The last assertion of the lemma follows from the observation that for all $m \le p^d$,

$$V_m = k[X]/(X-1)^m k[X] \simeq kP/(z-1)^m kP.$$

Theorem 8.3 Let G be a group with a cyclic Sylow p-subgroup P. Then each indecomposable kG-module is a homomorphic image of the kG-module kG (and so a homomorphic image of some principal indecomposable kG-module).

Proof Let V be an indecomposable kG-module. By §5.2 we know that there exists an indecomposable kP-module W such that $V \mid W^G$. By Lemma 8.3 we can find $w \in W$ such that $W = w(kP)$. Then $W^G = w(kP) \otimes_{kP} kG = w_0(kG)$ where $w_0 := w \otimes 1$. Hence the mapping $a \mapsto w_0 a$ shows that W^G is a homomorphic image of the kG-module kG. Since V is a direct summand of W^G, V is also a homomorphic image of the kG-module kG.

Corollary Under the hypotheses of the theorem there are only a finite number of nonisomorphic indecomposable kG-modules.

Proof By the proof above, each indecomposable kG-module V is isomorphic to a component of W^G for some indecomposable kP-module W. Then the corollary follows from Lemma 8.3 and the Krull–Schmidt theorem.

EXERCISES

1. Let $G = \langle a \rangle \times \langle b \rangle$, where a and b have orders p, and let k be any field of characteristic p. Let V be a vector space of dimension $2n + 1$, with a basis $\{u_0, u_1, \ldots, u_n; v_1, \ldots, v_n\}$. Define $u_i a = u_i b = u_i$, $0 \le i \le n$, $v_i(a - 1) = u_i$, $1 \le i \le n$; and $y_i(a - 1) = u_{i-1}$. Prove that V is an indecomposable kG-module.
2. Use Exercise 1 to show that if G is a group with a noncyclic Sylow p-subgroup, then there exist indecomposable kG-modules of arbitrarily large dimensions.

8.4 Some technical lemmas

In the next section we shall be describing properties of blocks of defect 1 in a group G of order $g = pg_0$, where $p \nmid g_0$. To prove these results we need information about characters that are sums of constituents of principal indecomposable modular characters of G (Lemma 8.4B). Lemma 8.4A will only be used for the proof of Lemma 8.4B.

Lemma 8.4A Let U_0 be a principal indecomposable AG-module and let V be a KG-module such that $V \mid U_0 \otimes_A K$. Then there exists an A-free AG-module U such that $U \otimes_A K \simeq V$ and $\bar{U}$ is an indecomposable kG-module.

Proof Since $U_0 \otimes_A K$ is completely reducible, we can write $U_0 \otimes_A K = V \oplus V'$ for some KG-submodule V' of $U_0 \otimes_A K$. Let $W := V \cap U_0$ and

8.4 SOME TECHNICAL LEMMAS

$W' := V' \cap U_0$ (where U_0 is assumed to be embedded in $U_0 \otimes_A K$). We define $U := U_0/W'$ and show that it satisfies the required conditions. First, U is A-free. Indeed, if $u \in U_0$ and $(u + W')a = 0$ for some $a \neq 0$ in A, then $ua \in W' \subseteq V'$, and so $u \in V'a^{-1} = V'$ as well as $u \in U_0$ and hence $u + W' = W'$. Hence U is A-torsion free and so by the structure theory of modules over a principal ideal domain (see Example 4 of §1.1), U is A-free as asserted.

Next, for any K-basis of V we can multiply by a suitable element of A to obtain a K-basis of V lying in U_0. Therefore $W \otimes_A K = V$; and similarly $W' \otimes_A K = V'$. Since U is A-free, this shows that $U \otimes_A K \simeq (U_0 \otimes_A K)/(W' \otimes_A K) \simeq V$.

Also since U is A-free, $\bar{U} \simeq \bar{U}_0/\bar{W}'$. By Theorem 3.4A, $\bar{U}_0$ is a principal indecomposable kG-module (since U_0 is a principal indecomposable AG-module) and so $\bar{U}_0$ has a unique maximal submodule by Theorem 3.4B. Therefore $\bar{U}$ must also have a unique maximal submodule and hence it is indecomposable. This proves the lemma.

Now let G be a group of order $g = pg_0$ with $p \nmid g_0$. Let P be a Sylow p-subgroup of G and put $C := C_G(P)$ and $N := N_G(P)$. Let B be a block of defect 1 in kG, and B_1 the unique block of kN such that $B_1^G = B$ (Theorem 6.3). With the notation of Theorem 8.2 we shall take $\chi_1, \chi_2, \ldots, \chi_q$ and χ_λ ($\lambda \in \Lambda$) as the irreducible ordinary characters that lie in the block B_1. The following result serves as a replacement for Theorem 6 of Dade [2].

Lemma 8.4B *Under the above hypotheses let V be a KG-module whose irreducible constituents lie in the block B of kG, and let ζ be the character afforded by V. Suppose that there exists a principal indecomposable AG-module U_0 such that $V \mid U_0 \otimes_A K$. Then for some i ($1 \leq i \leq q$) and some integers n_i, n_λ ($\lambda \in \Lambda$) each equal to 0 or 1,*

$$\zeta_N = n_i \chi_i + \sum_{\lambda \in \Lambda} n_\lambda \chi_\lambda + \theta$$

where θ is a sum of characters such that $\theta = 0$ on $N\setminus N^o$.

Proof By Lemma 8.4A we can find an A-free AG-module U which affords the character ζ and for which $\bar{U}$ is indecomposable. In particular, U is indecomposable. Since $P \subseteq N$, U is N-projective and so $U \mid (U_N)^G$, and Note 1 of §5.2 shows that for some indecomposable A-free AN-module W, $W \mid U_N$ and $U \mid W^G$. Write

(1) $$U_N = W_0 \oplus \cdots \oplus W_m,$$

where the W_i are indecomposable AN-modules and $W_0 \simeq W$. By Mackey's theorem (Theorem 2.1A), $(W^G)_N \simeq \bigoplus_{i=1}^n (V_{x_i})^N$, where $x_1 = 1, \ldots, x_n$ is a set of representatives of (N, N)-double cosets and V_{x_i} is an

$A(x_i^{-1}Nx_i \cap N)$-module. If $i \geq 2$, then $x_i \notin N$ and so $P \nsubseteq x_i^{-1}Nx_i \cap N$; hence in this case $x_i^{-1}Nx_i \cap N$ is a p'-group since $|P| = p$. Therefore for each $i \geq 2$, the vertex of V_{x_i} is 1. This shows that $(W^G)_N$ (and hence U_N) is a sum of indecomposable AN-modules such that all but one (namely $V_{x_1} \cong W$) is 1-projective. Hence in (1), the modules $W_1, \ldots, W_m$ are all 1-projective. Now for $i \geq 1$, each W_i is isomorphic to a principal indecomposable AN-module (Example 3 of §5.1); so the character afforded by W_i vanishes on $N \backslash N^o$ by the Example of §3.7. Thus, if θ is the character afforded by $W_1 \oplus \cdots \oplus W_m$, then $\theta = 0$ on $N \backslash N^o$. To complete the proof of the lemma it remains to show that the character afforded by $W \cong W_0$ has the form $n_i \chi_i + \sum_{\lambda \in \Lambda} n_\lambda \chi_\lambda$ (for some i) with each n_i, n_λ ($\lambda \in \Lambda$) equal to 0 or 1.

Since W is an A-free indecomposable AN-module, $\bar{W}$ is an indecomposable kN-module by Theorem 3.4B. Hence by Theorem 8.3 there is a surjective kN-homomorphism $kN \to \bar{W}$. Choose $w \in W$ such that $\bar{w}$ is the image of 1 under this homomorphism. Then $a \mapsto wa$ is an AN-homomorphism of $AN \to W$; and since $\overline{w(AN)} = \bar{W}$, this mapping is surjective by the structure theory of modules over principal ideal domains. Thus W is a homomorphic image of the AN-module AN. Since W is indecomposable this means that W is a homomorphic image of some principal indecomposable AN-module, say $e(AN)$. Let τ be the character afforded by $e(AN)$. We claim that $\tau = \chi_i + \sum_{\lambda \in \Lambda} \chi_\lambda$ for some i. Since the multiplicity of any irreducible character as a constituent of τ must be at least as great as its multiplicity in the character afforded by W, this will prove the lemma.

Thus (with the notation of Theorem 8.2) suppose that the principal indecomposable kN-module $\overline{e(AN)}$ corresponds to the modular irreducible character ψ_i, say. Then by Theorem 8.2(iii) and the example of §3.7 we have

$$\tau = \begin{cases} \chi_i + \sum_{\lambda \in \Lambda} \chi_\lambda & \text{on } G^\circ \\ 0 & \text{on } G \backslash G^\circ. \end{cases}$$

Thus, using Theorem 3.7B, we conclude that the multiplicity of χ_j in τ is (for $j = 1, \ldots, q$)

$$\langle \tau, \chi_j \rangle_G = \langle \tau, \psi_j \rangle_{G^\circ} = \delta_{ij} \quad \text{by (4) of §8.2;}$$

and the multiplicity of χ_λ in τ is (for all $\lambda \in \Lambda$)

$$\langle \tau, \chi_\lambda \rangle_G = \langle \tau, \psi_1 + \cdots + \psi_q \rangle_{G^\circ} = 1 \quad \text{by (3) of §8.2.}$$

Thus $\tau = \chi_i + \sum_{\lambda \in \Lambda} \chi_\lambda$ (on G) and the proof of the lemma is complete.

8.5 Groups of order $g = pg_0$ with $p \nmid g_0$

In this section let G denote a group of order $g = pg_0$ where $p \nmid g_0$. The purpose of this section is to describe the irreducible ordinary and modular

8.5 GROUPS OF ORDER $g = pg_0$ WITH $p \nmid g_0$

characters lying in a block with (cyclic) defect groups of order p. The general problem of a block, with a cyclic defect group, of an arbitrary group was solved by Dade [2]; but for our purpose a special case is sufficient. Moreover, a knowledge of this special case enables one to grasp the general situation better.

NOTATION Let P be a Sylow p-subgroup of G. As before put $C := C_G(P)$ and $N := N_G(P)$; note that $C^\circ = O_{p'}(C)$ and $C = P \times C^\circ$.

Let b be a block of kC. Then we know from Theorem 6.2B that there is a single irreducible modular character ϕ lying in b and the irreducible ordinary characters $\xi_1, \xi_2, \ldots, \xi_p$ (all of the same degree) lying in b have the form

$$\xi_j(zy) = \alpha_j(z)\phi(y) \quad (z \in P, y \in C^\circ)$$

where $\alpha_1 = 1_P, \alpha_2, \ldots, \alpha_p$ are the irreducible ordinary characters of P. With the notation of §8.2 there are $q + (p-1)/q$ irreducible ordinary characters of N lying in b^N, which we denote by $\chi_1, \chi_2, \ldots, \chi_q$ and χ_λ ($\lambda \in \Lambda$). These are related to the characters in b by

$$\xi_1^N = \chi_1 + \chi_2 + \cdots + \chi_q \quad \text{and} \quad \{\xi_j^N \mid j = 2, \ldots, p\} = \{\chi_\lambda \mid \lambda \in \Lambda\}$$

and moreover we have $\chi_\lambda = \chi_1 + \chi_2 + \cdots + \chi_q$ on N° for all $\lambda \in \Lambda$. We shall define $\tau := \chi_1 + \chi_2 + \cdots + \chi_q$.

Let us denote by $b_1 = b, b_2, \ldots, b_n$ the blocks of kC such that $b_i^N = b^N$. Then the description above holds for the blocks b_i and we shall denote the corresponding characters by $\xi_j^{(i)}, \alpha_j^{(i)}$ and $\phi^{(i)}$.

Theorem 8.5 Let G be a group of order $g = pg_0$, where $p \nmid g_0$, and let B be a block of defect 1 of kG. Let b be a block of kC such that $b^G = B$. Then with the notation above we have the following:

(i) There are exactly $q + (p-1)/q$ irreducible ordinary characters lying in B which we denote by $\zeta_1, \zeta_2, \ldots, \zeta_q$ and ζ_λ ($\lambda \in \Lambda$).

(ii) There exists $\varepsilon = 1$ or -1 such that $(\zeta_\lambda - \zeta_\mu) = \varepsilon(\chi_\lambda - \chi_\mu)^G$ for all $\lambda, \mu \in \Lambda$.

(iii) There exist $m_i = 1$ or -1 ($i = 1, 2, \ldots, q$) such that

$$(\zeta_i)_C = m_i \sum_{\xi^N = \tau} \xi + \theta_i,$$

where the sum is over all irreducible ordinary characters ξ of C with $\xi^N = \tau$ (namely, $\xi = \xi_1^{(i)}$ for $i = 1, 2, \ldots, n$), and $\theta_i \in \text{Char}(C)$ vanishes on $C \backslash C^\circ$. In particular ζ_i is constant on $P \backslash \{1\}$ because P lies in the kernel of each ξ (see Theorem 8.3).

(iv) For each $\lambda \in \Lambda$ we have

$$(\zeta_\lambda)_C = \varepsilon \sum_{\xi^N = \chi_\lambda} \xi + \theta_\lambda,$$

where the sum ranges over all irreducible ordinary characters ξ of C such that $\xi^N = \chi_\lambda$ (they all lie in the blocks $b_1, b_2, \ldots, b_n$ by Lemma 8.2), and $\theta_\lambda \in \text{Char}(C)$ vanishes on $C\backslash C^\circ$.

(v) For each $\lambda \in \Lambda$ we have

$$\zeta_\lambda = \varepsilon \sum_{i=1}^{q} m_i \zeta_i \quad \text{on} \quad G^\circ.$$

Remark The characters ζ_λ ($\lambda \in \Lambda$) will be called the *exceptional characters* of the block B and the characters $\zeta_1, \zeta_2, \ldots, \zeta_q$ the *nonexceptional characters*. The exceptional characters all have the same degree by (v).

Proof For convenience, the proof will be divided into several steps.

Step 1 *The irreducible characters of G determined by Λ.*

Let $S := N \backslash N^\circ$. We claim that

(1) $$y^{-1}Sy \cap S = \begin{cases} S & \text{if } y \in N \\ \varnothing & \text{if } y \in G \backslash N. \end{cases}$$

Thus S is a trivial intersection set with $N_G(S) = N$. Indeed it is easily seen that S is just the set of $x \in N$ whose p-part $x_p \in P\backslash\{1\}$, because $P \triangleleft N$. Let $y \in G$ and suppose $y^{-1}Sy \cap S$ contains some element x. Then x and yxy^{-1} lie in S, and so both x_p and $(y^{-1}xy)_p = y^{-1}x_p y$ lie in $P\backslash\{1\}$. Since $|P| = p$, this means that $P = \langle yx_p y^{-1} \rangle = y\langle x_p \rangle y^{-1} = yPy^{-1}$, and so $y \in N$. Since S is a normal subset of N, we conclude $y^{-1}Sy \cap S \ne \varnothing$ implies $y^{-1}Sy \cap S = S$ as asserted.

Now it follows (with the notation above) that for all $\lambda, \mu \in \Lambda$, $\chi_\mu - \chi_\lambda = (\chi_\mu - \tau) - (\chi_\lambda - \tau)$ is a generalized character of N which vanishes on $N\backslash S$. So using (1) it follows from Theorem 2.5B that if $|\Lambda| > 1$ then there exist distinct irreducible ordinary characters ζ_λ ($\lambda \in \Lambda$) such that for some $\varepsilon = 1$ or -1,

(2) $$(\chi_\lambda - \chi_\mu)^G = \varepsilon(\zeta_\lambda - \zeta_\mu) \quad \text{for all } \lambda, \mu \in \Lambda \text{ and } \lambda \ne \mu.$$

This gives $|\Lambda|$ irreducible ordinary characters ζ_λ ($\lambda \in \Lambda$) of G when $|\Lambda| > 1$. If $|\Lambda| = 1$, then we define ζ_λ as any irreducible constituent of $(\chi_\lambda - \tau)^G$ and put $\varepsilon = 1$. In particular, the condition (ii) holds.

Step 2 *Analysis of $(\chi_\lambda - \tau)^G$.*

We claim that there exist irreducible ordinary characters $\zeta_1, \zeta_2, \ldots, \zeta_t$ different from ζ_λ ($\lambda \in \Lambda$) such that for some integers $m_0, m_1, \ldots, m_t$ (independent of λ) with $m_i \ne 0$ for $i = 1, 2, \ldots, t$ we have

(3) $$(\chi_\lambda - \tau)^G = \varepsilon \zeta_\lambda + m_0 \sum_{\mu \in \Lambda} \zeta_\mu - \sum_{i=1}^{t} m_i \zeta_i .$$

8.5 GROUPS OF ORDER $g = pg_0$ WITH $p \nmid g_0$ 211

In the case $|\Lambda| = 1$ this is trivial, so suppose $|\Lambda| > 1$. Then for each $\mu \in \Lambda$, $\mu \neq \lambda$, we have by Theorem 2.5B that

(4) $\quad \langle (\chi_\lambda - \tau)^G, \zeta_\lambda - \zeta_\mu \rangle = \langle (\chi_\lambda - \tau)^G, \varepsilon(\chi_\lambda - \chi_\mu)^G \rangle$
$$= \varepsilon \langle \chi_\lambda - \tau, \chi_\lambda - \chi_\mu \rangle = \varepsilon$$

since both $\chi_\lambda - \tau$ and $\chi_\lambda - \chi_\mu$ vanish on $N \backslash S \, (= N^\circ)$. Moreover, if $\lambda \neq \mu \neq v$, then

$$\langle (\chi_\lambda - \tau)^G, \zeta_\mu \rangle - \langle (\chi_v - \tau)^G, \zeta_\mu \rangle = \langle (\chi_\lambda - \chi_v)^G, \zeta_\mu \rangle$$
$$= \langle \varepsilon(\zeta_\lambda - \zeta_v), \zeta_\mu \rangle = 0.$$

Thus for $\lambda \neq \mu$ we have that

$$\langle (\chi_\lambda - \tau)^G, \zeta_\mu \rangle = m_0 ,$$

say, is independent of λ and μ; and so by (4), $\langle (\chi_\lambda - \tau)^G, \zeta_\lambda \rangle = m_0 + \varepsilon$. This shows that $(\chi_\lambda - \tau)^G$ has the form described in (3). The fact that $m_1, m_2, \ldots, m_t$ are independent of λ follows from the equation

$$\langle (\chi_\lambda - \tau)^G, \zeta_i \rangle - \langle (\chi_\mu - \tau)^G, \zeta_i \rangle = \langle \varepsilon(\zeta_\lambda - \zeta_\mu), \zeta_i \rangle = 0.$$

Using the definition of $\tau = \sum_{i=1}^{q} \chi_i$ and Theorem 2.5A we obtain from (3) the relation

$$1 + q = \langle \chi_\lambda - \tau, \chi_\lambda - \tau \rangle_N = \langle (\chi_\lambda - \tau)^G, (\chi_\lambda - \tau)^G \rangle_G$$
$$= (m_0 + \varepsilon)^2 + (|\Lambda| - 1)m_0^2 + \sum_{i=1}^{t} m_i^2.$$

Hence

(5) $\quad q = 2m_0 \varepsilon + |\Lambda| m_0^2 + \sum_{i=1}^{t} m_i^2.$

Step 3 *The restriction of the characters ζ_i to C.*

By the Frobenius reciprocity theorem (Theorem 2.5A), for all $j > 1$,

$$\langle (\zeta_i)_C, \xi_j - \xi_1 \rangle_C = \langle \zeta_i, (\chi_\lambda - \tau)^G \rangle = -m_i$$

by (3), where $\chi_\lambda = \xi_j^N$. Thus if we define $c := \langle (\zeta_i)_C, \xi_1 \rangle - m_i$, then the contribution to $(\zeta_i)_C$ from the characters in the block b is

(6) $\quad c \sum_{j=1}^{p} \xi_j + m_i \xi_1.$

We also note that since the ξ_j all have the same degree, it follows by

Theorem 4.2C that

(7) $$\sum_{j=1}^{p} \xi_j(x) = \frac{1}{\xi_j(1)} \sum_{j=1}^{p} \xi_j(x)\xi_j(1) = 0 \quad \text{for all} \quad x \in C\backslash C^\circ.$$

If $b_1 = b$, $b_2, \ldots, b_n$ are the blocks of kC such that $b_l^N = b^N$, then we can apply the above analysis to each block b_l. Since the constants m_i depend only on the characters in b^N, they are independent of the blocks b_l. Hence from (6) and (7) we conclude that

(8) $$(\zeta_i)_C = m_i \sum_{j=1}^{n} \xi_1^{(j)} + \beta_i + \gamma_i,$$

where β_i is a sum of characters from the blocks $b_1, b_2, \ldots, b_n$ with the property $\beta_i = 0$ on $C\backslash C^\circ$, and γ_i is a **Z**-linear combination of characters from blocks different from $b_1, b_2, \ldots, b_n$. Since $\xi_1^{(l)}(zy) = \phi_1^{(l)}(y)$ for all $z \in P$ and $y \in C^\circ$, it follows from (8) that the generalized decomposition number of ζ_i at $\phi^{(l)}$ is m_i. Since $m_i \neq 0$ by hypothesis, the Second Main Theorem (Theorem 6.5) shows that ζ_i lies in the block B. Again Theorem 6.5 shows that γ_i in (8) is 0 on $C\backslash C^\circ$. This proves (iii).

Step 4 *The restriction of the characters ζ_λ to C.*

Let $m := \langle \zeta_\lambda, \tau^G \rangle$. Then if $\xi_i^N = \chi_\mu$, $\langle (\zeta_\lambda)_C, \xi_i \rangle = \langle \zeta_\lambda, (\chi_\mu - \tau)^G \rangle + m = \varepsilon \, \delta_{\lambda, \mu} + m_0 + m$ by (3), and $\langle (\zeta_\lambda)_C, \xi_1 \rangle = \langle \zeta_\lambda, \tau^G \rangle = m$. Therefore the contribution of b to $(\zeta_\lambda)_C$ is

$$(m + m_0) \sum_{i=2}^{p} \xi_i + m\xi_1 + \varepsilon \sum_{\xi_j^N = \chi_\lambda} \xi_j$$

$$= (m + m_0) \sum_{i=1}^{p} \xi_i - m_0 \xi_1 + \varepsilon \sum_{\xi_j^N = \chi_\lambda} \xi_j.$$

As in Step 3, we obtain

$$(\xi_\lambda)_C = (m + m_0) \sum_{l=1}^{n} \sum_{i=1}^{p} \xi_i^{(l)} - m_0 \sum_{l=1}^{n} \xi_1^{(l)} + \varepsilon \sum_{l=1}^{n} \sum_{(\xi_j^{(l)})^N = \chi_\lambda} \xi_j^{(l)} + \gamma_\lambda,$$

where γ_λ is the sum of characters from blocks different from $b_1, b_2, \ldots, b_n$. Then

(9) $$(\zeta_\lambda)_C = -m_0 \sum_{l=1}^{n} \xi_1^{(l)} + \varepsilon \sum_{l=1}^{n} \sum_{(\xi_j^{(l)})^N = \chi_\lambda} \xi_j^{(l)} + \beta_\lambda + \gamma_\lambda,$$

where β_λ is a sum of characters from the blocks $b_1, b_2, \ldots, b_n$ with $\beta_\lambda = 0$ on $C\backslash C^\circ$ and γ_λ is a sum of characters from other blocks. Consider the generalized decomposition number $d_{\lambda l}^z$ of ζ_λ with respect to $\phi^{(l)}$ at $z \in P\backslash\{1\}$. From (9), we get

$$d_{\lambda l}^z = -m_0 + \varepsilon \sum_{(\xi_j^{(l)})^N = \chi_\lambda} \alpha_j(z).$$

8.5 GROUPS OF ORDER $g = pg_0$ WITH $p \nmid g_0$

Thus we have

$$(10) \quad \sum_{\lambda \in \Lambda} d_{\lambda l}^{\bar{z}} = -m_0|\Lambda| + \varepsilon \sum_{j=2}^{p} \alpha_j(z) = -m_0|\Lambda| - \varepsilon,$$

because $\sum_{j=1}^{p} \alpha_j(z) = 0$ by the ordinary character relations. Now $-m_0|\Lambda| - \varepsilon \neq 0$; this is obvious for $|\Lambda| \geq 2$, and for $|\Lambda| = 1$ it follows from (3) and the definition of ζ_λ in this case. Thus at least one ζ_λ lies in the block B.

For any $\lambda, \mu \in \Lambda$, $\chi_\lambda - \chi_\mu$ vanishes outside S. Therefore by (2), $\zeta_\lambda - \zeta_\mu$ vanishes outside $\bigcup_{x \in G} x^{-1}Sx$, and in particular it is 0 on G°. If η is any principal indecomposable modular character of G, then $\langle \eta, \zeta_\lambda - \zeta_\mu \rangle_{G^\circ} = 0$. Therefore if one of the ζ_λ appears in η, then all ζ_λ appear in η as constituents. But at least one ζ_λ lies in B and so all ζ_λ ($\lambda \in \Lambda$) lie in B. Hence by Theorem 6.4, $\gamma_\lambda = 0$ on $C\backslash C^\circ$. Thus (9) may be written

$$(\zeta_\lambda)_C = -m_0 \sum_{\xi^N = \tau} \xi + \varepsilon \sum_{\xi^N = \chi_\lambda} \xi + \theta_\lambda,$$

where $\theta_\lambda := \beta_\lambda + \gamma_\lambda$ is 0 on $C\backslash C^\circ$. To complete the proof of (iv) it remains to show that $m_0 = 0$.

Step 5 *If the characters χ_λ are suitably labeled, then $m_0 = 0$, $m_i = 1$ or -1 for $i = 1, \ldots, t$, and $t = q$.*

Choose e as the block idempotent of AG lying in B. Then (see §4.2) each irreducible KG-module lying in B is isomorphic to a KG-composition factor of the KG-module $e(AG) \otimes_A K$. This latter module is completely irreducible by Maschke's theorem; therefore, if V is a direct sum of nonisomorphic irreducible KG-modules lying in B, then $V | e(AG) \otimes_A K$. Let α be the character that is afforded by such a KG-module V. Then by Lemma 8.4B there exists an integer j ($1 \leq j \leq q$) and some n_j, n_λ ($\lambda \in \Lambda$) equal to 0 or 1 such that for some $\theta \in \text{Char}(N)$ vanishing on $N\backslash N^\circ$ we have

$$\alpha_N = n_j \chi_j + \sum_{\lambda \in \Lambda} n_\lambda \chi_\lambda + \theta.$$

Now suppose that $\mu \in \Lambda$ and choose the character ξ_j of C so that $\xi_j^N = \chi_\mu$. Since $\xi_j = \xi_1$ on C° (see the beginning of this section) and $\xi_1^N = \tau$, therefore $0 = \langle \theta_C, \xi_j - \xi_1 \rangle_C = \langle \theta, (\xi_j - \xi_1)^N \rangle_N = \langle \theta, \chi_\mu - \tau \rangle$. Hence from the above expression for α_N we conclude that $\langle \alpha_N, \chi_\mu - \tau \rangle = n_\mu - n_j = 0, 1,$ or -1. Thus we have shown that for all characters α of the above type

$$(11) \quad \langle \alpha, (\chi_\mu - \tau)^G \rangle_G = \langle \alpha_N, \chi_\mu - \tau \rangle_N = 0, 1, \text{ or } -1.$$

We first apply (11) with $\alpha = \zeta_i$ ($i = 1, \ldots, t$). Then from (3) we see that $-m_i = 0, 1,$ or -1. Since ζ_i is a constituent of $(\chi_\mu - \tau)^G$ by the way in which it was chosen, we conclude that $m_i = 1$ or -1 when $i \geq 1$.

We now apply (11) with $\alpha = \sum_{\lambda \in \Lambda} \zeta_\lambda$. Then from (3) we obtain $m_0 |\Lambda| + \varepsilon = 0, 1,$ or -1. Since $\varepsilon = 1$ or -1, this shows that $m_0 = 0$ when $|\Lambda| > 2$. If $|\Lambda| = 2$, then either $m_0 = 0$ or $m_0 = 1$, $\varepsilon = -1$; in the latter case, if we interchange the labels on the two χ_λ ($\lambda \in \Lambda$), then the equations (3) still hold with the new values $m_0 = 0$ and $\varepsilon = 1$. Finally, if $|\Lambda| = 1$, then (3) shows that $m_0 + \varepsilon \neq 0$ (since ζ_λ is a constituent of $(\chi_\lambda - \tau)^G$); hence replacing m_0 by 0 and ε by $m_0 + \varepsilon$ gives a valid equation (3) in which $m_0 = 0$. This shows that in all cases we may take $m_0 = 0$. The relation $t = q$ then follows immediately from (5).

This completes the proof of (iv), and (v) now follows from (3) because $(\chi_\mu - \tau)^G = 0$ on G° (Theorem 2.5B).

Step 6 $\zeta_1, \zeta_2, \ldots, \zeta_q$ *and* ζ_λ ($\lambda \in \Lambda$) *are the only irreducible ordinary characters lying in B.*

It is enough to show that any irreducible ordinary character ζ lying in B is a constituent of $(\chi_\mu - \tau)^G$ for some $\mu \in \Lambda$. Suppose the contrary. Then for all $j > 1$,

$$\langle \zeta_C, \xi_j - \xi_1 \rangle = \langle \zeta, (\xi_j - \xi_1)^G \rangle = \langle \zeta, (\chi_\lambda - \tau)^G \rangle = 0,$$

where $\chi_\lambda = \xi_j^N$. This shows that $\langle \zeta_C, \xi_j \rangle = \langle \zeta_C, \xi_1 \rangle = c$, say, and so the contribution to ζ_C from characters lying in b is given by $c \sum_{j=1}^p \xi_j$, which is 0 on $C \backslash C^\circ$ by (7). Making a similar analysis at the other blocks $b = b_1, b_2, \ldots, b_n$ we find that

$$\zeta_C = \beta + \gamma,$$

where β is a sum of characters from the blocks $b_1, b_2, \ldots, b_n$ with $\beta = 0$ on $C \backslash C^\circ$, and γ is a sum of characters from other blocks. By the Second Main Theorem (Theorem 6.4), $\gamma = 0$ on $C \backslash C^\circ$. But this implies $\langle \zeta_P, 1_P \rangle = \zeta(1)/p$, and so $p | \zeta(1)$. Since B is a block of defect 1, this is impossible by Theorem 5.5B. Hence all irreducible ordinary characters of G lying in B are constituents of $(\chi_\lambda - \tau)^G$ ($\lambda \in \Lambda$). On the other hand, Steps 3 and 4 show that these constituents lie in B. Hence (i) is proved, and the proof of the theorem is complete.

EXAMPLE 1 Let G be a group of order $g = pg_0$, where $p \nmid g_0$, and let B_0 be the principal block of kG. Then B_0 has defect 1 (see the example of §4.5), and by the Third Main Theorem the principal block b_0 of kC is the only block of kC with $b_0^G = B$. In the notation of §8.2 this shows that $F = N$ and $q = n$, where $n := |N : C|$. Moreover, the unique irreducible modular character of b_0 is 1_C, so $\xi_1 = 1_C$ and $C^\circ \subseteq \text{Ker } \xi_i$ for each i. Then the theorem shows that

$$(\zeta_i)_C = m_i 1_C = \theta_i \qquad (i = 1, 2, \ldots, n).$$

8.6 GROUPS WITH REPRESENTATION OF DEGREE $d < \frac{1}{2}(p-1)$

In particular, $\zeta_i - m_i 1_C = 0$ outside of G° since if $x \notin G^\circ$, then x is conjugate to an element of $C\backslash C^\circ$. Now (see §8.2) N ($= F$) acts on the set $\{\xi_2, \ldots, \xi_p\}$ of characters in b_0 permuting them in orbits of length n ($= q$). Any two characters in the same orbit induce to the same character of N. If $\{\xi_{i_1}, \ldots, \xi_{i_n}\}$ is one orbit, then $\xi_{i_1}^N = \xi_{i_2}^N = \cdots = \xi_{i_n}^N = \chi_\mu$, say, and so $\xi_{i_1}, \ldots, \xi_{i_n}$ are n distinct constituents of $(\chi_\mu)_C$. Counting degrees we see that $(\chi_\mu)_C = \xi_{i_1} + \cdots + \xi_{i_n}$. Hence we conclude from the theorem that for each of the $(p-1)/n$ exceptional characters ζ_μ ($\mu \in \Lambda$) we have

$$(\zeta_\mu)_C = \varepsilon \sum_{\xi^N = \chi_\mu} \xi + \theta_\mu = \varepsilon(\chi_\mu)_C + \theta_\mu.$$

EXAMPLE 2   In the case where $|\Lambda| = 1$, it follows from the theorem that

$$(\varepsilon \zeta_\lambda + m_i \zeta_i)_C = \sum_\xi \xi + \varepsilon \theta_\lambda + m_i \theta_i,$$

where ξ ranges over all characters in the block $b_1, \ldots, b_n$. It follows from (7) and the conditions of (iii) and (iv) that this implies that

$$\varepsilon \zeta_\lambda + m_i \zeta_i = 0 \quad \text{on} \quad C\backslash C^\circ.$$

In particular, ζ_λ is constant on $P\backslash\{1\}$.

EXERCISES

[In the following exercises suppose that the hypotheses of Theorem 8.5 hold.]

1. Show that G has exactly q irreducible modular characters which lie in B.
2. Prove that the exceptional characters ζ_λ in B form a single p-conjugate set of characters, while each nonexceptional character ζ_i is p-conjugate to itself (see §4.6).
3. Determine the decomposition matrix of B and the Cartan matrix of B.
4. Prove that there are at least two irreducible ordinary characters whose restrictions to G° are irreducible modular characters lying in block B.
5. Let $\alpha = \sum_{\lambda \in \Lambda} \zeta_\lambda$. Then prove that $\langle \alpha_C, \phi^y \rangle_{C^\circ} = 1$ for any y in N.

8.6 Groups with a faithful representation of degree $d < \frac{1}{2}(p-1)$

We shall apply the results of Theorem 8.5 to show that no simple group of order $g = pg_0$ with $p \nmid g_0$ can have a nontrivial character of degree $d < \frac{1}{2}(p-1)$. This result is due to Brauer [3] and will be used in the next section to prove a more general theorem about groups with a faithful representation of degree $d < \frac{1}{2}(p-1)$.

Lemma 8.6 Let G be a group of order $g = pg_0$ with $p \nmid g_0$. Let P be a Sylow p-group of G and put $C := C_G(P)$ and $N := N_G(P)$. Let ψ be an irreducible ordinary character of G of degree $d < \frac{1}{2}(p - 1)$ such that $P \nsubseteq \operatorname{Ker} \psi$. Then

(i) ψ lies in a block B of defect 1, ψ is an exceptional character, and at least two exceptional characters lie in B. Let b be a block of kC such that $b^G = B$.

(ii) For each exceptional character ζ lying in B, there is an ordinary irreducible character ξ lying in b such that

$$\zeta_C = \sum_{i=1}^{t} \xi^{u_i}$$

where $\xi^{u_1}, \ldots, \xi^{u_t}$ are the distinct conjugates of ξ in N.

(iii) We can find an exceptional character ζ lying in the block B such that $\zeta \neq \psi$, $\langle \zeta_P, \psi_P \rangle = 0$, and $\langle \zeta_{C^\circ}, \psi_{C^\circ} \rangle = |N : C|q$, where q is defined as in Theorem 8.5.

(iv) $q \leq d$.

Proof (i) Since $p \nmid d = \psi(1)$, ψ lies in a block B of defect 1 by Theorem 4.5B. We next claim that ψ is not constant on $P\backslash\{1\}$. Indeed if it were, then we should have the integer $\langle \psi_P, 1_P \rangle = (1/p)\{\psi(1) + (p - 1)\psi(z)\}$ for each $z \in P\backslash\{1\}$. But then $\psi(z)$ is an integer and $\psi(z) \equiv \psi(1) \pmod{p}$. Since $|\psi(z)| \leq \psi(1) < \frac{1}{2}(p - 1)$, this implies that $\psi(z) = \psi(1)$ for all $z \in P$; that contradicts the hypothesis $P \nsubseteq \operatorname{Ker} \psi$. Thus ψ is not constant in $P\backslash\{1\}$. This shows that ψ is an exceptional character (Theorem 8.5(iii)), and that $|\Lambda| > 1$ (Example 2 of §8.5).

(ii) In the notation of Theorem 8.5 ζ_C has the form

(1) $$\pm \sum_{\xi^N = \chi} \xi + \theta$$

for some irreducible ordinary character χ of N and some $\theta \in \operatorname{Char}(C)$ with $\theta = 0$ on $C\backslash C^\circ$. The ξ are irreducible ordinary characters of C and by Theorem 6.2B we know that ξ_{C° is equal to the unique modular irreducible character of the block b in which ξ lies. Since ζ is an exceptional character, there are exactly q of the ξ in the sum (1) that lie in a given block b (see Theorem 6.2B). Thus we have

$$\langle \zeta_{C^\circ}, \xi_{C^\circ} \rangle = \pm q + \langle \theta_{C^\circ}, \xi_{C^\circ} \rangle = \pm q + p\langle \theta, \xi \rangle$$

because $\theta = 0$ on $C\backslash C^\circ$ and $|C : C^\circ| = p$. Since the left-hand side lies between 0 and d, and $q = (p - 1)/|\Lambda| \leq \frac{1}{2}(p - 1)$ by (i), we conclude that $\langle \theta, \xi \rangle = 0$ and the positive sign holds in (1). Since θ is 0 on $C\backslash C^\circ$, $\langle \theta_P, 1_P \rangle = \theta(1)/p$, and so $\theta(1) \equiv 0 \pmod{p}$. Since the positive sign holds in (1) and $\langle \theta, \xi \rangle = 0$ for each ξ in the sum (1), θ is a sum of irreducible characters of C with nonnegative coefficients. Because ζ is an exceptional character,

8.6 GROUPS WITH REPRESENTATION OF DEGREE $d < \frac{1}{2}(p-1)$

$\zeta(1) = d < p$, and so $\theta(1) = 0$, which shows that $\theta = 0$. Hence we have shown that $\zeta_C = \sum_{\xi^N = \chi} \xi$ for some irreducible character χ of N. By the Frobenius reciprocity theorem the sum is over all irreducible constituents of χ_C, and by Clifford's theorem these are the conjugates in N of any one of the ξ. This proves (ii).

(iii) Let ξ be an irreducible ordinary character of C. By Theorem 6.2B, ξ_P is a multiple of a single irreducible character of P. Thus it follows from (ii) that the irreducible constituents of ψ_C form a single class of irreducible characters of P conjugate under N. Since $\psi(1) < \frac{1}{2}(p-1)$, there exists an irreducible character ξ_0 lying in b such that $\langle \psi_P, (\xi_0)_P \rangle = 0$ and $\langle (\xi_0)_P, 1_P \rangle = 0$. Let ζ be the exceptional character of B associated with the character ξ_0^N of N. Then by (ii)

$$\zeta_C = \sum_{i=1}^{s} \xi_0^{v_i}$$

for suitable $v_i \in N$. (Actually $s = t$. Why?) By the choice of ξ_0 no irreducible constituent of ζ_P equals an irreducible constituent of ψ_P, and so $\langle \zeta_P, \psi_P \rangle = 0$.

Finally, as we noted in the proof of (ii), exactly q of the irreducible constituents ξ_{C° of ζ are equal to the unique modular irreducible character in b. This is true for each of the $|N:F| = |N:C|/q$ blocks b_l such that $b_l^G = B$ (see the notation of §8.2). The same is true of the character ψ. Therefore

$$\langle \zeta_{C^\circ}, \psi_{C^\circ} \rangle = q^2 \cdot |N:C|/q = q|N:C|.$$

(iv) The t characters ξ^{u_i} in (ii) are the irreducible constituents of ξ^N. By Theorem 8.2, q of these constituents lie in b, and so $d = \zeta(1) \geq t \geq q$.

Theorem 8.6 Let G be a group of order $g = pg_0$ with $p \nmid g_0$, and let P be a Sylow p-subgroup. If $G = G'$ (the commutator group), then each irreducible ordinary character ψ of G with degree $d < \frac{1}{2}(p-1)$ has $P \subseteq \text{Ker } \psi$.

Proof We shall suppose that G has such a character ψ of degree $d < \frac{1}{2}(p-1)$ with $P \nsubseteq \text{Ker } \psi$ and obtain a contradiction. We shall continue with the notation introduced in §8.2 and §8.5. Note that all the information from Lemma 8.6 is available. Let B_0 be the principal block of kG. Then by Example 1 of §8.5 we know that B_0 has $n := |N:C|$ nonexceptional characters $\zeta_1, \zeta_2, \ldots, \zeta_n$ and $(p-1)/n$ exceptional characters ζ_λ ($\lambda \in \Lambda$). These satisfy

(2) $\qquad (\zeta_i)_C = m_i 1_C + \theta_i \qquad$ and $\qquad (\zeta_\lambda)_C = \varepsilon(\chi_\lambda)_C + \theta_\lambda,$

where each $m_i = \pm 1$, $\varepsilon = \pm 1$ and $\theta_i, \theta_j \in \text{Char}(C)$ vanish on $C \backslash C^\circ$. We shall not use the corresponding invariants for the block B in which ψ lies, but the invariant q will refer to B. The character ζ will be chosen to satisfy (iii) of Lemma 8.6.

Step 1 For each i, ζ_i is a constituent of $\psi\bar\psi$ if $m_i = 1$ and a constituent of $\psi\bar\zeta$ if $m_i = -1$ (the bar denotes complex conjugation).

Each element of $G\backslash G°$ is conjugate in G to an element of $C\backslash C°$. Indeed if $x \in G\backslash G°$ has p-part x_p, then there exists $y \in G$ such that $y^{-1}x_p y \in P\backslash\{1\}$ and hence $y^{-1}xy \in C\backslash C°$. Thus it follows from (2) that $\zeta_i = m_i$ on $G\backslash G°$. On the other hand, since ψ and ζ are exceptional characters in the same block, $\psi - \zeta = 0$ on $G°$ by Theorem 8.5 (v). Therefore $(\psi - \zeta)(\zeta_i - m_i 1_G) = 0$ on G. If $m_i = 1$, then $\psi\bar\zeta_i + \zeta = \psi + \zeta\bar\zeta_i$ and so $\langle\psi\bar\psi, \zeta_i\rangle = \langle\psi, \psi\zeta_i\rangle \geq 1$. On the other hand, if $m_i = -1$, then $\psi\bar\zeta_i + \psi = \zeta + \zeta\bar\zeta_i$, and so $\langle\psi\bar\zeta, \zeta_i\rangle = \langle\psi, \zeta\bar\zeta_i\rangle \geq 1$.

Step 2 For some $\lambda \in \Lambda$, ζ_λ is a constituent of $\psi\bar\psi$ if $\varepsilon = -1$ and a constituent of $\psi\bar\zeta$ if $\varepsilon = 1$.

Since the exceptional characters in a given block all have the same degree, Theorem 4.2C shows that for all $x \in G\backslash G°$

$$\zeta_\lambda(1) \sum_{\mu \in \Lambda} \zeta_\mu(x) + \sum_{i=1}^{n} \zeta_i(x)\zeta_i(1) = 0.$$

Since $\zeta_i(x) = m_i$ by (2), it follows from Theorem 8.5(v) that

$$\zeta_\lambda(1) = \varepsilon \sum_{i=1}^{n} \zeta_i(x)\zeta_i(1).$$

Hence we obtain $\sum_{\mu \in \Lambda} \zeta_\mu(x) = -\varepsilon$ for any $x \in G\backslash G°$. As we saw above, $\psi - \zeta = 0$ on $G°$; and so $(\psi - \zeta)(\sum_{\mu \in \Lambda} \zeta_\lambda + \varepsilon) = 0$ on G. If $\varepsilon = 1$, then $\psi \sum_{\mu \in \Lambda} \zeta_\lambda + \psi = \zeta + \zeta \sum_{\mu \in \Lambda} \zeta_\mu$. Hence

$$\langle\psi\bar\zeta, \sum_{\mu \in \Lambda} \zeta_\mu\rangle = \langle\psi, \zeta \sum_{\mu \in \Lambda} \zeta_\mu\rangle \geq 1,$$

and so $\langle\psi\bar\zeta, \zeta_\lambda\rangle \geq 1$ for some $\lambda \in \Lambda$. Similarly, if $\varepsilon = -1$, then $\langle\psi\bar\psi, \sum_{\mu \in \Lambda} \zeta_\mu\rangle \geq 1$, and so $\langle\psi\bar\psi, \zeta_\lambda\rangle \geq 1$ for some λ.

Step 3 The degrees of the characters ζ_i ($i = 1, 2, \ldots, n$) and ζ_λ.

By Lemma 8.6(iii), $\langle(\psi\bar\zeta)_P, 1_P\rangle = \langle\psi_P, \zeta_P\rangle = 0$, and so by Step 1, $\langle(\zeta_i)_P, 1_P\rangle = 0$ whenever $m_i = -1$. Thus it follows from (2) that if $m_i = -1$, then $1 = \langle(\theta_i)_P, 1_P\rangle = \theta_i(1)/p$ and so $\theta_i(1) = p$. Thus $\zeta_i(1) = p - 1$ whenever $m_i = -1$. On the other hand it follows from (2) that if $m_i = 1$, then $\zeta_i(1) = 1 + \theta_i(1)$. Since θ_i is 0 on $C\backslash C°$, $\langle(\theta_i)_P, 1_P\rangle = \theta_i(1)/p$ so $p \mid \theta_i(1)$. The hypothesis on G implies that only the trivial character has degree 1. When $m_i = 1$, then $\zeta_i(1) \geq p + 1$ except for the trivial character ζ_1.

From (2) we have $\zeta_\lambda(1) = \varepsilon\chi_\lambda(1) + \theta_\lambda(1)$. Again, since $\theta_\lambda = 0$ on $C\backslash C°$, $p \mid \theta_\lambda(1)$. On the other hand, $\chi_\lambda(1) = n$ by the example of §8.2. Hence $\zeta_\lambda(1) \geq p - n$ if $\varepsilon = -1$ and $\zeta_\lambda(1) \geq n$ if $\varepsilon = 1$. In fact, if $\varepsilon = 1$ then $(\zeta_\lambda)_C =$

8.6 GROUPS WITH REPRESENTATION OF DEGREE $d < \frac{1}{2}(p-1)$

$(\chi_\lambda)_C$, and so $\zeta_\lambda(1) = \chi_\lambda(1)$. Indeed, we showed above that $\langle(\psi\bar\zeta)_P, 1_P\rangle = 0$, and so by Step (2), $0 = \langle(\zeta_\lambda)_P, 1_P\rangle = \langle(\chi_\lambda)_P, 1_P\rangle + \theta_\lambda(1)/p$. By Theorem 8.3, $P \nsubseteq \operatorname{Ker}\chi_\lambda$ and so by Clifford's theorem, 1_P is not an irreducible constituent of $(\chi_\lambda)_P$. Therefore $\langle(\chi_\lambda)_P, 1_P\rangle = 0$ and so $\theta_\lambda(1) = 0$; in particular, $\langle(\chi_\lambda)_P, (\theta_\lambda)_P\rangle = 0$. Thus we conclude that χ_λ and θ_λ have no common irreducible constituent, and so by (2), $\theta_\lambda = 0$ and $(\zeta_\lambda)_C = (\chi_\lambda)_C$.

Step 4   The case when $\varepsilon = 1$.

Without loss in generality we may suppose $m_i = 1$ for $i = 1, 2, \ldots, l$ and $m_i = -1$ for $i = l+1, \ldots, n$. Then by Theorem 8.5(v) we have

(3) $$\varepsilon\zeta_\lambda(1) + \sum_{i=l+1}^{n} \zeta_i(1) = \sum_{i=1}^{l} \zeta_i(1).$$

For $\varepsilon = 1$ the information from Step 3 gives

$$n + (n-l)(p-1) = \sum_{i=1}^{l} \zeta_i(1) \geq (l-1)(p+1) + 1.$$

Hence $n(p+1) \geq 2pl$. On the other hand, Lemma 8.6 shows that

$$nq = \langle(\zeta)_{C^\circ}, (\psi)_{C^\circ}\rangle = \langle(\psi\bar\zeta)_{C^\circ}, 1_{C^\circ}\rangle$$

and so

(4) $$nq \geq \left\langle \left(\zeta_\lambda + \sum_{i=l+1}^{n} \zeta_i\right)_{C^\circ}, 1_{C^\circ}\right\rangle$$

by Steps 1 and 2. Now by Step 3 we have for $\varepsilon = 1$ that

$$\langle(\zeta_\lambda)_{C^\circ}, 1_{C^\circ}\rangle = \langle(\chi_\lambda)_{C^\circ}, 1_{C^\circ}\rangle = n \quad \text{by the example of §8.2.}$$

On the other hand, for $m_i = -1$,

$$0 \leq \langle(\zeta_i)_C, 1_C\rangle = -1 + \langle\theta_i, 1_C\rangle = -1 + (1/p)\langle(\theta_i)_{C^\circ}, 1_{C^\circ}\rangle$$

because $\theta_i = 0$ on $C\backslash C^\circ$. Thus $\langle(\theta_i)_{C^\circ}, 1_{C^\circ}\rangle \geq p$ and hence

(5) $$\langle(\zeta_i)_{C^\circ}, 1_{C^\circ}\rangle \geq p - 1 \quad \text{whenever} \quad m_i = -1.$$

Thus we conclude from (4) that

$$nq \geq n + (n-l)(p-1)$$

and so $l(p-1) \geq n(p-q)$. This combined with the inequality above gives

$$n(p^2 - 1) \geq 2pl(p-1) \geq 2p(p-q)n$$

and so $p - q < \frac{1}{2}p$. But $q \leq d$ by Lemma 8.6, and so we have a contradiction. Thus the case $\varepsilon = 1$ is impossible.

Step 5 The case $\varepsilon = -1$.

In this case the equation (3) and the information from Step 3 gives

$$(n - l)(p - 1) = \zeta_\lambda(1) + \sum_{i=1}^{l} \zeta_i(1) \geq (p - n) + (l - 1)(p + 1) + 1$$

and so $n \geq 2l$. On the other hand, we obtain analogously to (4) the relation

$$nq \geq \left\langle \left(\sum_{i=l+1}^{n} \zeta_i \right)_{C^\circ}, 1_{C^\circ} \right\rangle \geq (n - l)(p - 1)$$

by (5). Therefore

$$nq \geq (n - \tfrac{1}{2}n)(p - 1) = \tfrac{1}{2}n(p - 1),$$

which implies $q \geq \tfrac{1}{2}(p - 1)$. But again this is impossible because $q \leq d$ by Lemma 8.6. Thus in either case we have a contradiction. This proves the theorem.

Remark It has been shown in Feit [3] that if G is a group of order $g = pg_0$ with $p \nmid g_0$, and G has a faithful ordinary representation of degree $\tfrac{1}{2}(p - 1)$, then either G has a normal Sylow p-subgroup or $G/Z(G) \cong PSL(2, p)$.

EXERCISE

Let G be a group of order $g = pg_0$ with $p \nmid g_0$ and let B_0 be the principal p-block of G. Suppose that G has an irreducible ordinary character of degree $\tfrac{1}{2}(p - 1)$ and that $\zeta^1 := 1_G$ is the only irreducible ordinary character of degree 1 lying in B_0. Then prove the following:

(i) If $p \equiv 1 \pmod 4$, the set of irreducible ordinary characters lying in B_0 consists of ζ^1, $\tfrac{1}{4}(p - 5)$ characters of degree $p + 1$, two p-conjugate characters of degree $\tfrac{1}{2}(p + 1)$, and $\tfrac{1}{4}(p - 1)$ characters of degree $p - 1$.

(ii) If $p \equiv 3 \pmod 4$, the set of irreducible ordinary characters lying in B_0 consists of ζ^1, $\tfrac{1}{4}(p - 1)$ characters of degree $p + 1$, two p-conjugate characters of degree $\tfrac{1}{2}(p - 1)$, and $\tfrac{1}{4}(p - 3)$ characters of degree $p - 1$.

8.7 Criteria for normal Sylow p-groups

In the present section we give a proof (based on Brauer's result in §8.6) of the theorem that a group with a faithful ordinary representation of degree $d < \tfrac{1}{2}(p - 1)$ has a normal abelian Sylow p-subgroup.

Note If G is a group with a faithful KG-module V of dimension $d \leq p - 1$, then its Sylow p-subgroups are always abelian. Indeed, if P is a

8.7 CRITERIA FOR NORMAL SYLOW p-GROUPS

Sylow p-subgroup, then the restriction V_P is completely reducible and the dimensions of the irreducible constituents of V_P divide $|P|$ (Theorem 2.4B). Since $d < p$, this means that the irreducible constituents of V_P all have dimension 1. Since V is faithful, this shows that P is abelian.

In proving the general result we shall need to use the following (simpler) result for solvable groups. This result (due to Itô) shows that when G is solvable we can prove a slightly stronger result.

Theorem 8.7A Let G be a solvable group which has a faithful ordinary representation of degree $d < p - 1$. Then G has a normal abelian Sylow p-subgroup.

Proof Suppose the theorem is not true and let G be a minimal counterexample. Let ζ be a character of a faithful ordinary representation of G with smallest possible degree d. Then $d < p - 1$ and clearly $d > 1$. Let P be a Sylow p-subgroup of G; P is abelian by the note above.

Step 1 ζ *is irreducible.*

Otherwise we could write $\zeta = \zeta_1 + \zeta_2$ as a sum of characters. By the choice of ζ, ζ_1 and ζ_2 are not faithful, and so by the choice of G, $O_p(G/\text{Ker } \zeta_i)$ is the Sylow p-subgroup of $G/\text{Ker } \zeta_i$ ($i = 1, 2$). Then $O_p(G/\text{Ker } \zeta_1 \times G/\text{Ker } \zeta_2)$ is the Sylow p-subgroup of $G/\text{Ker } \zeta_1 \times G/\text{Ker } \zeta_2$. Since $\text{Ker } \zeta_1 \cap \text{Ker } \zeta_2 = \text{Ker } \zeta = 1$, there is an injective homomorphism of G into this latter group given by $x \mapsto (x \text{ Ker } \zeta_1, x \text{ Ker } \zeta_2)$. But that implies that $O_p(G)$ is the Sylow p-subgroup of G contrary to the hypothesis.

Step 2 *For each normal subgroup H of G with $H \neq G$, $H \cap P \subseteq Z(G)$.*

Since G is a minimal counterexample, $O_p(H)$ is the Sylow p-subgroup of H. Put $L = C_G(O_p(H)) \triangleleft G$. Since P is abelian, $P \subseteq L$. If $L \neq G$, then $P = O_p(L)$ by the minimality of G, and so $P \triangleleft G$ since $L \triangleleft G$. This is contrary to hypothesis, so $L = G$; hence $P \cap H \subseteq O_p(H) \subseteq Z(G)$.

Step 3 $G = PG'$, $P \cap G' = 1$, *and* $G'' \subseteq Z(G)$.

Since PG' is a normal subgroup of G, and $P \nsubseteq Z(G)$, $PG' = G$ by Step 2. Next let T be a representation of G which affords ζ. Let $z \in P \cap G'$. Then $z \in Z(G)$ by Step 2 and so $T(z)$ is a scalar $\alpha 1$, say. However, since $z \in G'$ it is a product of commutators and so $\alpha^d = \det T(z) = 1$. On the other hand, since z is a p-element, α is a p-power root of 1. Since $d < p$, this shows that $\alpha = 1$, and so $z = 1$. Thus $P \cap G' = 1$.

Finally, since $P \cap G' = 1$ and $G'' \neq G'$, therefore $G''P \neq G$. By the minimality of G, $P \triangleleft G''P$. Since $G'' \triangleleft G''P$, this shows that $G''P = G'' \times P$. Hence $P \subseteq C_G(G'') \triangleleft G$. Since $P \nsubseteq Z(G)$, Step 2 now shows that $C_G(G'') = G$ and so $G'' \subseteq Z(G)$.

Step 4 $Z(G') = G' \cap Z(G)$ and $|G' : Z(G')| = d^2$.

Let λ be an irreducible constituent of $\zeta_{G'}$. Since $G = PG'$ and $G' \triangleleft G$, the group P acts transitively on the set of conjugates of λ. Since $p \nmid \zeta(1)$, it follows that $\lambda^x = \lambda$ for all $x \in G$. By Theorem 8.1, λ is the restriction of an irreducible character of G to G', and so $\zeta_{G'} = \lambda$, which shows that $\zeta_{G'}$ is irreducible. Let T be a representation which affords ζ. Then using Schur's lemma, we have $T(x)$ a scalar for each $x \in Z(G')$ and hence $x \in Z(G)$. Thus $Z(G') \subseteq Z(G) \cap G'$ and the reverse inequality is trivial. We next show that $\zeta = 0$ on $G' \backslash Z(G')$. Indeed if $x \in G' \backslash Z(G')$ then there exists $y \in G' \backslash C_{G'}(x)$. Since $G'' \subseteq Z(G)$, Schur's lemma shows that $T(x^{-1}y^{-1}xy)$ is a scalar, $\alpha 1$ say, and $\alpha \neq 1$ because $xy \neq yx$. Thus $\alpha T(x) = T(y^{-1}xy)$, which yields $\alpha \zeta(x) = \zeta(x)$. Since $\alpha \neq 1$, $\zeta(x) = 0$ as asserted. Thus $\zeta = 0$ on $G' \backslash Z(G')$. On the other hand, $|\zeta(x)| = d$ for $x \in Z(G')$ because there $T(x)$ is a scalar $\alpha 1$ with $|\alpha| = 1$. Therefore

$$1 = \langle \zeta_{G'}, \zeta_{G'} \rangle = \left\{ \sum_{x \in Z(G')} |\zeta(x)|^2 \right\} / |G'| = d^2 |Z(G')| / |G'|$$

and $|G' : Z(G')| = d^2$ as asserted.

Step 5 Contradiction.

Let $S := C_{G'}(P)$. Then $S \supseteq Z(G) \cap G' \supseteq G''$ by Step 3, and so $S \triangleleft G'$. Then $S \triangleleft G'P = G$. Since $P \subseteq C_G(S) \triangleleft G$, and $P \nsubseteq Z(G)$, Step 2 shows that $C_G(S) = G$, so $S \subseteq Z(G)$. Hence by Step 4 $Z(G') = S = C_{G'}(P)$. Now P acts by conjugation on G' and S is the union of its orbits of length 1. All other orbits under P have lengths that are powers of p greater than 1, so p divides

$$|G'| - |S| = |S|\{|G' : Z(G')| - 1\} = |S|(d^2 - 1)$$

(using Step 4). Since $p \nmid |G'|$ by Step 3, this implies $p \mid (d-1)(d+1)$. Hence $d + 1 \geq p$ contrary to the choice of d.

Lemma 8.7 *Let H be a normal subgroup of the group G, let ζ be an irreducible ordinary character of G such that $H \nsubseteq \mathrm{Ker}\, \zeta$, and suppose some $x \in G$ satisfies $C_H(x) = 1$. Then $\zeta(x) = 0$.*

Proof Enumerate the irreducible ordinary characters of G so that $\zeta_1, \ldots, \zeta_r$ contain H in their kernels and $\zeta_{r+1}, \ldots, \zeta_s$ do not. The condition $C_H(x) = 1$ shows that $|C_G(x)| = |C_{G/H}(Hx)|$; indeed each Hy centralizing Hx in G/H contains exactly one element of $C_G(x)$. The characters $\zeta_1, \ldots, \zeta_r$ correspond to a complete set of irreducible characters for G/H, so the usual orthogonality relations give

$$\sum_{i=1}^{r} |\zeta_i(x)|^2 = |C_{G/H}(Hx)| = |C_G(x)| = \sum_{i=1}^{s} |\zeta_i(x)|^2.$$

Hence $\zeta_{r+1}(x) = \cdots = \zeta_s(x) = 0$.

8.7 CRITERIA FOR NORMAL SYLOW p-GROUPS

Theorem 8.7B Let G be a group with a faithful ordinary representation of degree $d < \frac{1}{2}(p-1)$. Then G has a normal abelian Sylow p-subgroup.

Proof We shall suppose that the theorem is false and obtain a contradiction by reducing to the situation of Theorem 8.6. Let G be a minimal counterexample. Then G has a faithful ordinary character ζ of degree $d < \frac{1}{2}(p-1)$, and clearly G is not abelian so $d > 1$. Let P be Sylow p-subgroup of G; by the note above, P is abelian.

Step 1 For each proper normal subgroup H of G, $P \cap H \subseteq Z(G)$.

The proof of Step 2 of Theorem 8.7A applies.

Step 2 $G = G'$.

Since $P \nsubseteq Z(G)$, it follows from Step 1 that $PG' = G$. Suppose $G \neq G'$. Then the argument of Step 3 of Theorem 8.7A applies to show that $G' \cap P = 1$. Since P is not normal in G we can choose a prime q so that $q \mid |G : N_G(P)|$. Let Q be a Sylow q-subgroup of G'. For each $x \in G$, $x^{-1}Qx$ is also a Sylow q-subgroup of G', so $x^{-1}Qx = y^{-1}Qy$ for some $y \in G'$. This shows that $x \in N_G(Q)y \subseteq N_G(Q)G'$. This holds for all $x \in G$, so $G = N_G(Q)G'$. Hence $N_G(Q)$ contains a Sylow p-subgroup P_1, say, of G. Consider the subgroup QP_1. Then $Q \triangleleft QP_1$, so QP_1 is solvable. On the other hand, since P_1 is conjugate to P, the choice of q shows that $Q \nsubseteq N_G(P_1)$, so P_1 is not normal in QP_1. This contradicts Theorem 8.7A. Thus $G = G'$ as asserted.

Step 3 ζ is irreducible.

The proof of Step 1 of Theorem 8.7A applies.

Step 4 $Z(G)$ is the unique maximal normal subgroup of G. It is a p'-group and $G/Z(G)$ is simple and noncyclic.

Let H be a maximal normal subgroup of G. Then the subgroup $PH \neq G$; for otherwise $G/H \simeq P/P \cap H$ is abelian and so $H = G$ by Step 2. By Step 1, $P \cap H \subseteq Z(G) = Z(G) \cap G'$; so we can apply the first part of the argument of Step 3 of Theorem 8.7A to show that $P \cap H = 1$. Since PH is a proper subgroup of G, the minimality of G shows that $P \triangleleft PH$. But $H \triangleleft PH$, so $PH = P \times H$. This shows that $P \subseteq C_G(H) \triangleleft G$. Since $P \nsubseteq Z(G)$, Step 1 now shows that $C_G(H) = G$, so $H \subseteq Z(G)$. Since G is nonabelian and H is a maximal normal subgroup, $H = Z(G)$. Since $H \cap P = 1$, $Z(P)$ is a p'-group. Since H is a maximal normal subgroup, $G/Z(G)$ is simple. Finally, $G/Z(G)$ is noncyclic because otherwise G would be abelian.

Step 5 Let $N := N_G(P)$ and $N_0 := \{x \in N \mid C_P(x) \neq 1\}$. Then

(1) $x^{-1}Px \cap P = 1$ for all $x \in G \backslash N$

(2) $$x^{-1}N_0 x \cap N_0 = \begin{cases} N_0 & \text{if } x \in N \\ Z(G) & \text{if } x \in G \setminus N. \end{cases}$$

In particular, both $P \setminus \{1\}$ and $N_0 \setminus Z(G)$ are trivial intersection sets.

First suppose $x^{-1}Px \cap P \neq 1$ for some $x \in G$. Then $x^{-1}Px \cap P \not\subseteq Z(G)$ by Step 4, and so $H := C_G(x^{-1}Px \cap P)$ is a proper subgroup of G. By the minimality of G, H has a (unique) normal Sylow p-subgroup. Both P and $x^{-1}Px$ are Sylow p-subgroups of H because P is abelian; and so $P = x^{-1}Px$ and $x \in N$. This proves (1).

It is easily verified that N_0 is a normal subset of N, so to prove (2) we must show that $x^{-1}N_0 x \cap N_0 \not\subseteq Z(G)$ implies $x \in N$. By hypothesis there exists y such that xyx^{-1}, y both lie in $N_0 \setminus Z(G)$. Put $P_1 := C_P(xyx^{-1})$ and $P_2 := C_P(y)$; these are nontrivial by the definition of N_0. Now $x^{-1}P_1 x$ and P_2 are p-subgroups of $C_G(y)$ and $C_G(y) \neq G$ by the choice of y. By the minimality of G, $C_G(y)$ has a unique Sylow p-subgroup, which will contain both $x^{-1}P_1 x$ and P_2. Since the p-subgroups of G are all abelian, this means that $x^{-1}P_1 x \subseteq C_G(P_2)$. But $P \subseteq C_G(P_2)$ and $C_G(P_2) \neq G$ by Step 5. Therefore P is the unique Sylow p-subgroup of $C_G(P_2)$ and $x^{-1}P_1 x \subseteq P$. Hence $1 \neq P_1 \subseteq xPx^{-1} \cap P$, and so $x \in N$ by (1). This proves (2).

We now introduce the following notation. Put $p^m := |P|$, $n := |Z(G)|$ and $p^m nh := |N|$. Let $\zeta_1 = 1_G, \zeta_2, \ldots, \zeta_s$ be the irreducible ordinary characters of G and write

$$(\zeta_i)_N = \alpha_i + \beta_i \quad (i = 1, \ldots, s),$$

where α_i is the sum of the irreducible constituents of $(\zeta_i)_N$, none of which contain P in their kernels, and β_i is the sum of the remaining irreducible constituents. Since the representation affording ζ_i is scalar on $Z(G)$, $|\beta_i(x)| = \beta_i(1)$ for all $x \in PZ(G)$.

Step 6 For each $i > 1$, $\beta_i(1)^2 p^m < \zeta_i(1)^2$.

It follows from (2) that $N_0 \setminus Z(G)$ has $|G : N|$ distinct conjugates that are mutually disjoint. Therefore for $i > 1$

$$1 = \langle \zeta_i, \zeta_i \rangle = \frac{1}{g} \sum_{x \in G} |\zeta_i(x)|^2$$

$$\geq \frac{|G : N|}{g} \sum_{x \in N_0 \setminus Z(G)} |\zeta_i(x)|^2 + \zeta_i(1)^2/g.$$

For each irreducible character ζ_i, $|\zeta_i(x)| = \zeta_i(1)$ whenever $x \in Z(G)$. Therefore

$$1 > \frac{1}{|N|} \left\{ \sum_{x \in N_0} |\zeta_i(x)|^2 - |Z(G)| \cdot \zeta_i(1)^2 \right\}$$

8.7 CRITERIA FOR NORMAL SYLOW p-GROUPS

$$= \frac{1}{|N|} \sum_{x \in N_0} \{|\alpha_i(x)|^2 + \alpha_i(x)\overline{\beta_i(x)} + \overline{\alpha_i(x)}\beta_i(x) + |\beta_i(x)|^2\} - \zeta_i(1)^2/hp^m$$

$$\geq \langle \alpha_i, \alpha_i \rangle + 2\langle \alpha_i, \beta_i \rangle + \beta_i(1)^2/h - \zeta_i(1)^2/hp^m$$

since $\alpha_i = 0$ on $N \backslash N_0$ by Lemma 8.7 and $|\beta_i(x)| = \beta_i(1)$ on $PZ(G)$ by the definition of β_i. However, the definition of α_i and β_i shows that $\langle \alpha_i, \beta_i \rangle = 0$. Moreover since $\zeta_i \neq \zeta_1 = 1_G$, Step 4 shows that $P \not\subseteq \text{Ker } \zeta_i$, so $\alpha_i \neq 0$. Therefore we conclude

$$\zeta_i(1)^2 = p^m\{\beta_i(1)^2 + h\langle \alpha_i, \alpha_i \rangle - h\} \geq p^m \beta_i(1)^2.$$

Step 7 $(1_P)^G = \sum_{i=1}^{s} \beta_i(1)\zeta_i$.

By Frobenius reciprocity, $\langle 1_P^G, \zeta_i \rangle = \langle 1_P, (\zeta_i)_P \rangle = \langle 1_P, (\alpha_i)_P + (\beta_i)_P \rangle$. By the definition of α_i and β_i and Clifford's theorem we have $\langle 1_P, (\alpha_i)_P \rangle = 0$ and $\langle 1_P, (\beta_i)_P \rangle = \beta_i(1)$, and so the result follows.

Step 8 *The conclusion.*

We first show that $d \leq 1 + \sum_{i=2}^{s} \langle \zeta_i, \zeta\bar\zeta \rangle \beta_i(1)$. Indeed by Step 7,

$$\sum_{i=1}^{s} \langle \zeta_i, \zeta\bar\zeta \rangle \beta_i(1) = \langle (1_P)^G, \zeta\bar\zeta \rangle$$

$$= \langle 1_P, (\zeta\bar\zeta)_P \rangle = \langle \zeta_P, \zeta_P \rangle$$

$$\geq \zeta(1) = d$$

since ζ_P is a sum of $\zeta(1)$ characters of degree 1. Since $\langle \zeta_1, \zeta\bar\zeta \rangle \beta_1(1) = \langle \zeta, \zeta \rangle = 1$, the assertion follows.

Therefore, from Step 6 it follows that

$$d \leq 1 + \sum_{i=2}^{s} \langle \zeta_i, \zeta\bar\zeta \rangle \zeta_i(1)/p^{\frac{1}{2}m}$$

$$= 1 + \{\langle 1^G, \zeta\bar\zeta \rangle - 1\}/p^{\frac{1}{2}m}$$

because $1^G = \sum_{i=1}^{s} \zeta_i(1)\zeta_i$ (see the Example of §2.5). Since $\langle 1^G, \zeta\bar\zeta \rangle = \zeta(1)^2 = d^2$, this shows that $p^{\frac{1}{2}m} \leq (d^2 - 1)/(d - 1) = d + 1$. But $d + 1 < \frac{1}{2}(p + 1)$ by hypothesis, and so $m = 1$. However by Step 2 this means that the hypotheses of Theorem 8.6 are satisfied; and that shows that our counterexample is impossible. This completes the proof of the theorem.

EXERCISES

1. Let G be a group with a normal p-complement for some prime factor p of $|G|$. If G has a faithful irreducible ordinary character of degree $d < p - 1$, show that G has a normal Sylow p-subgroup.

2. Let G be a group. If every irreducible ordinary character ζ of G has degree $d \leq p - 1$, prove that G has a normal Sylow p-subgroup.

3. Let G be a group and ζ a faithful ordinary character of degree d. If Ω is the set of all prime factors p of $|G|$ such that $d < p - 1$, then show that G has an abelian Hall Ω-subgroup.

4. Let P be a Sylow p-subgroup of a group G such that $C_G(u) = P$ for every $u \neq 1$ in P and let p^n be the order of P. If G has a faithful ordinary representation of degree $d < \frac{1}{2}(p^n - 1)$, prove that either $P \triangleleft G$ or no proper subgroup of P is normal in $N_G(P)$, $N_G(P)/U$ is nonabelian and $d^2 - d \geq 2(p^n - 1)$.

8.8 Notes and comments

The earliest attempt to analyze the characters in a p-block of a group goes back to a paper of Brauer [2], who successfully determined the structure of blocks of defect 0 and 1. In [3] Brauer investigated the p-blocks of a group G whose order is divisible by p only to the first power and then applied this information to prove that such a group with a faithful representation of degree $d < \frac{1}{2}(p - 1)$ is not simple. In fact, in [3] Brauer proves Theorem 8.6 without the hypothesis that $G = G'$. There are further applications of the results of Brauer [3] on the blocks of defect 1. For example, the following result of Brauer [13] leads to some important theorems on simple groups whose orders are divisible by a prime to the first power only.

Theorem 8.8A Let G be a group with $G = G'$ and let z be an element of order p such that $C_G(z) = \langle z \rangle$. Then the following are true.

(i) There are integers n and t with $t \mid p - 1$ such that

$$g = (p - 1)p(1 + np)/t,$$

where $g := |G|$, $1 + np$ is the number of subgroups of order p, and t is the number of conjugate classes of elements of order p.

(ii) Assume that there are no positive integers u and h such that $n(u + 1) = u(u + 1) + h(pu + 1)$. Then one of the following holds

(a) $n = 1$, $t = 2$, and $G \simeq PSL(2, p)$, $p > 3$;
(b) $n = (p - 3)/2$, $t = (p - 1)/2$, $p = 2^m + 1 > 3$, and $G \simeq PSL(2, p - 1)$;

(iii) if $n < (p + 3)/2$, then $G \simeq PSL(2, p)$, $p > 3$, or $G \simeq PSL(2, p - 1)$, where $p = 2^m + 1 > 3$ is a Fermat prime.

Using this result, Brauer and Tuan [1] proved the following result.

Theorem 8.8B If G is a noncyclic simple group of order $g = pq^b g_0$, where p and q are distinct primes and $g_0 < p - 1$, then $G \simeq PSL(2, p)$ where $p = 2^m + 1$ or $G \simeq PSL(2, 2^m)$ where $p = 2^m + 1 > 3$.

8.8 NOTES AND COMMENTS

As a consequence of this result, the only simple groups of order prq^b, where p, q, r are distinct primes, are $PSL(2, 5)$ and $PSL(2, 7)$. As another application of the above ideas we have a result of Brauer and Reynolds [1].

Theorem 8.8C If the order g of a simple noncyclic group G has a prime factor $p > g^{1/3}$, then G is isomorphic to $PSL(2, p)$ where $p > 3$ or to $PSL(2, p - 1)$ where $p > 3$ is a Fermat prime.

After the two papers of Brauer [3] on the structure of blocks with defect groups of order p, the theory remained stagnant until Thompson [1] generalized the result of Brauer in a special case. Using this idea of Thompson, Dade [2] generalized Brauer's results [3] to blocks with cyclic defect groups.

Theorem 8.8D Let p be a fixed prime, B a p-block of G with a cyclic defect group D of order p^d, b the unique p-block of $N := N_G(D)$ such that $b^G = B$, c a fixed p-block of $C := C_G(D)$ such that $c^N = b$, and E the set of all elements x of N such that $c^x = c$. Then

(i) $q := |E : C|$ is a factor of $p - 1$.
(ii) E/C acts on the set of nontrivial irreducible ordinary characters of D such that each orbit has $(p^d - 1)/q$ elements.
(iii) If Λ is a set of representatives of the orbits in (ii), then B has q irreducible modular characters $\phi^1, \phi^2, \ldots, \phi^q$ and $q + (p^d - 1)/q$ irreducible ordinary characters $\zeta^1, \zeta^2, \ldots, \zeta^q$ and $\zeta^\lambda (\lambda \in \Lambda)$.
(iv) The decomposition numbers associated with B are 0 and 1. The decomposition numbers $d_{\lambda j}$ associated with (ζ^λ, ϕ^j) are all equal for a fixed j. If $d_{\lambda j} = 1$, then there is exactly one ζ^i such that $d_{ij} = 1$. If $d_{\lambda j} = 0$, then there are exactly two ζ^i such that $d_{ij} = 1$.

Moreover the generalized decomposition numbers are also determined. The situation bears a close resemblance to that of blocks of defect 1 (see Theorem 4.6B). The proof of Theorem 8.8D given in Dade [2] is based on a complicated induction argument, but the paper should be accessible to someone who has followed the arguments of the present chapter. There are two other proofs of this result, one due to Peacock [1] and the other to Feit [5]. These results of Brauer and Dade do not seem to extend to other types of blocks.

The theorems of §8.3 are due to Green [1,2]. Theorem 8.7B was proved by Feit and Thompson [1]. For further results in the latter direction see Blau [2], Lindsey [1], and Winter [1]. In Feit [3] Theorem 8.7B is extended as follows.

Theorem 8.8E Let G be a group of order g which has a faithful representation of degree n over the complex field. Let $g = g_1 h$ where $(g_1, h) = 1$ and $p > n + 1$ for every prime p dividing h. Then G contains a normal abelian subgroup of order h or h/p for some prime p.

There are other applications of Theorem 8.6 and Theorem 8.7B. For example see Brauer [9,11,12], Brauer and Fowler [1], Hall [2], Herzog [1], and Wales [1,2,3,4,5]. An interesting problem in finite group theory has to do with the determination of all simple groups with orders $p^a q^b r^c$ divisible by exactly three distinct primes p, q, r. By Thompson [2] the primes involved in such groups are 2, 3, 5, 7, 13 and 17. No example of simple group is known in which all three exponents a, b, c are greater than 1. The first step is the determination and classification of all simple groups of order $2^a 3^b r$, r prime. In this case Brauer's method is powerful enough to give a complete solution. Leon and Wales [1] have extended the result to the case $2^a 3^b r^c$ where the Sylow r-subgroups are cyclic. The results of Brauer, Leon, and Wales can be stated in the following combined form.

Theorem 8.8F The only nonabelian simple groups of order $2^a 3^b r^c$, with cyclic Sylow r-subgroups, are A_6, $Sp(4, 3)$, $U_3(3)$, $PSL(3, 3)$, and $PSL(2, r)$ for $r = 5, 7, 13$, and 17.

The cases $r = 5$, $c = 1, 2$ were settled by Brauer; the cases $r = 7, 13, 17$ with $c = 1$ were treated by Wales; and the general case was given by Leon and Wales.

References

Gorenstein [1] consists of a compilation of 3052 reviews of papers and books in finite group theory from the journal *Mathematical Reviews* (1940–1970). This gives an excellent coverage of the literature up to about 1969 classified by topics. Dornhoff [1] also offers an extensive bibliography of papers in representation theory of finite groups that appeared before the end of 1971. In view of these sources, we have included in the bibliography below only items that have been quoted in the text or that have appeared rather recently.

Alex, L. J.
1. Simple groups of order $2^a 3^b 5^c 7^d p$. *Trans. Amer. Math. Soc.* **173** (1972), 389–399.
2. On simple groups of order $2^a 3^b 7^c p$. *J. Algebra* **25** (1973), 113–124.
3. Index two simple groups. *J. Algebra* **31** (1974), 262–275.

Alperin, J. L.
1. Up and down fusion. *J. Algebra* **28** (1974), 206–209.

Alperin, J. L., Brauer, R., and Gorenstein, D.
1. Finite groups with quasi-dihedral and wreathed Sylow 2-subgroups. *Trans. Amer. Math. Soc.* **151** (1970), 1–261.
2. Finite simple groups of 2-rank two. *Scripta Math.* **29** (1971), 191–214.
3. The extended ZJ-theorem. *Proc. Gainesville Conf. (1972)* 6–7. "Math. Studies" 7, North-Holland Publ., Amsterdam (1973).

Artin, E., Nesbitt, C. J., and Thrall, R. M.
1. "Rings with Minimum Condition." Univ. of Michigan Press, Ann Arbor, 1944.

Basmaji, B. G.
1. Modular representations of metabelian groups. *Trans. Amer. Math. Soc.* **196** (1972), 389–399.

Bender, M.
1. The Brauer–Suzuki–Wall theorem. *Illinois J. Math.* **18** (1974), 229–235.

Berman, S. D.
1. The number of irreducible representations of a finite group over an arbitrary field. *Dokl. Akad. Nauk SSSR* **106** (1956), 767–769.
2. Characters of linear representations of a finite group over an arbitrary field. *Mat. Sb.* **44** (86), (1958), 409–456.
3. Generalized characters of finite groups. *Dopovīdī Akad. Nauk Ukraïn. RSR* (1957), 112–115.
4. Modular representations of finite supersolvable groups. *Dopovīdī Akad. Nauk Ukraïn. RSR* (1960), 586–589.

Blau, H. I.
1. Indecomposable modules for direct products of finite groups. *Pacific J. Math.* **46** (1974), 39–44.
2. On linear groups of degree $p - 2$. *J. Algebra* **36** (1975), 495–498.

Brauer, R.
1. On the connection between the ordinary and modular characters of groups of finite order. *Ann. of Math.* **42** (2), (1941), 926–935.
2. Investigations on group characters. *Ann. of Math.* **42** (2), (1941), 936–958.
3. On groups whose order contains a prime number to the first power I, II. *Amer. J. Math.* **64** (1942), 401–420, 421–440.
4. On the representation of a group of order g in the field of gth roots of unity. *Amer. J. Math.* **67** (1945), 461–471.
5. On blocks of characters of groups of finite order I, II. *Proc. Nat. Acad. Sci. U.S.A.* **32** (1946), 182–186, 215–219.
6. Applications of induced characters. *Amer. J. Math.* **69** (1947), 709–716.
7. A characterization of the characters of groups of finite order. *Ann. of Math.* **57** (2), (1953), 357–377.
8. Zur Darstellungstheorie der Gruppen endlicher Ordnung I, II. *Math. Z.* **63** (1956), 406–444; **72** (1959/60), 25–46.
9. Some applications of the theory of blocks and characters of finite groups I–V. *J. Algebra* **1** (1964), 152–167, 307–334; **3** (1966), 225–255; **17** (1971), 489–521; and **28** (1974), 433–460.
10. On the first main theorem on blocks of characters of finite groups. *Illinois J. Math.* **14** (1970), 183–187.
11. Types of blocks of representations of finite groups. *Proc. Symp. Pure Math.* **21** (*Amer. Math. Soc.* 1971) pp. 7–11.
12. Character theory of finite groups with wreathed Sylow 2-subgroups. *J. Algebra* **19** (1971), 547–592.
13. On permutation groups of prime degree and related classes of groups. *Ann. of Math.* **44** (2), (1943) 57–79.
14. On 2-blocks with dihedral defect groups. *Symp. Math. Roma* **23** (1972), 367–397. Academic Press, London (1974).
15. On the structure of blocks of characters of finite groups. *In* "Lecture Notes in Math.," **372**, pp. 103–130. Springer-Verlag, Berlin and New York, 1974.

Brauer, R., and Feit, W.
1. On the number of irreducible characters of finite groups in a given block. *Proc. Nat. Acad. Sci. U.S.A.* **45** (1959), 361–365.
2. An analogue of Jordan's theorem in characteristic p. *Ann. of Math.* **84** (2), (1966), 119–131.

Brauer, R., and Fong, P.
1. On the centralizers of *p*-elements in finite groups. *Bull. London Math. Soc.* **6** (1974), 319–324.

Brauer, R., and Fowler, K. A.
1. On groups of even order. *Ann. of Math.* **62** (1955), 565–583.

Brauer, R., and Nesbitt, C. J.
1. "On Modular Representations of Groups of Finite Order." Univ. of Toronto Press, Toronto, 1937.
2. On the modular characters of groups. *Ann. of Math.* **42** (2), (1941), 556–590.

Brauer, R., and Reynolds, W. F.
1. On a problem of E. Artin. *Ann. of Math.* **68** (2), (1958), 713–720.

Brauer, R., and Suzuki, M.
1. On finite groups of even order whose 2-Sylow group is a quaternion group. *Proc. Nat. Acad. Sci. U.S.A.* **45** (1959), 1757–1759.

Brauer, R., and Tate, J.
1. On the characters of finite groups. *Ann. of Math.* **62** (2), (1955), 1–7.

Brauer, R., and Tuan, H. F.
1. On simple groups of finite order I. *Bull. Amer. Math. Soc.* **51** (1945), 756–766.

Brauer, R., and Wong, W. J.
1. Some properties of finite groups with wreathed Sylow 2-subgroups. *J. Algebra* **19** (1971), 263–273.

Bryant, R. M., and Kovacs, L.
1. Tensor products of representations of finite groups. *Bull. London Math. Soc.* **4** (1972), 133–135.

Burnside, W.
1. "The Theory of Groups of Finite Order," 2nd ed. Cambridge Univ. Press, London and New York, 1911, (reprinted Dover, New York, 1955).

Carleson, J. F.
1. Block idempotents and the Brauer correspondence. *Bull. Austral. Math. Soc.* **5** (1971), 337–340.
2. The modular representation ring of a cyclic 2-group. *J. London Math. Soc.* **11** (2), (1975), 91–92.

Carter, R. W.
1. "Simple Groups of Lie Type." Wiley, London, 1972.

Carter, R. W., and Lusztig, G.
1. On the modular representations of general linear and symmetric groups. *Math. Zeit.* **136** (1974), 193–242.

Chevalley, C.
1. Theorie des blocs. *Sem. Bourbaki 1972/73. In* "Lecture Notes in Math.," **383**, pp. 34–39. Springer-Verlag, Berlin and New York, 1974.

Clifford, A. H.
1. Representations induced in an invariant subgroup. *Ann. of Math.* **38** (2), (1937), 533–550.

Curtis, C. W., and Reiner, I.
1. "Representation Theory of Finite Groups and Associative Algebras." Wiley (Interscience), New York, 1962.

Dade, E. C.
1. On Brauer's second main theorem. *J. Algebra* **2** (1965), 299–311.
2. Blocks with cyclic defect groups. *Ann. of Math.* **84** (2), (1966), 20–48.
3. Block extensions. *Illinois J. Math.* **17** (1973), 198–272.
4. Degrees of modular irreducible representations of p-solvable groups. *Math. Z.* **104** (1968), 141–143; correction, **105** (1968), 172.
5. Character theory of finite simple groups. *In* "Finite Simple Groups" (M. B. Powell and G. Higman, eds.), pp. 249–327. Academic Press, New York, 1971.

Dagger, S. W.
1. On the blocks of Chevalley groups. *J. London Math. Soc.* **3** (2), (1971), 21–29.

Dixon, J. D.
1. "The Structure of Linear Groups." Van Nostrand-Reinhold, London, 1971.

Donovan, P., and Freislich, M. R.
1. Representable functions on the category of modular representations of a finite group with cyclic Sylow subgroup. *J. Algebra* **32** (1974), 356–364.
2. Representable functions on the category of modular representations of a finite group with Sylow subgroup $C_2 \times C_2$. *J. Algebra* **32** (1974), 365–369.

Dornhoff, L.
1. "Group Representation Theory" (Parts A and B). Dekker, New York, 1971–72.

Feit, W.
1. "Characters of finite groups." Benjamin-Addison Wesley, New York, 1967.
2. "Representations of finite groups" (mimeographed notes). Math. Dept., Yale, Univ., New Haven, Connecticut, 1969.
3. Groups with a faithful representation of degree less than $p - 1$. *Trans. Amer. Math. Soc.* **112** (1964), 287–303.
4. On finite linear groups. *J. Algebra* **5** (1967), 378–400.
5. "Blocks with cyclic defect groups" (mimeographed notes). Math. Dept., Yale Univ., New Haven, Connecticut, 1975.

Feit, W., and Thompson, J. G.
1. Groups which have a faithful representation of degree less than $\frac{1}{2}(p - 1)$. *Pacific J. Math.* **11** (1961), 1257–1262.
2. Solvability of groups of odd order. *Pacific J. Math.* **13** (1963), 775–1029.

Fong, P.
1. On the characters of p-solvable groups. *Trans. Amer. Math. Soc.* **98** (1961), 263–284.
2. On decomposition numbers of J_1 and $R(q)$. *Symp. Math. Convegni, Roma* **13** (1972), pp. 415–422. Academic Press, London (1974).

Fong, P., and Gaschütz, W.
1. A note on the modular representations of solvable groups. *J. Reine Angew. Math.* **208** (1961), 73–78.

Gagola, S. M.
1. A note on lifting Brauer characters. *Proc. Amer. Math. Soc.* **53** (1975), 295–300.

REFERENCES

Glauberman, G.
1. Central elements in core-free groups. *J. Algebra* **4** (1966), 403–420.
2. On groups with a quaternion Sylow 2-subgroup. *Illinois J. Math.* **18** (1974), 60–65.
3. Direct factors of Sylow 2-subgroups, I, II. *J. Algebra* **28** (1974), 133–161, 162–173.

Goldschmidt, D. M.
1. An application of Brauer's second main theorem. *J. Algebra* **20** (1972), 72–77.
2. 2-fusion in finite groups. *Ann. of Math.* **99** (2), (1974), 70–117.

Goldschmidt, D. M., and Isaacs, I. M.
1. Schur indices in finite groups. *J. Algebra* **33** (1975), 191–199.

Gorenstein, D.
1. "Reviews on Finite Groups." Amer. Math. Soc., Providence, Rhode Island, 1974.
2. Finite simple groups and their classification. *Israel J. Math.* **19** (1974), 5–66.

Gorenstein, D., and Harada, K.
1. Finite groups whose Sylow 2-subgroups are the direct product of two dihedral groups. *Ann. of Math.* **95** (2), (1972), 1–54.
2. Finite groups with Sylow 2-subgroups of type $PSp(3, q)$, q odd. *J. Fac. Sci. Univ. Tokyo Sect. I A Math.* **20** (1973), 341–372.

Gorenstein, D., and Walter, J. H.
1. On finite groups with dihedral Sylow 2-subgroups. *Illinois J. Math.* **6** (1962), 553–593.
2. The characterization of finite groups with dihedral Sylow 2-subgroups, I, II, and III. *J. Algebra* **2** (1965), 85–151, 218–270, and 354–393.

Gow, R.
1. Schur indices and modular representations. *Math. Zeit.* **144** (1975), 97–99.

Gow, R., and Humphreys, J. F.
1. Normal p-complements and irreducible representations. *J. London Math. Soc.* **11** (2), (1975), 308–312.

Green, J. A.
1. On the indecomposable representations of a finite group. *Math. Z.* **70** (1958), 430–445.
2. Blocks of modular representations. *Math. Z.* **79** (1962), 100–115.
3. Axiomatic representation theory of finite groups. *J. Pure Appl. Algebra* **1** (1971), 41–77.
4. Walking around the Brauer tree. *J. Austral. Math. Soc.* **17** (1974), 197–213.
5. Vorlesungen über Modulare Darstellungstheorie endlicher Gruppen. *Vorlesungen aus dem Math. Inst. Giessen, Heft* **2** (1974).

Green, J. A., and Hill, R.
1. On a theorem of Fong and Gaschütz. *J. London Math. Soc.* **1** (2), (1969), 573–576.

Hall, M., Jr.
1. "The Theory of Groups." Macmillan, New York, 1959.
2. On the number of Sylow subgroups in a finite group. *J. Algebra* **7** (1967), 363–371.
3. Simple groups of order less than one million. *J. Algebra* **20** (1972), 98–102.

Hamernik, W.
1. The linear character of an indecomposable module of a group algebra. *J. London Math. Soc.* **7** (2), (1973), 220–224.
2. Induced modules with cyclic vertex. *Proc. Int. Conf. on Representation Theory.* Carleton Univ. (1974) Ottawa, Canada.

Hamernik, W., and Michler, G. O.
1. On Brauer's main theorem on blocks with normal defect groups. *J. Algebra* **22** (1972), 1–11.

Harada, K.
1. On some 2-groups of normal 2-rank 2. *J. Algebra* **20** (1972), 90–93.

Herstein, I. N.
1. "Topics in Algebra." Ginn (Blaisdell), Boston, Massachusetts, 1964.

Herzog, M.
1. On finite groups with a cyclic Sylow subgroup. *Illinois J. Math.* **14** (1970), 188–193.
2. Simple groups with cyclic central 2-Sylow intersection. *J. Algebra* **25** (1973), 307–312.
3. Central 2-Sylow intersection. *Pacific J. Math.* **45** (1973), 535–538.

Herzog, M., and Schult, E.
1. Groups with central 2-Sylow intersection of rank at most one. *Proc. Amer. Math. Soc.* **38** (1973), 465–470.

Higman, D. G.
1. Modules with a group of operators. *Duke Math. J.* **21** (1954), 369–376.
2. Indecomposable representations at characteristic p. *Duke Math. J.* **21** (1954), 377–381.

Humphreys, J. E.
1. Representations of $SL(2, p)$. *Amer. Math. Monthly* **82** (1), (1975), 21–39.

Humphreys, J. F.
1. Groups with modular irreducible representations of bounded degree. *J. London Math. Soc.* **5** (2), (1972), 233–234.
2. Defect groups for finite groups of Lie type. *Math. Zeit.* **119** (1971), 149–152.

Hunt, D. C.
1. Character tables of certain finite simple groups. *Bull. Austral. Math. Soc.* **5** (1971), 1–42.

Huppert, B.
1. "Endliche Gruppen I." Springer, Berlin, 1967.
2. Bemerkungen zur modularen Darstellungstheorie I. *Absolut unzerlegbare moduln. Arch. der Mitt.* **26** (1975), 242–249.

Iizuka, K.
1. On Brauer's theorem on sections in the theory of blocks of group characters. *Math. Z.* **75** (1960–61), 299–304.
2. A note on blocks of characters of a finite group. *J. Algebra* **20** (1972), 196–201.

Iizuka, K., and Itô, Y.
1. A note on blocks and defect groups of finite groups. *Kumamoto J. Sci. (Math.)* **9** (1972), 25–32.

Isaacs, I. M.
1. Character degrees and derived length of a solvable group. *Canad. J. Math.* **27** (1975), 146–151.
2. Complex p-solvable linear groups. *J. Algebra* **24** (1973), 513–530.
3. "Character Theory of Finite Groups." Academic Press, New York, 1976.
4. Lifting Brauer characters of p-solvable groups. *Pacific J. Math.* **53** (1974), 171–188.

Isaacs, I. M., and Scott, L.
1. Blocks and subgroups. *J. Algebra* **20** (1972), 630–636.

REFERENCES

Itô, N.
1. On the characters of soluble groups. *Nagoya Math. J.* **3** (1951), 31–48.
2. Note on the characters of soluble groups. *Nagoya Math. J.* **39** (1970), 23–28.

James, G. D.
1. The modular characters of Mathieu groups. *J. Algebra* **27** (1973), 57–111.

Jeyakumar, A. V.
1. Principal indecomposable representations for the group $SL(2, q)$. *J. Algebra* **30** (1974), 444–458.

Keown, R.
1. "An Introduction to Group Representation Theory." Academic Press, New York, 1975.

Kerber, A.
1. Representations of permutation groups, I, II. *In* "Lecture Notes in Math.," **240, 495**. Springer-Verlag, Berlin and New York, 1971, 1976.

Kerber, A., and Peel, M. H.
1. On the decomposition numbers of symmetric and alternating groups. *Mitt. Math. Sem. Giessen Heft* **91** (1971), 45–81.

Landrock, P.
1. A counterexample to a conjecture on the Cartan invariants of a group algebra. *Bull. Amer. Math. Soc.* **5** (1973), 223–224.

Lang, S.
1. "Algebra." Addison Wesley, Reading, Massachusetts, 1965.

Leon, J. S.
1. On simple groups of order $2^a 3^b 7^c$ containing a cyclic Sylow subgroup. *J. Algebra* **28** (1974), 326–341.
2. On simple groups of order $2^a 3^b 5^c$ containing a cyclic Sylow subgroup. *J. Algebra* **21** (1972), 450–457.

Leon, J. S., and Wales, D. B.
1. Simple groups of order $2^a 3^b p^c$ with cyclic Sylow p-groups. *J. Algebra* **29** (1974), 246–254.

Leonard, H. S., Jr.
1. Finite linear groups having an abelian Sylow subgroup, I, II. *J. Algebra* **20** (1972), 57–69; **26** (1973), 368–382.

Lindsey, J. H.
1. Complex linear groups of degree less than $4p/3$. *J. Algebra* **23** (1972), 452–475.

Martineau, R. P.
1. On 2-modular representations of the Suzuki group. *Amer. J. Math.* **94** (1972), 55–72.

Mason, D. R.
1. Finite simple groups with Sylow 2-subgroup dihedral wreath Z_2. *J. Algebra* **26** (1973), 10–68.

Michler, G. O.
1. "Blocks and centers of group algebras" (mimeographed notes). Tulane Univ., 1970–71.
2. The kernel of a block of a group algebra. *Proc. Amer. Math. Soc.* **37** (1973), 47–49.
3. The blocks of p-nilpotent groups over arbitrary fields. *J. Algebra* **24** (1973), 303–315.

Nagao, H.
1. On a conjecture of Brauer for *p*-solvable groups. *J. Math. Osaka City Univ.* **13** (1962), 35–38.
2. A proof of Brauer's theorem on generalized decomposition numbers. *Nagoya Math. J.* **22** (1963), 73–77.

Newman, M. F., ed.
1. *Proc. Second Int. Conf. on Theory of Groups* (1973). *In* "Lecture Notes in Math." **372**, Springer, Berlin, 1974.

Osima, M.
1. Notes on blocks of group characters. *Math. J. Okayama Univ.* **4** (1955), 175–188.
2. On block idempotents of modular group rings. *Nagoya Math. J.* **27** (1966), 429–433.
3. On the generalized decomposition numbers of the alternating group. *Proc. Japan Acad.* **47** (1971), 757–760.

Passman, D. S.
1. "Infinite Group Rings." Dekker, New York, 1971.
2. Blocks and normal subgroups. *J. Algebra* **12** (1969), 569–575.
3. Central idempotents in group rings. *Proc. Amer. Math. Soc.* **22** (1969), 555–556.

Peacock, R. M.
1. Blocks with a cyclic defect group. *J. Algebra* **34** (1975), 232–259.

Powell, M. B., and Higman, G.
1. "Finite Simple Groups." Academic Press, New York, 1971.

Puttaswamaiah, B. M.
1. Determination of Brauer characters. *Canad. J. Math.* **26** (1974), 746–752.
2. Brauer characters and Grothendieck rings. *Canad. J. Math.* **27** (1975), 1025–1028.

Reiner, I.
1. On the number of irreducible modular representations of a finite group. *Proc. Amer. Math. Soc.* **15** (1964), 810–812.

Reynolds, W. F.
1. Blocks and normal subgroups of finite groups. *Nagoya Math. J.* **22** (1963), 15–32.
2. Fields related to Brauer characters. *Math. Z.* **135** (1973–74), 363–367.

Richen, F.
1. Blocks of defect zero of split (B, N) pairs. *J. Algebra* **21** (1972), 275–279.

Robinson, G. de B.
1. "Representation Theory of the Symmetric Group." Univ. of Toronto Press, Toronto, 1961.

Rosenberg, A.
1. Blocks and centers of group algebras. *Math. Z.* **76** (1961), 206–216.

Rothschild, B.
1. Degrees of irreducible modular characters of blocks with cyclic defect groups. *Bull. Amer. Math. Soc.* **73** (1967), 102–104.

Rukolaine, A. V.
1. Degrees of modular representations of *p*-solvable groups. *Vestnik Leningrad. Univ.* **17** (1962), 41–48.

Scott, L.
1. Modular permutation representations. *Trans. Amer. Math. Soc.* **175** (1973), 101–121.
2. The modular theory of permutation representations. *Proc. Symp. Pure Math.* **21** (*Amer. Math. Soc.*, 1971), pp. 137–144.

Serre, J. P.
1. "Représentations Linéaires des Groupes Finis." Hermann, Paris, 1967.

Smith, F. L.
1. Finite groups whose Sylow 2-subgroups are the direct product of a dihedral and a semi-dihedral group. *Illinois J. Math.* **17** (1973), 352–386.
2. Groups whose Sylow subgroups are the direct product of two semidihedral groups. *Illinois J. Math.* **17** (1973), 387–396.

Spiegel, H.
1. Blockkorrespondenzen und p'-Normalteiler. *Arch. der Math.* **25** (1974), 483–487.

Solomon, L.
1. The representation of finite groups in algebraic number fields. *J. Math. Soc. Japan* **13** (1961), 144–164.

Solomon, R.
1. Finite groups with Sylow 2-subgroups of type A_{12}. *J. Algebra* **24** (1973), 346–378.

Suzuki, M.
1. Application of group characters. *Proc. Symp. Pure Math.* **6** (*Amer. Math. Soc.*, 1962), pp. 101–105.

Swan, R. G.
1. The Grothendieck ring of a finite group. *Topology* **2** (1963), 85–110.

Thompson, J. G.
1. Vertices and sources. *J. Algebra* **6** (1967), 1–6.
2. Nonsolvable finite groups all of whose local subgroups are solvable, I, II, III, IV, and V. *Bull. Amer. Math. Soc.* **74** (1968), 383–407; *Pacific J. Math.* **33** (1970), 451–537; **39** (1971), 483–534; **48** (1973), 511–592; and **50** (1974), 215–297.

Tsushima, Y.
1. On the blocks of defect 0. *Nagoya Math. J.* **44** (1971), 57–59.

Wales, D.
1. Defect groups in p-constrained groups. *J. Algebra* **14** (1970), 572–574.
2. Simple groups of order $p \cdot 3^a \cdot 2^b$. *J. Algebra* **16** (1970), 183–190.
3. Simple groups of order $7 \cdot 3^a \cdot 2^b$. *J. Algebra* **16** (1970), 575–596.
4. Simple groups of order $17 \cdot 3^a \cdot 2^b$. *J. Algebra* **17** (1971), 429–433.
5. Simple groups of order $13 \cdot 3^a \cdot 2^b$. *J. Algebra* **20** (1972), 124–143.

Winter, D. L.
1. Finite groups having a faithful representation of degree less than $(2p + 1)/3$. *Amer. J. Math.* **86** (1964), 608–618.
2. On the structure of certain p-solvable linear groups. *J. Algebra* **33** (1975), 170–190.

Index

A

Absolutely indecomposable module, 24
Absolutely irreducible module, 25
Action of a group, 1
Afforded characters, 40
Algebra, 3
Algebraic conjugate character, 60
Algebraic integer, 46
Alperin, J. L., 196
Annihilator, 3
Artin, E., 88
Artinian module, 5

B

Basis, 3
Berman, S. D., 32, 61, 88
Binomial theorem, 16
Blau, H. I., 227
Blichfeldt's theorem, 39
Block, 89, 90
   with cyclic defect groups, 197
   with cyclic Sylow p-subgroups, 205
   of a group algebra, 89, 90
   idempotent, 90

   with normal p-subgroups, 144
   orthogonality relations, 98
   of small defect, 112
Brauer, R., 32, 50, 57, 61, 78, 88, 113, 115, 133,
   141, 168, 196, 215, 226, 227, 228
Brauer character, 77
Brauer correspondence, 149, 150, 152
Brauer homomorphism, 142
Brauer's theorem on induced characters, 52, 54
Brauer's theorem on splitting fields, 57
Burnside, W., 11, 40, 60, 70, 88, 166, 168
Burnside, Frobenius, and Schur theorem, 40

C

Carleson, J. F., 115
Cartan invariant, 79
Cartan matrix, 79
Central character, 91
Central primitive idempotent, 90
Character, 39
Characterization of characters, 55
Class function, 40, 42
Class sum, 42
Clifford's theorem, 37
Complete orthogonal set of idempotents, 23
Completely reducible module, 11

Composition factor, 7
Composition series, 6
Conjugate blocks, 153
Conjugate modules, 37
Conjugate representations, 84
Coordinate functions, 41
Curtis, C. W. and Reiner, I., 32, 60, 61, 88, 133

D

Dade, E., 88, 115, 141, 168, 209, 227
Decomposable module, 20
Decomposition
  matrix, 80
  number, 80
Defect group, 100, 101, 102
  of a block, 102
  of a class, 101
Degree
  of a character, 43
  of a representation, 4
Derivative of a module, 131
Dixon, J. D., 133
Dornhoff, L., 88
Dot product, 175
Dual of a module, 14

E

Elementary subgroup, 53
Equivalent representations, 11
Exponent of a group, 56
Extension of the First Main Theorem, 152

F

Faithful action, 1
Faithful representation, 4
Feit, W., 61, 115, 133, 141, 168, 196, 220, 227
Field of p-adic numbers, 63
First Main Theorem, 149, 151
Fitting's lemma, 6
Fong, P., 84, 88, 168
Fong–Swan theorem, 85
Fowler, K. A., 115, 228
Free, 3
Frobenius, G. A., 40

Frobenius character, 39
Frobenius reciprocity, 49
Full defect, 109
Full set of irreducible modules, 11
Fusion, 169

G

Gaschütz, W., 88
General linear group, 4
Generalized character, 52
Generalized decomposition numbers, 157, 158, 169, 170
Generalized quaternion group, 184
Glauberman, G., 61, 192, 196
Glauberman's Z^*-theorem, 191, 192
Goldschmidt, D. M., 61, 196
Gorenstein, D., 184, 196
Green, J. A., 32, 88, 116, 121, 140, 168, 227
Green's theorem, 123, 126
Group
  algebra, 2
  characters, 39
  of type $(2^m, 2^m)$, 178

H

Hall, M., 168, 184, 228
Harada, K., 196
Herzog, M., 227
Higman, D. G., 117
Hill, R., 88
Homogeneous components, 38
Huppert, B., 60

I

Idempotent, 21
  refinement, 64
Iizuka, K., 168
Imprimitive module, 34
Imprimitivity, system of, 34
Indecomposable module, 20
Induced characters, 48
Induced modules, 14, 33
Induction, 33
Inequivalent representations, 11

INDEX

Inner product, 44
Integral closure, 66
Integral elements, 66
Intertwining number, 29, 30
Irreducible character, 43, 77, 81
Irreducible constituent, 7
Irreducible modular character, 77, 81
Isaacs, M., 61, 80
Ito, N., 88, 221

J

Jacobson radical, 8
Jordan–Hölder theorem, 6
Jordan's theorem, 113

K

Kernel
  of a character, 43
  of a module, 4
  of a representation, 4
Krull–Schmidt theorem, 22, 74
Krull–Schmidt–Azumaya theorem, 22

L

Lang, S., 22
Leon, J. S., 228
Lifting idempotents, 73
Lindsey, J. H., 227
Linear representation, 4

M

Mackey's theorem, 35
Maschke's theorem, 12
Matrix representation, 5
Michler, G. O., 114, 168
Module, 2
Modular character, 62, 77
  of a p-solvable group, 84
Modular representation, 62, 69
Monic polynomial, 46
Monomial module, 38
Multiplicity, 7, 46

N

Nagao, H., 88, 168
Nesbitt, C. J., 78, 88, 115
Nil ideal, 9
Noetherian module, 5
Nonexceptional characters, 210

O

Orbit, 2
Ordinary character, 43
Ordinary representation, 69
Orthogonal class functions, 44
Orthogonal idempotents, 23, 79
Orthogonality relations, 45, 82, 98
Orthonormal, 44
Osima, M., 97, 115, 168

P

p-adic algebras, 65
p-adic field, 63
p-adic integers, 62, 63
p-adic numbers, 63
p-conjugate characters, 113
  element, 17
  part of an element, 17
  regular element, 17
  section, 162
  singular element, 17
  solvable group, 84
p'-element, 17
p'-part of an element, 17
Passman, D. S., 32, 168
Peacock, R. M., 227
Primitive idempotent, 23
Primitive module, 34
Principal block, 94
Principal character, 40, 43
Principal indecomposable module, 26
Principal indecomposable modular character, 81
Projective (H-projective), 117

Q

Quasi-elementary subgroup, 53

R

Rank of a module, 3
Reduced decomposition matrix, 113
Reducible character, 43
Reducible module, 6
Reducible representation, 5, 6
Reduction modulo, 70
Regular character, 44
Regular module, 3
Reiner, I., 32, 60, 61, 88, 133
Relatively projective modules, 116, 117
Representation, 4
Representation module, 1
Restriction, 33
Reynold's theorem, 147, 168
Rosenberg, A., 168
Rukolaine, A. V., 88

S

Schur, I., 40, 133
Schur's lemma, 7
Scott, L., 168
Second Main Theorem, 157, 160
Semisimple ring, 11
Serre, J. R., 84
Short exact sequence, 116, 117
Source, 120
Split exact sequence, 117
Splitting field, 11, 56
Structure theorem for modules over principal ideal domain, 3
Suzuki, M., 50, 196
Swan, R. G., 84
System of imprimitivity, 34

T

Tate, J., 61
Tensor products, 13
Third Main Theorem, 162, 163
Thompson, J. G., 141, 227, 228
Torsion-free module, 3
Transitive, 2
Trivial intersection set, 50
Tuan, H. F., 226

V

Vertex, 120

W

Wales, D., 228
Walter, J. H., 184, 196
Wedderburn, J. H. M., 57
Wedderburn structure theorem, 9
Winter, D. L., 88, 227